THE CYTOSKELETON IN CELL DIFFERENTIATION AND DEVELOPMENT

EDITED BY

RICARDO B MACCIONI AND JUAN ARECHAGA

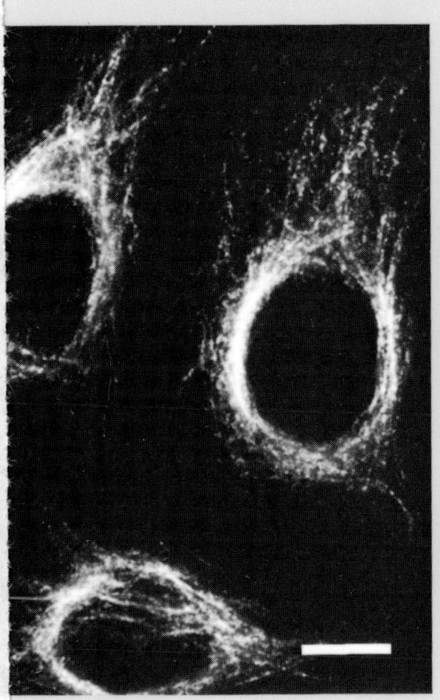

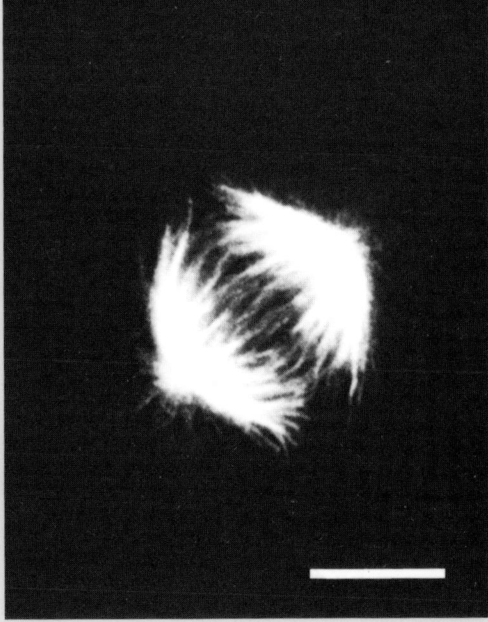

PUBLISHED FOR THE ICSU PRESS BY
IRL PRESS
OXFORD · WASHINGTON DC

ICSU Symposium Series

Volume 8

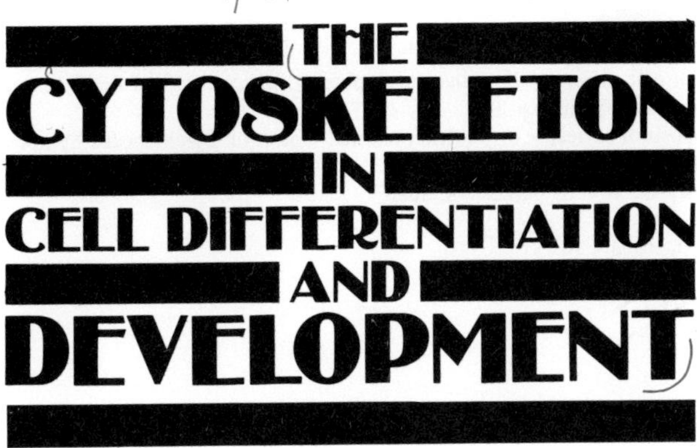

Proceedings of the First International Symposium

Granada, Spain

April 21–25, 1987

Edited by

Ricardo B.Maccioni and Juan Aréchaga

Published for the ICSU Press by

International Union of Biochemistry Symposium no. 160

Published by IRL Press Limited
PO Box 1, Eynsham, Oxford OX8 1JJ, UK
PO Box Q, McLean, Virginia 22101-0850, USA

Printed in Great Britain by Information Printing Limited

Library of Congress catalog number: 87-26251

British Library Cataloguing in Publication Data

The Cytoskeleton in cell differentiation and development : proceedings of the first international symposium, Granada, Spain, April 21–25, 1987.
—(ICSU symposium series; v.8)—(International Union of Biochemistry Symposium; no. 160).
1. Cytoskeleton
I. Maccioni, Ricardo B. II. Aréchaga, Juan
III. Series IV. Series
574.87'34 QH603.C96

ISBN 1-85221-093-1

Also in this series:
ICSU Press Symposium no. 1
Genetic Manipulation: Impact on Man and Society
1984: Published for the ICSU Press by Cambridge University Press
ISBN 0 521 26417 0

ICSU Press Symposium no. 2
Striga: Biology and Control
1984: Published by the ICSU Press and I.D.R.C.
ISBN 0-947946-17-9

ICSU Press Symposium no. 3
H^+-ATPase (ATP Synthase): Structure, Function, Biosynthesis
1984: Published for the ICSU Press by Adriatica Editrice, Bari, Italy
ISBN 0-930-357-00-0

ICSU Press Symposium no. 4
Progress in Bio-organic Chemistry and Molecular Biology
1984: Published for the ICSU Press by Elsevier Science Publishers
ISBN 0 444 80643 1

ICSU Press Symposium no. 5
Global Change
1985: Published for the ICSU Press by Cambridge University Press
ISBN 0 521 30670 1 hard cover ISBN 0 521 31499 2 paperback

ICSU Press Symposium no. 6
Membranes and Muscle
1985: Published for the ICSU Press by IRL Press
ISBN 0-947946-40-3

ICSU Press Symposium no. 7
Integration and Control of Metabolic Processes: Pure and Applied Aspects
1987: Published for the ICSU Press by Cambridge University Press
ISBN 0 521 34273 2

Cover illustration: Intermingling of Mβ5 between interphase and spindle microtubules in HeLa cells. Photomicrograph of HeLa interphase (left) and spindle (right) microtubules identified using antibody specific for Mβ5. Bar = 10 microns.

Contents

MICROTUBULE ASSOCIATED PROTEINS IN CELL DIFFERENTIATION

CELLULAR AND MOLECULAR BIOLOGY OF CYTOSKELETON DURING DEVELOPMENT

CYTOSKELETON IN DEVELOPMENTAL NEUROBIOLOGY

INTERMEDIATE AND ACTIN FILAMENTS IN CELL DIFFERENTIATION AND CANCER BIOLOGY

INTERACTION OF THE CYTOSKELETON COMPONENTS DURING DEVELOPMENT

CYTOPLASMIC ORGANIZATION DURING DEVELOPMENT

Speakers and Editors

ARÉCHAGA,J. – Department of Biochemistry, Biophysics and Genetics, University of Colorado Health Sciences Center, Denver, CO 80262, USA

AVILA,J. – Centro de Biología Molecular (C.S.I.C.-U.A.M.), Canto Blanco, 28049 Madrid, Spain

BAJER,A.S. – Department of Biology, University of Oregon, Eugene, OR 97403, USA

BERLIN,R.D. – University of Connecticut Health Center, Farmington, CT 06032, USA

CAPLOW,M. – Department of Biochemistry, University of North Carolina at Chapel Hill, Chapel Hill, NC 27514–7231, USA

CELIS,J.E. – Department of Medical Biochemistry, Aarhus University, DK-8000, Aarhus C, Denmark

COHEN,W.D. – Department of Biological Sciences, Hunter College of the University of New York, NY 10021, USA

COWAN,N.J. – New York University School of Medicine, New York, NY 10016, USA

EVANS,R. – Department of Pathology, University of Colorado Health Sciences Center, Denver, CO 80262, USA

GUNDERSEN,G.G. – Department of Biology, University of California, Los Angeles, CA 90024, USA

JORCANO,J.L. – Division of Membrane Biology and Biochemistry, Institute of Cell and Tumor Biology, German Cancer Research Center, and Center of Molecular Biology, University of Heidelberg, D-6900, Heidelberg, FRG

KRYSTOSEK,A. – University of Colorado Health Sciences Center, Denver, CO 80262, USA

MACCIONI,R.B. – Department of Biochemistry, Biophysics and Genetics, University of Colorado Health Sciences Center, Denver, CO 80262, USA

MARCO,R. – Instituto de Investigaciones Biomédicas del CSIC and Departamento de Bioquímica, UAM Facultad de Medicina, Universidad Autonoma de Madrid, Madrid 28029, Spain

NUÑEZ,J. – INSERM U282, Hôpital Henri Mondos, 94010 Créteil, France

PAULIN,D. – Unité de Génétique Cellulaire du Collège de France et de l'Institut Pasteur, Université Paris 7, 25 rue du Dr Roux, Paris 75724, Cedex 15, France

PIERCE,G.B. – Department of Pathology, University of Colorado School of Medicine, Denver, CO 80262, USA

RUCH,J.V. – Institut de Biologie Médicale, UER Médecine 11, rue Humann, Strasbourg, France

SEEDS,N.W. – Department of Biochemistry, University of Colorado Health Sciences Center, Denver, CO 80262, USA

STENT,G.S. – Department of Molecular Biology, University of California, Berkeley, CA 94720, USA

TAMM,S.L. – Boston University Marine Program, Marine Biological Laboratory, Woods Hole, MA 02543, USA

VALDIVIA,M.M. – Department of Biochemistry, Facultad de Ciencias, Universidad de Cadiz, Puerto Real, Cadiz, Spain

VERA,J.C. – Department of Biochemistry, Biophysics and Genetics, University of Colorado Health Sciences Center, Denver, CO 80262, USA

WICHE,G. – Institute of Biochemistry, University of Vienna, 1090 Vienna, Austria

WILSON,L. – Department of Biological Sciences, University of California, Santa Barbara, CA 93106, USA

Preface

This volume contains the invited and contributed presentations of the 1st International Symposium **The Cytoskeleton in Cell Differentiation and Development** which took place under the auspices of the Department of Biochemistry, Biophysics and Genetics of the University of Colorado Health Sciences Center and the Instituto F.Oloriz for Developmental Biology of the University of Granada. The Symposium was held at the Medical School of the University of Granada, Spain on April 21–25, 1987, and was sponsored by the International Union of Biochemistry (IUB), the International Society for Developmental Biologists, and the Spanish Societies of Biochemistry and Cell Biology.

The understanding of the structural and functional aspects of the early development and differentiation of the cell is one of the most stimulating problems in modern biology. A striking characteristic of developmental processes is the continuous increase in structural complexity leading to cellular asymmetry. The weight of the evidence which has been accumulated over the years indicates very strongly that the cytoskeletal network is directly involved in determining the shape and internal organization of most eukaryotic cells, but the detailed molecular and cellular aspects of the roles that the different components of cytoskeleton play in embryogenesis and cell differentiation remain to be elucidated.

The main purpose of this Symposium was to bring together scientists with different experimental approaches and points of view, to explore in depth the complex relationships between the cytoskeleton and developmental processes. The collection of research articles included in this book is an overview of the recent advances in the field and shows the remarkable progress that has taken place in our understanding of the subject.

In retrospect, the Symposium has been very fruitful and we want to acknowledge the stimulating input of the invited speakers and the contributors to the poster sessions. Thanks are also due to the various participants who acted as moderators and to the members of the Organizing Committee. We are also grateful to the sponsors of this Conference and the several institutions whose support and encouragement have been essential to the success of the Symposium. Last, but not least, we would like to thank the staff of the ICSU Press for their endeavors in publishing this volume.

Ricardo B.Maccioni   Juan Aréchaga

This is ICSU

This monograph, the printed record of the IUB Symposium 160, is published by the ICSU Press in partnership with the IRL Press.

The ICSU Press is the publishing house of the International Council of Scientific Unions, an international non-governmental scientific organization, whose principal objective is to encourage international scientific activity for the benefit of all. ICSU's membership is composed of 20 Scientific Unions, representing various disciplines and national bodies such as academies or research councils in approximately 70 countries.

Since its creation in 1931, ICSU has adopted a policy of non-discrimination, affirming the rights of all scientists throughout the world – without regard to race, religion, political philosophy, ethnic origin, citizenship, sex or language – to join in international scientific activities.

To fulfill its objectives, ICSU initiates, designs and coordinates international interdisciplinary research programs. ICSU acts as a focus for the exchange of ideas, the communication of scientific information, and the development of scientific standards, nomenclature, units, etc.

For programs in multi- or transdisciplinary fields, such as Antarctic, Oceanic, Space Research, or Genetic Experimentation, which are not under the aegis of one of the Scientific Unions, and for activities in areas common to all Unions, such as Teaching of Science, Data, Science and Technology in Developing Countries, ICSU creates Scientific Committees or Commissions.

Members of the ICSU family organize in many parts of the world scientific conferences, congresses, symposia, summer schools and meetings of experts, as well as meetings to decide policies and programs.

ICSU members produce a wide range of publications, including newsletters, handbooks, proceedings of meetings, congresses and symposia, professional scientific journals, data, standards, etc. Some of these are published or managed by the ICSU Press, including BioEssays, the monthly current-awareness journal, sponsored by ICSU's Biological Unions, and BioFactors, a new journal, published for the IUB by IRL Press.

ICSU cooperates with a number of international governmental and non-governmental organizations, including UNESCO, UNDP and WHO.

Further information about ICSU may be obtained from the ICSU Secretariat, 51 Boulevard de Montmorency, 75016 Paris, France.

PLENARY LECTURES

THE ROLE OF CELL LINEAGE IN DEVELOPMENT

G.S. Stent
Department of Molecular Biology, University of California, Berkeley, California 94720, USA

Origins of Cell Lineage Studies. Studies of developmental cell lineage -- that is of the fate of individual cells, or blastomeres, that arise in the early embryo -- were begun in the 1870's, in the context of the controversy then raging about Ernst Haeckel's "biogenetic" law. The biogenetic law seemed to imply that cells of the metazoan blastula recapitulate the non-differentiated tissues of a remote sponge-like ancestor. Only after gastrulation would the germ layers -- ectoderm, mesoderm, endoderm -- be destined to form the highly differentiated tissues characteristic of more recent metazoan ancestors. This implication was tested by the founder of American experimental embryology, Charles O. Whitman (1887). By observing the cleavage pattern of early leech embryos -- which is also the main experimental material of this brief review article -- Whitman traced the fate of individual cells from the uncleaved egg to the germ-layer stage and concluded that, contrary to the implication of the biogenetic law, a characteristic postembryonic fate can be assigned to each identified blastomere and to the clone of its descendant cells. These findings suggested, moreover, that the differentiated properties that characterize a given cell of the mature animal are somehow determined by its genealogical line of descent from the egg.

Despite its highly promising beginnings, the study of developmental cell lineage went into decline after the turn of this century. It remained a biological backwater for the next 50 years, probably because the discovery of regulative and inductive phenomena in the development of echinoderms and chordates focussed the attention of the embryologists on cell interactions rather than on cell lineage as causal factors in cell differentiation. The first dramatic development in this direction was Hans Driesch's discovery in 1891, that upon separation of the two cells produced by the first cleavage of the sea urchin egg, each cell is capable of developing into a whole, albeit smaller embryo. This finding showed that, in accord with the implications of the biogenetic law and contrary to the view of cell lineage as a determinant of cell fate, individual blastomeres contain the entire developmental potential of the uncleaved egg -- that is, they are totipotent.

When Hans Spemann and Hilde Mangold showed in 1924 that grafting an exogenous dorsal blastoporal lip on the ventral aspect of an amphibian gastrula induces the development of a second, supernumerary central nervous system, the attention of embryologists became focussed on the mechanism by which the cells in one part of the embryo induce the developmental fate of pluripotent cells in another part of the embryo.

For the next thirty years, the search for inducers formed the core project of experimental embryology. However, despite intensive efforts, not one substance was ever identified for which the role of a specific inducer could be convincingly demonstrated. Now, in retrospect, the reason for this failure is quite apparent: the experimental embryologists of the 1930's, 1940's and 1950's lacked the modern molecular biological knowledge which we now know to be necessary to account for the chemical basis of the determination of developmental cell fate.

Revival of Cell Lineage Studies. It may have been the disappointment over the lack of progress in uncovering the molecular basis of morphogenetic gradients and inducers that brought a revival of interest in the developmental role of cell lineage about 20 years ago. This revival was accompanied by the introduction of analytical techniques more precise and far-reaching than those available to Whitman and other 19th century pioneers. This recent renaissance of developmental cell lineage analysis reawoke interest in the study of embryos of protostomes, such as nematodes (Sulston et al., 1983), leeches (Stent et al., 1982) and insects (Garcia-Bellido and Meriam, 1969), in which the inductive aspects of development are much less prominent than in embryos of deuterostomes, such as echinoderms and chordates. And with this renewed interest in protostomal development Whitman's old idea of the determinative role of cell lineage came back into favor. Among the reasons which led to the renewed belief in a causal nexus between the line of descent of a cell and its fate is that in the embryogenesis of nematodes, leeches and insects serially and bilaterally homologous cell types are, on the whole, generated via homologous genealogical pathways. Indeed, as we shall see presently, this generative homology can be thought to account for the evolution of the segmented body plan of some metazoa in the first place. Another reason is that certain gene mutations or other manipulations which lead to changes in cell lineage also lead to changes in cell fate.

Agents of Commitment. There are two kinds of commonly considered agents which may commit embryonic cells to their fate. One kind is represented by _intracellular determinants_ of cell differentiation, which account for the commitment of sister cells to different fates in terms of their unequal partition in successive cell divisions. Here cell lineage would play a crucial role in cell commitment by consigning particular subsets of intracellular determinants to particular cells. The other commonly considered kind of agent consists of a set of _intercellular inducers_, which are anisotropically distributed over the volume of the embryo. Here cell lineage would play a crucial role in cell commitment by placing particular cells at particular sites within the inductive field, and hence governing the pattern of their exposure to inducers. In some cases cell lineage was found to play its determinative role in cell commitment by bringing about the orderly, unequal partitioning of intracellular determinants over daughter cells in successive cell divisions (Whittaker, 1973) and in other cases, by bringing about an orderly topographic cell placement (Shankland, 1984).

Typological and Topographic Commitment Hierarchies. Regardless of which of these alternative agents might happen to be responsible for commitment to developmental cell fate, it had been generally expected that such commitment proceeds stepwise, according to a _typologically_ hierarchic sequence. For instance, it was thought that in the developmental line of ancestry of a cholinergic motor neuron there would occur a commitment first to ectoderm rather than to mesoderm, then to nervous tissue rather than skin, then to neuron rather than glial cell, then to motor neuron rather than sensory neuron, and finally to synthesis of choline acetyltransferase rather than glutamic acid decarboxylase.

One important, albeit negative, insight brought by the modern cell lineage studies is that development does _not_ generally proceed according to that expected typologically hierarchic commitment sequence. Instead it transpired that the commitment to differential cell fates, manifest in developmental cell lineage, appears to be typologically arbitrary. For instance, of two differentiated sister cells, one may be a neuron and the other an epidermal cell, whereas of two anatomically similar neurons, one may have arisen on the ectodermal branch and the other on the genealogically very remote mesodermal branch of the lineage tree. Instead of being typologically hierarchic, in nematodes and leeches, the commitment sequence turns out to be largely _topographically_ hierarchic. That is to say, in these embryos, where cell migration plays a relatively minor (though definite) role, it is the position of two cells rather than their phenotype which tends to be correlated with the closeness of their genealogical relation. Or, in other words, the spatially ordered sequence of cell divisions represented by that genealogical relation is so arranged that most differentially committed postmitotic cells arise at, or very close to, the sites where they are actually needed (Stent, 1985).

Segmentation. Another important insight brought by modern cell lineage studies pertains to the developmental origin of body segments, which provide, in fact, a particularly important example of the topographically hierarchic character of the commitment sequence. Several protostomal as well as deuterostomal phyla, such as the annelids, arthropods and vertebrates, share the general structural feature of their bodies being composed of a periodic series of bilaterally symmetric segments. Each segment corresponds to a module, or *metamere*, of regularly iterated morphological elements, such as appendages, skin specializations, muscles or nerve cell ganglia. This basic morphological segmentation pattern is often obscured, however, because the metameres, rather than being exactly alike, usually differ at various positions along the longitudinal body axis, from head to tail.

An important concept relating to the embryonic origin of body segments dating back to William Bateson is that each segment is poised to generate the basic metamere pattern, and that the position-specific departures from the basic pattern, or segmental *heterosis*, arises from a specific deflection of the local tissues from their basic developmental pathway. In the fruit fly *Drosophila* a set of genes has been identified, and are presently under intensive molecular-biological study, whose products are necessary, and in some cases sufficient, for generating the position-specific heterotic departures from the basic metamere pattern. These genes are designated as *homeotic* genes, because, in line with the tradition of classical genetics, according to which genes are named, not after their normal function, but after their mutant phenotype, in homeotic *Drosophila* mutants segments manifest the basic metamere pattern rather than their position-specific *heterotic* departure from the basic pattern.

Leeches. Here I concentrate on the role of cell lineage in the development of the body segments of leeches. The bilaterally symmetric, tubular body of leeches, which form the class Hirudinea of the phylum Annelida, consists of 32 segments and a non-segmental prostomium. The metameric morphological elements include the segmental ganglion of the ventral nerve cord, a bilateral pair of excretory organs, or nephridia, and three subdivisions of the skin, or annuli. One of these three annuli lies in register with the segmental ganglion and includes circumferentially distributed sensory organs, or sensilla. The frontmost four and the rearmost seven body segments are fused, and, reflecting the position-dependent departures from the basic, midbody metamere pattern, i.e. heterosis, contain the specialized morphological features of the head and tail regions.

The location of the segment border, i.e. the interface between successive midbody (i.e. unfused) metameres, had long been the subject of controversy. At the turn of the century, both William Castle and J. P. Moore claimed that the ganglion and its in-register sensillar annulus mark the middle of the segment. Accordingly, Castle and Moore chose the furrow separating two successive non-sensillar annuli as the location of the segment border. The Castle-Moore view of the centrality of the segmental ganglion came to be generally accepted, not only for segmentation in leeches but also in insects. By contrast, Whitman had proposed earlier, on the basis of his embryological finding that the ganglion lies astride the septal margin of the somite, that the middle of the sensillar annulus marks the segment border rather than the middle of the segment, i.e. that the ganglion is an intersegmental rather than segmental structure. Now, with the wisdom of hindsight, it is obvious that this controversy was largely futile, since there is no objective way of fixing the beginnings and ends of the morphological modules of a periodically repeated structure. As we shall see, however, developmental cell lineage analyses do permit an objective definition of the spatial limits of individual _generative_ (rather than morphological) metameres. These analyses showed that in this controversy, as in several others in which he had been involved, Whitman was closer to the truth than his adversaries.

Leech Embryogenesis. Development of the fertilized leech egg proceeds via a stereotyped sequence of (holoblastic) cleavages, giving rise to an embryo whose cells can be identified individually by various criteria, including size, position, birth order and cytoplasmic specializations (Figure 1). The ectodermal and mesodermal segmental tissues arise from bilateral sets of five large blastomeres, designated _teloblasts_. By a series of several iterated, highly unequal divisions, each teloblast generates a bandlet of several dozen much smaller primary blast cells. The bandlet designated m will give rise to the mesodermal and the bandlets designated n, o, p and q to the ectodermal components of the 32 segments. Ipsilateral bandlets merge to form left and right germinal bands and the two bands migrate over the surface of the embryo and eventually coalesce in rostrocaudal sequence to form a sheet of cells, the germinal plate, along the future ventral midline. At the time of its formation, the n, o, p and q bandlets lie in mediolateral alphabetical order on the superficial aspect of the germinal plate, while the m bandlet lies on its deep aspect.

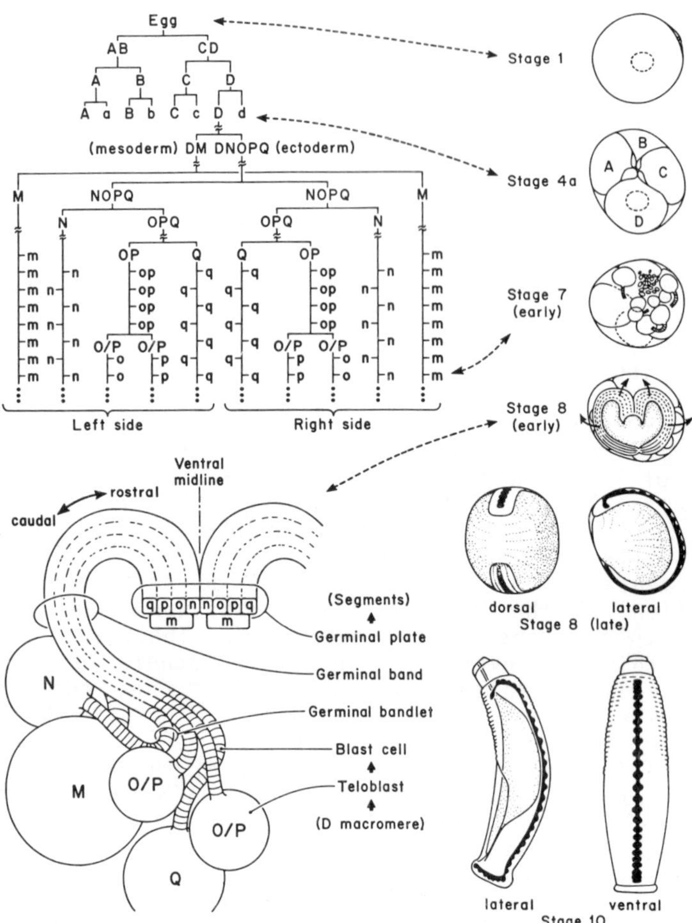

Fig.1. Schematic summary of the development of the leech. Upper left: cell pedigree leading from the uncleaved egg to the teloblast pairs, M, N, O/P, O/P, and Q; and the paired primary blast cell bandlets. Lower left: hemilateral disposition of the teloblasts and their primary blast cell bandlets within the germinal band and germinal plate. Right margin: diagrammatic views of the embryo at various stages. In the stage 8 (early) embryo, the heart-shaped germinal bands migrate over the surface of the embryo in the directions indicated by the arrows. In the stage 8 (late) embryo the germinal plate is shown to lie on the ventral midline, with the nascent central nervous system and its ganglia indicated in black. In the stage 10 embryo shown, body closure is nearly complete (Weisblat et al., 1984).

Proliferation and differentiation of the m, n, o, p and q blast cell clones gives rise to a morphological periodicity of the germinal plate, reflecting formation of the 32 body segments. Within the germinal plate, the clones founded by the older blast cells of each bandlet lie anterior to the clones founded by the younger blast cells. Hence throughout most of its development, the germinal plate embodies a caudorostral developmental progression, according to which the development of a given segment is more advanced than that of its next posterior segment. Cell proliferation also causes the germinal plate to expand laterally around the circumference of the embryo, until the right and left leading edges of the plate meet along the future dorsal midline. At this point, formation of the body tube of the leech is complete.

The coordinates of the morphological periodicity pattern appear to be provided by an orthogonal network of longitudinal and circular muscle fibers, as visualized by immunofluorescent staining with a monoclonal anti-muscle antibody. In particular the rectangular territory of the ganglionic rudiment is enclosed by two serially successive circular muscle fibers at its anterior and posterior margins and by a bilateral pair of longitudinal muscle fibers at its lateral margins (Torrence and Stuart, 1986).

Cell Lineage Tracers. To refine and extend Whitman's century-old genealogical studies, we developed a novel lineage tracer technique. This technique consists of injecting a tracer molecule into an identified embryonic cell, which is passed on exclusively to the lineal descendants of that cell. These descendants can then be identified at a later developmental stage, by observing the distribution pattern of the tracer within the tissues. One type of tracer we have used is the enzyme horseradish peroxidase (HRP), which, upon a particular histochemical treatment of the embryonic tissues, causes formation of a black precipitate in the cells containing it (Weisblat et al., 1978; 1984). Another type of tracer of our design are adducts of the fluorescent dyes fluorescein and rhodamine and carrier molecules, such as dextran (Weisblat et al., 1980). The distribution of these fluorescent tracers can be observed in living tissues under the fluorescence microscope. The fluorescein labeled tracer has an additional, very useful property: it can serve also as a specific photosensitizer. Thus illuminating the embryonic tissues with light of a particular wavelength leads to death by photo-oxidation of all descendants of the tracer-injected precursor cell, but not of any other, genealogically unrelated cells with which the labeled cells may be intermingled (Shankland, 1984). This type of tracer makes it possible, therefore, to examine the developmental effects of the selective ablation of cells of particular lines of descent.

By use of the cell lineage tracer techniques to mark the differentiated post-mitotic descendants of individual blast cell bandlets we found that each bandlet gives rise to a particular hemilateral _kinship group_ of segmentally iterated identified cells (Weisblat et al., 1984). Hence these findings extended Whitman's inference of the determinative role of genealogical relations for developmental fate to the ultimate level of identified post-mitotic cells. They showed moreover, that there is little mixing of cells across the ventral and dorsal midlines of the embryo.

Segmental Founder Cells. The cell lineage tracer technique was adapted also for ascertaining the number of primary blast cells from each bandlet which found the kinship group of ectodermal and mesodermal cells of each morphological segment (Weisblat and Shankland, 1985). For this purpose, the teloblast of origin of a blast cell bandlet is injected with the lineage tracer after it has already produced a few unlabeled primary cells. In the resulting embryo, a boundary is then observed between anterior, unlabeled segments, derived from primary blast cells born prior to the injection, and posterior labeled segments born after the injection. At first, this boundary is rather sharp, suggesting that there is little longitudinal intermixing of blast cell clones. Now if each hemilateral segmental kinship group represents a clone descended from a single primary blast cell, then the antero-posterior label boundary should always coincide with the segment boundary. However, if each hemilateral segmental kinship group is founded by two successively born primary blast cells, then in about a half of the embryos, the label boundary should coincide with the segment boundary and in the other half of the embryos it should course midway between two segment boundaries.

The results of this experiment were as follows: In the case of the ectodermal n and q bandlets, the two types of label boundaries were observed with equal frequency. This showed that each hemilateral n kinship group is derived from two primary n blast cells (designated as n_s and n_f) which are serial successors in the blast cell bandlet, and the domains of the descendant clones which alternate rostrocaudally. The same is true for the two q segmental founder cells (designated q_f and q_s) and their descendant clones. However, in the case of the mesodermal m and ectodermal o and p bandlets, only a single label boundary was observed, and this single boundary coursed midway through the segment (as defined by Castle and Moore) i.e. midway through two successive segmental ganglia. This showed that each hemilateral m, o and p kinship group is derived from a single m, o and p primary blast cell and that, moreover, the generative metamere corresponds to Whitman's rather than to Castle and Moore's definition of the segment. A similar phase shift between the traditionally accepted morphological segment and of that of the generative metamere has been recently observed in insects, where the generative metamere has been designated as "parasegment" (Martinez-Arias and Lawrence, 1985).

Thus we conclude that in the leech each hemilateral segment arises as seven distinct cell clones, six ectodermal clones and one mesodermal clone. Each of the seven heterologous, hemisegmental clones comprises a few dozen characteristic cells, which have arisen from their founder blast cell via a clone-specific, stereotyped cell lineage pattern. By contrast, bilaterally and serially homologous primary blast cells generate homologous cell lineage patterns. As specific ablation experiments using the photosensitizing tracer methods have shown, the periodicity of that pattern is (largely) an autonomous property of individual blast cells, rather than being the product of intercellular interactions. An interesting exception is formed by the o and p blast cells, of which one can take on the fate of the other and whose commitment to one of two alternative fates does depend on intercellular interactions (Weisblat and Blair, 1984; Shankland and Weisblat, 1984).

In the course of growth and development of the hemisegmental founder blast cell sextet there occurs extensive (but stereotyped rather than random) interclonal cell mixing among the members of both heterologous clones of the same metamere and of homologous clones belonging to adjacent metameres. The latter kind of cell mixing eventually leads to a developmental interdigitation of serially successive generative metameres, and hence to a lack of isomorphism between generative metameres and morphological segments. For instance, the nephridial tissues and the muscle fibers derived from the same primary m blast cell clone must be assigned to different morphological segments of the postembryonic leech, no matter how its segment borders are defined. Moreover, the transegmental distribution of members of a single primary blast cell clone can be enhanced experimentally by photoablating an adjacent, serially homologous blast cell. These findings imply that in leech development the borders of morphological segments do not correspond to borders of clonal restriction of generative metameres.

The differences in segment morphology, or heterosis, along the longitudinal axis of the leech body are reflected in corresponding modifications of the cell pattern to which some serially successive generative metameres give rise. These modifications do not seem to be attributable, however, to intrinsic differences between serially homologous primary blast cells -- for instance to the rank of their birth -- as can be inferred from experiments in which slippage of a blast cell bandlet has been displaced to an inappropriate segment. Under these conditions, the displaced primary blast cell will develop in accord with its ectopic position rather than its birth rank (Shankland, 1984).

Role of the Cytoskeleton. There can be little doubt that, in the developmental phenomena described here, cell lineage functions in the commitment of cells to their fate via a stereotyped pattern of cell divisions. Regardless of whether the agents of commitment are intracellular determinants or intercellular inducers, it must be the orderly sequence of cleavages, meridional or equatorial, spiral or radial, symmetric or asymmetric, which causes particular branches of the genealogical tree of developmental cell descent to be committed to particular fates. This stereotyped cell division pattern is, of course, a reflection of an orderly sequence of mitotic spindle orientations, which, in turn reflects the orderly movement of the centrosomes to predestined intracellular positions during each mitotic cycle. It is in the dynamic positioning of centrosomes that the cytoskeleton, particularly the microtubules, is likely to play one of its most important roles in development. For instance, the asymmetric positioning of the centrosomes near one pole of an embryonic cell about to enter a highly unequal cleavage is probably attributable to a lower net rate of elongation of microtubules from the centrosomes towards that pole than toward the other pole of the cell. Thus one of the most critical processes now in want of understanding in the commitment of cells to their developmental fate is the governance of the determinate succession of cytoskeletal rearrangements from one cell division to the next that underlies the stereotyped embryonic cleavage pattern.

REFERENCES

Garcia-Bellido, A. and Merriam, J.R. (1969). Cell lineage of the imaginal disk in *Drosophila* gynadromorphs. J. exp. Zool. 170, 61-76.

Martinez-Arias, A. and Lawrence, P.A. (1985). Parasegments and compartments in the *Drosophila* embryo. Nature, Lond. 313, 639-642.

Shankland, M. (1984). Positional control of supernumerary blast cell death in the leech embryo. Nature, Lond. 307, 541-543.

Shankland, M. and Weisblat, D.A. (1984). Stepwise commitment of blast cell fates during the positional specification of the O and P cell lines in the leech embryo. Dev. Biol. 106, 326-342.

Stent, G.S. (1985). The role of cell lineage in development. Phil Trans. R. Soc. Lond. B312, 3-19.

Stent, G.S., Weisblat, D.A., Blair, S.S. and Zackson, S.L. (1982). Cell lineage in the development of the leech nervous system. In Neuronal Development (ed. N. Spitzer), pp. 1-44. Plenum, New York.

Sulston, J.E., Schierenberg, E., White, J.G. and Thomson, J.N. (1983). The embryonic cell lineage of the nematode _Caenorhabditis elegans_. Dev. Biol. 100, 64-119.

Torrence, S.A. and Stuart, D.K. (1986). Gangliogenesis in leech embryos: Migration of neural precursor cells. J. Neurosci. 6, 2736-2746.

Weisblat, D.A. and Blair, S.S. (1984). Developmental indeterminacy in embryos of the leech _Helobdella triserialis_. Dev. Biol. 101, 326-335.

Weisblat, D.A., Kim, S.Y. and Stent, G.S. (1984). Embryonic origins of cells in the leech _Helobdella triserialis_. Dev. Biol. 104, 65-85.

Weisblat, D.A., Sawyer, R.T. and Stent, G.S. (1978). Cell lineage analysis by intracellular injection of a tracer enzyme. Science, Wash. 202, 1295-1298.

Weisblat, D.A. and Shankland, M. (1985). Cell lineage and segmentation in the leech. Phil. Trans. R. Soc. Lond. B312, 39-56.

Weisblat, D.A., Zackson, S.L., Blair, S.S. and Young, J.D. (1980). Cell lineage analysis by intracellular injection of fluorescent tracers. Science, Wash. 209, 1538-1541.

Whitman, C.O. (1887). A contribution to the history of germ layers in Clepsine. J. Morphol. 1, 105-182.

Whittaker, J.R. (1973). Segregation during ascidian embryogenesis of egg cytoplasmic information for tissue specific enzyme development. Proc. natn. Acad. Sci. U.S.A. 70, 2096-2100.

THE EMBRYOLOGIC BASIS OF CARCINOMA

G. B. Pierce

Department of Pathology, University of Colorado School of Medicine, Denver, Colorado 80262

Carcinoma does not have a base in embryology in the sense that Cohnheim postulated. Cohnheim believed that cancer arose from embryonic cells that escaped maturation and became cancerous in adult life (Cohnheim, 1889). Although proof for or against the idea has not been generated it is now clear that carcinoma cells arise from the undifferentiated stem cells of a lineage (Stevens, 1967), and that the features of the neoplastic phenotype are superimposed on the determination of the normal lineage (Pierce and Speers, 1987). Features of the malignant phenotype taken individually are not harmful to the host; all are traits of embryonic cells expressed as various phases of development. It was thus possible to conceptualize the cancer cell as one out of context in terms of time (Pierce and Johnson, 1971). While, true, this idea did not put the cellular facts of neoplasia in order, nor did it have predictive value. It did, however, lead to an extremely valuable working concept of cancer that states: carcinoma is a caricature of the process of tissue renewal (Pierce, 1974; Pierce, et al. 1978; Pierce, 1983). This concept has predictive value and brings order to the facts of neoplasia (Fig. 1). The data establishing the validity of the concept have been reviewed recently in detail (Pierce and Speers, 1987) so only a few points will be briefly outlined as an introduction to the embryologic experiments that we are performing on carcinomas of various types.

Carcinomas are derived from determined undifferentiated normal stem cells, and give rise to malignant stem cells (Stevens, 1967). Whereas division of the normal stem cells is regulated to replace cells as they become senescent, the malignant stem cells produce many copies of themselves for each one that matures and expresses the differentiated features of the lineage (Pierce, 1974). The essential feature of the caricature is this overproduction of undifferentiated malignant stem cells, which imparts an undifferentiated appearance to the tumor. This appearance is not the result of dedifferentiation because the normal stem cell is no more nor less differentiated than the malignant stem cell to which it gives rise (Pierce et al. 1967).

When malignant stem cells differentiate they do so according to the original determination of the lineage (Pierce and Wallace, 1971; Pierce et al. 1978; Pierce, 1983). Thus, carcinogenesis does not change the original histiotypic determination of the normal stem cell; yet, when the malignant stem cell differentiates, the malignant phenotype is usually abrogated. The reason is not clear.

From the foregoing it is clear that it should be possible to cure cancer patients with metastasis by directing the differentiation of cancer cells to benign cells (Pierce, 1961).

Clinical oncologists now realize that at least two factors are important in cytotoxic chemotherapy (Bloch, 1984; Frisney, 1985; Carr et al. 1981). Obviously, many cancer cells are killed by therapy, but many are also made to differentiate, or as illustrated in Fig. 1, they are shifted to the right. Search is now underway for nontoxic agents suitable for directing differentiation as an alternative to cytotoxic therapy (Bloch, 1984; Frisney, 1985).

A series of naturally occurring growth and differentiation factors have been identified for cells of the leukocyte lineages (Metcalf, 1985; Sachs, 1986). Interestingly, these agents when applied to cultures of leukemia cells _in vitro_ direct the differentiation of the highly malignant stem cells into mature monocytes and granulocytes. It is not known whether this is accomplished by converting the malignant population of cells into postmitotic senescent cells (shift to the right in Fig. 1) or by reversing the malignant process (converting S' to S in Fig. 1).

The idea of reversing the malignant process, which is usually overlooked in most attempts to direct differentiation by chemicals, forms the basis of this paper.

The precedent for this approach to tumor therapy

stems from the Brinster experiment (Brinster, 1974), which has been extended and confirmed in many laboratories (Papaioannou et al. 1975; Mintz and Illmensee, 1975). The rationale for the Brinster experiment lay in the demonstration that embryonal carcinoma cells were multipotent and differentiated into the primitive cells of the three germ layers, ectoderm, mesoderm and endoderm (Pierce and Dixon, 1959; Pierce et al. 1960; Kleinsmith and Pierce, 1964). This led to the idea that embryonal carcinoma was the neoplastic equivalent of the inner cell mass of the blastocyst, which also forms the three germ layers (Pierce, 1967). This was proved true when inner cell mass cells spontaneously transformed into embryonal carcinoma cells after culture in vitro (Evans and Kaufman, 1981; Martin, 1981). Brinster injected some embryonal carcinoma cells from black mice into the blastocysts of white mice, put the injected blastocyst into pseudopregnant females and acquired a chimeric mouse as evidenced by its black and white coat color (Brinster, 1974). Such animals are chimeric in all tissues as evidenced by glucose phosphate isoenzyme analyses (Papaioannou et al. 1975; Mintz and Illmensee, 1975).

It is to be stressed that little is known about the biology of chimeric animals or their cancer derived normal tissues. Not all cell lines produce chimeras. Some chimeras are born with tumors (Bradley et al. 1984; Hanaoka et al. 1986). The fate of cancer derived normal cells in the host is not known: for example, are they more susceptible to carcinogenesis or are they already initiated cells and only require promotion for the expression of the malignant phenotype? These are all important questions if direction of differentiation of malignant to benign cells is to replace cytotoxic therapy for cancer.

Little is known of induction in the blastocyst, but it appeared from the Brinster experiment (Brinster, 1974) that the internal environment of the blastocyst could induce a malignant stem cell to become an apparently normal stem cell. The resulting normal stem cell and its offspring responded to homeostatic control eventuating chimeric mice whose tissues were populated by embryo and cancer derived and apparently normal cells.

Assays were made to determine the mechanism of blastocyst regulation of embryonal carcinoma, and to see if the mechanism could be adapted clinically as a noncytotoxic cure for cancer. These assays have indicated that not all lines of embryonal carcinoma are regulated by the blastocyst (Pierce et al. 1982). Of those that do, one is experimentally derived and one arose spontaneously, each was aneuploid. The regulation of tumor formation is specific, the blastocyst does not regulate tumors that have no normal counterpart in the blastocyst (Pierce et al. 1982). There is a restriction point in the cell cycle at which the embryonal carcinoma cell is regulated (Wells, 1982). Regulation is mediated by a soluble factor in blastocele fluid plus contact with trophectoderm (Pierce et al. 1984). Because of the paucity of blastocele fluid, it was impossible until recently to determine if the inner cell mass could also regulate embryonal carcinoma cells in the presence of blastocele fluid. The technical problem was overcome in the following manner: emptied and washed zonae pellucidae were used as carriers for the inner cell mass and cancer cells to be tested. These preparations were placed in giant blastocysts (made by fusion of 8 eight-cell eggs) and incubated in tissue culture for 24 hr (Pierce, et al. 1987). The methods employed for injecting the giant blastocysts with these zonae pellucidae were adapted from the work of Gardner (1972) who put inner cell masses into normal blastocysts and from that of Pedersen and Spindle (1980) who put normal blastocysts into giant blastocysts.

The zonae pellucidae prevented contact of the cancer cells with the cells of the giant blastocyst, yet allowed the inner cell mass cells and cancer cells to be bathed by blastocele fluid. Under these circumstances the cancer cells were regulated and could not form colonies of cancer cells when rescued from the giant blastocyst and cultured *in vitro* (Pierce et al. 1987).

The presence of an inhibitor of growth of embryonal carcinoma cells by blastocele fluid was demonstrated by placing two embryonal carcinoma cells in empty zonae pellucidae and determining the number of cells after 24 hr of culture in giant blastocysts in relationship to controls, which were cultured in media. The cells resumed normal growth upon rescue from blastocele fluid suggesting that contact with trophectoderm or inner cell mass is required for the irreversible loss of malignant attributes (Pierce, et al. 1987). It is not known if cell contact provides an inductive event or is only required for cell attachment and thereby for polarity of the embryonal carcinoma cells.

Because of the logistical problems involved with acquiring sufficient blastocele fluid, it was decided to see if other embryonic fields could regulate their closely related cancers and whether or not these fields might be more amenable to investigation.

Neuroblastoma. Neuroblastoma (C-1300) is regulated in terms of tumor formation in the neural crest migratory route (Podesta et al. 1984) and in the adrenal anlagen (Wells and Miotto, 1986), but this system was abandoned for further study because of its complexity. Suffice it to say that tumor formation was regulated, and in parallel with the experience with the embryonal carcinoma it appeared likely that regulation was by directing differentiation of the cells. It must be added, however, that this has not been established.

Melanoma. Tumor formation of B16 melanoma cells was not regulated when the cells were placed in the neural crest migratory route but it was regulated in the skin of the embryo on or about the day normal pigment cell precursors arrived in that skin (Gerschenson et al. 1986). Pigment cell precursors arrive in the skin of the back by the 10th day and in the limb bud on the 14th day of fetal life. To demonstrate this effect, it was necessary to develop techniques for operating upon 14 day old mouse embryos *in utero*. Briefly, incisions were made through purse-string sutures over the embryos. When the embryos were visible through the yolk sacs a dam of transparent kitchen plastic wrap was pressed over the embryo to prevent herniation through the incision. The requisite number of cells were then injected into the skin at the appropriate site. Melanoma cells injected into the skin of the back at 14 days grew at control levels, but those injected into the limb bud on the same day produced fewer tumors than anticipated. In addition, tumor cells placed in the skin of the back at the 10th day of fetal life did not produce tumors in expected numbers. Thus, it appeared that growth of melanoma cells was inhibited on or about the day normal pigment cell precursors arrived in the embyronic skin. On the basis of the experiments with embryonal carcinoma in the blastocyst, it was suspected that a diffusible factor might be involved in the regulation of melanoma cells in the embryonic skin. Accordingly, media were conditioned by growing equal amounts of skin of the back or limb bud in Eagles minimal essential medium plus 10% fetal calf serum and the resultant conditioned media were added to cultures of C-1300 melanoma cells. The melanoma cells were inhibited in their growth by the media from 14 day limb bud in relationship to that of the skin of the back (Gerschenson et al. 1986).

Leukemia. Gootwine, *et al*. and Webb *et al*. (Gootwine et al. 1982; Webb et al. 1984) injected many leukemia cells into the placentas of 10 day old mouse embryos. Although the majority of the embryos died, a small number survived and were interesting because they were chimeric in their leukopoietic tissues.

From the above studies, it is apparent that if four of four tumors injected into the appropriate embryonic sites failed to produce tumors in expected numbers. In two of the instances, the lack of growth of the malignant cells could be attributed to differentiation of the malignant cells in the embryonic environment. The differentiations resulted in normal cells in the sense that they responded to homeostatic regulation (Brinster 1974, Gootwine et al. 1982). In the other two situations, proof that the cells differentiated has not been obtained as yet (Podesta et al. 1984; Gerschenson, et al. 1986). However, it is postulated that if the closely corresponding environment can regulate the malignant phenotype in terms of tumor or colony formation in four of four cases, that there is an embryonic environment capable of regulating every malignant type of tumor (Pierce and Speers, 1987).

The mechanism of regulation is not known. In the embryonal carcinoma a soluble factor plus cell contact is required. In the case of melanoma a soluble factor is acquired in the absence of cell contact.

The idea that carcinomas are caricatures of the process of tissue renewal has predictive value. It suggests that factors made by tumors will be made by the corresponding normal cell lineage during its development. The evidence comes from studies of plant teratomas in which the tumors produced two growth stimulants that were produced by embryonic plant cells but which were repressed by differentiation and were not demonstrable in the adult plant. The gene loci for these factors are derepressed by carcinogenesis (Braun, 1956). In addition, transforming growth factor beta is produced by tumors (Todaro et al. 1980) and it is present in the embryo (Rizzino, 1985). Finally, platelet derived growth factor is made by embryonal carcinoma (Gudas et al. 1983; Rizzino and Bowen-Pope, 1985) and by the the blastocyst (Rizzino, personal communication). It is important to note that PDGF is believed to have a regulatory role for endothelium in the adult, but it is produced in an embryonic environment devoid of endothelium. Thus, it can be assumed that molecules produced in the embryo may subserve functions that will be completely different from those subserved in the adult. It will be interesting to determine the roles that these molecules play in embryogenesis.

Finally, in the experiments discussed today, it should be noted that we are employing cancer cells to probe developmental events in the hope of finding cures for cancer that will be nontoxic. Note the obverse of the experiments, we are employing tumor forming ability and colony forming ability of neoplastic cells as unique markers for probing the mechanisms of induction and expression of the neoplastic phenotype in the embryo. In the appropriate embryonic environment these tumor markers are lost. The use of such cells should give important information concerning embryonic induction.

ACKNOWLEDGEMENTS

This work was supported in part by a gift from RJR Nabisco, Inc. and National Institutes Grant CA-15823, CA-35367 and CA-36069

REFERENCES

Bloch, A. (1984). Induced cell differentiation in cancer therapy. Cancer Treatment Reports 68(1), 199-205.

Bradley, A., Evans, M., Kaufman, M., and Robertson, E. (1984). Formation of germ-line chimeras from embryo-derived teratocaricnoma cell lines. Nature 309, 255-257.

Braun, A.C. (1956) '2-growth-substance systems accompanying the inversion of normal to tumor cells in crown gall. Cancer Res 16, 53-56.

Brinster, R.L. (1974). Effect of cells transferred into the mouse blastocyst on subsequent development. J. Exp. Med. 140, 1049-1056.

Carr, B.I., Gilchrist, K.W. and Carbone, P.P. (1981). The variable transformation in mestastases from

testicular germ cell tumors: the need for selective biopsy. J. Urol 126, 52-54.
Cohnheim, J. (1889). Lectures on General Pathology Vol. II, p. 789 (translated by A.B. McKee) New Sydenham Society, London.
Evans, M.J. and Kaufman, M.H. (1981). Establishment in culture of pluripotential cells from mouse embryos. Nature 292, 154-156.
Frisney, R.I. (1985). Induction of differentiation in neoplastic cells. Anti Cancer Research 5, 111-130.
Gardner, R.L. (1972). An investigation of inner cell mass and trophoblast tissues following their isolation from the mouse blastocyst. J. Embryol. Exp. Morphol. 28, 279-312.
Gerschenson, M., Graves, K., Carson, S.D., Wells, R.S. and Pierce, G.B. (1986). Regulation of melanoma by the embryonic skin. Proc. Natl. Acad. Science 83, 7307-7310.
Gootwine, E., Webb, C.G., and Sachs, L. (1982) Participation of myeloid leukemia cells injected into embryos in hematopoietic differentiation in adult mice. Nature 299, 63-65.
Gudas, L.J., Singh, J.P. and Stiles, C.D. (1983). Secretion of growth regulatory molecules by teratocarcinoma stem cells. In Teratocarcinoma Stem Cells (ed. L.M. Silver, G.R. Martin, & S. Strickland), pp. 229-236, Cold Spring Harbor Laboratory.
Hanaoka, K., Kato, Y., and Noguchi, T. (1986) Comparative study on the ability of various teratocarcinomas to form chimeric mouse embryos. Dev. Growth Differ 28, 223-231.
Kleinsmith, L.J. and Pierce, G.B. (1964). Multipotentiality of single embryonal carcinoma cells. Cancer Res. 24, 1544-1551.
Martin, G.R. (1981). Isolation of a pluripotential cell line from early mouse embryos cultured in medium conditioned by teratocarcinoma stem cells. Proc. Natl. Acad. Sci. USA 78, 7634-7638.
Metcalf, D., (1985). The granulocyte-macrophage colony-stimulating factors. Science 229, 16-22.
Mintz, B., and Illmensee, K. (1975). Normal genetically mosaic mice produced from malignant teratocarcinoma cells. Proc. Natl. Acad. Sci USA 72, 3585-3589.
Papaioannou, V.E., McBurney, M.W., Gardner, R.L., and Evans, R.L. (1975). Fate of teratocarcinoma cells injected into early mouse embryos. Nature (London) 258, 70-73.
Pedersen, R.A. and Spindle, A.I. (1980). Role of the

blastocele microenvironment in early mouse differentiation. Nature 284, 550-552.

Pierce, G.B. (1961). Teratocarcinomas. A Problem in Developmental Biology. Fourth Canadian Cancer Conference (ed. R. Begg) pp. 119-137, Vol. 4, Academic Press, New York.

Pierce, G.B. (1967). Teratocarcinoma: Model for a devleopmental concept of cancer. In Current Topics in Developmental Biology (ed. A.A. Moscona and A. Monroy), pp. 223-246, Vol. 2, Academic Press, New York.

Pierce, G.B. (1974). Neoplasms, differentiations and mutations. Am. J. Path. 77, 103-118.

Pierce, G.B. (1983). The cancer cell and its control by the embryo. Am. J. Path. 113, 117-124.

Pierce, G.B., Aguilar, D., Hood, G., Wells, R.S. (1984). Trophectoderm in control of murine embryonal carcinoma. Cancer Res. 44, 2987-3996.

Pierce, G.B. and Dixon, F.J. (1959). Testicular teratomas. I. The demonstration of teratogenesis by metamorphosis of multipotential cells. Cancer 12, 573.

Pierce, G.B., Dixon, F.J., and Verney, E.L. (1960). Teratocarcinogenic and tissue forming potentials of the cell types comprising neoplastic embryoid bodies. Lab. Invest. 9, 583-602.

Pierce, G.B., and Johnson, L.D. (1971). Differentiation and cancer. In Vitro 7, 140-145.

Pierce, G.B., Lewellyn, A., Williams-Skip, C., and Parchment, R. in preparation, 1987.

Pierce, G.B., Pantazis, C.G., Caldwell, J.E. and Wells, R.S. (1982). Specificity of the control of tumor formation by the blastocyst. Cancer Res. 42, 1082-1087.

Pierce, G.B. Shikes, R.H., and Fink, L.M. (1978). Cancer: A Problem of Developmental Biology. Prentice-Hall, Inc., Englewood Cliffs, New Jersey.

Pierce, G.B. and Speers, W.C. (1987 in press). Tumors are caricatures of the process of tissue renewal: prospects for therapy by directing differentiation. Cancer Res.

Pierce, G.B., Stevens, L.C., and Nakane, P.K. (1967). Ultrastructural analysis of the early development of teratocarcinomas. J. Nat. Cancer Inst. 39, 755-773.

Pierce, G.B. and Wallace, C. (1971). Differentiation of malignant to benign cells. Cancer Res. 31, 127-134.

Podesta, A.H., Mullins, J., Pierce, G.B. and Wells, R.S. (1984). The neurula-stage mouse embryo in

control of neuroblastoma. Proc. Natl. Acad. Sci USA 81(23), 7608-7611.
Rizzino, A. (1985). Early mouse embryos produce and release factors with transforming growth factor activity. In Vitro, Cellular and Dev. Biol. 21, 531-536.
Rizzino, A., Bowen-Pope, D.F. (1985). Production of PDGF-like growth factors by embryonal carcinoma cells and binding of PDGF to their endoderm-like cells. Dev. Biol. 110, 15-22.
Sachs, L. (1986). Growth, differentiation and reversal of malignancy. Scientific American 254, 30-37.
Stevens, L.C. (1967). Origin of testicular teratomas from primordial germ cells in mice. J. Natl. Cancer Inst. 38, 549-552.
Todaro, G.J., Fryling, C. and DeLarco, J.E. (1980). Transforming growth factors produced by certain human tumor cells: polypeptides that interact with epidermal growth factor receptors. Proc. Natl. Acad. Sci USA 77, 5258-5262.
Webb, C.W., Gootwine, E., and Sachs, L. (1984). Developmental potential of myeloid leukemia cells injected rats midgestation embryos. Dev. Biol. 101, 221-224.
Wells, R.S. (1982). An in vitro assay for regulation of embryonal carcinoma by the blastocyst. Cancer Res. 42, 2736-2741.
Wells, R.S. and Miotto, K.A. (1986). Widespread inhibition of neuroblastoma cells in the 13- to 17-day old mouse embryo. Cancer Res. 46, 1659-1662.

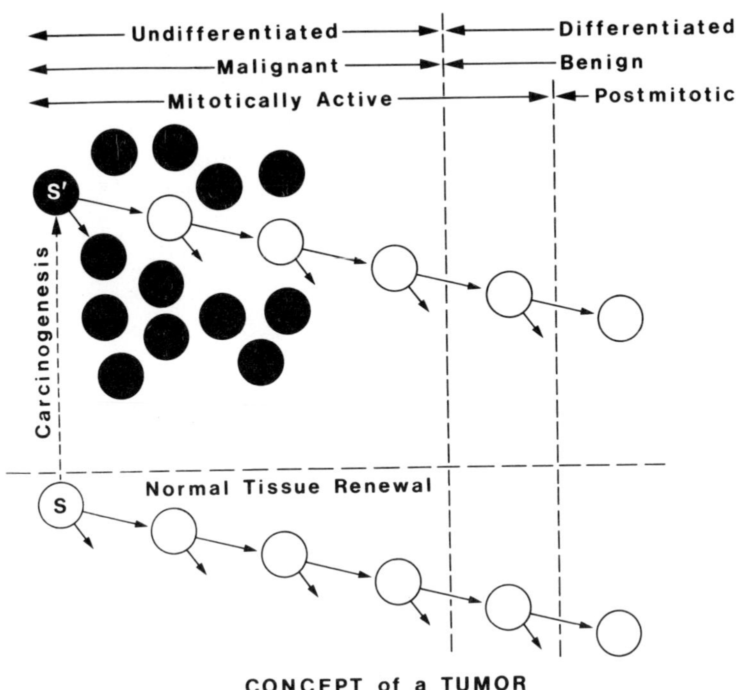

CONCEPT of a TUMOR

1

Normal tissue renewal with its stem cell S is illustrated in the lower cell lineage. The neoplastic lineage, originates from S by carcinogenesis, and the overproduction of undifferentiated malignant cells (black circles) in relationship to the ones that differentiate (white circles) is the caricature.

DYNAMICS OF MICROTUBULES AND F-ACTIN IN HIGHER PLANT ENDOSPERM MITOSIS ANALYZED WITH IMMUNO-GOLD AND VIDEO MICROSCOPY.

Bajer A. S[1]., M. Vantard[2], C. Schmit[2], C. Cypher[3], P. C. Hewitt[1], T. T. Huynh[1] and J. Molè-Bajer[1].

[1]Department of Biology, University of Oregon, Eugene, OR 97403, USA.

[2]Laboratoire de Biologie Cellulaire Vegetale, Universite L. Pasteur, 67083 Strasbourg-cedex, France.

[3]Hutchison Cancer Research Center, Seattle, WA 98104, USA.

ABSTRACT

Microtubules and F-actin were visualized in endosperm of the higher plant Haemanthus with immunogold or silver enhanced immunogold, and were analyzed by conventional and video enhanced microscopy in intact cells and cell fragments. Timed fixations demonstrated that the preparatory procedure perturbs instantly, but transiently, microtubule arrangements. In whole cells mitosis proceeds from very early prophase, and in fragments a characteristic spontaneous reorganization of microtubules is triggered. During the reorganization, converging microtubule centers appear to act as nucleating centers and are instrumental in the microtubule redistributions that occur. In large cell fragments, microtubule rearrangement invariably results in the formation of chromosome-free spindles and phragmoplasts with all the basic features of "non-kinetochore" mitotic transport which distributes spindle inclusions during normal mitosis (Bajer et al. 1987).

Preliminary results on the redistribution of F-actin during mitosis indicate that these microfilaments are also subject to non-kinetochore transport. During mitosis F-actin filaments form a cage around the spindle and phragmoplast. The distribution of F-actin depends on the presence of microtubules. The F-actin framework collapses when microtubules are disassembled by the drugs amiprophos methyl and oryzalin. Taxol, which stabilizes microtubules, appears to stretch the F-actin framework as it promotes microtubule assembly. Cytochalasin-D causes disruption of actin filaments without a noticeable effect on mitosis or cell plate formation. These results suggest that actin filaments are kept under tension by microtubules or microtubule related transport. Although our data point to microtubule/F-actin interaction, they argue against the direct involvement of F-actin in the mechanisms of chromosome segregation and cytokinesis in higher plants.

INTRODUCTION

Abbreviations: APM - O-methyl-O-(4-methyl-6-nitrophenyl)-N-isopropyl-phosphorothioamidate, DIC - Nomarski differential interference contrast system, IGS - immuno-gold stain, IGSS - immuno-gold silver enhanced, MAP - microtubule associated protein, MT(s) microtubule(s), VEM - video analog/digital enhanced microscopy.

Recent progress in our understanding of MT dynamics in vitro, and in our ability to visualize MTs in vivo and in model systems warrants the revision of several basic concepts of MT function during mitosis. Some of the new features of MT organization reported here, were detected using VEM of MTs and F-actin permanently labeled with IGS and IGSS in endosperm tissue of a higher plant, Haemanthus katherinae.

MT BEHAVIOR IN VITRO. Recent in vitro shearing and dilution experiments (Kristofferson et al. 1986, Mitchison and Kirschner 1984) demonstrated an unexpected redistribution of mean MT length after perturbation of the MT steady state [called dynamic instability (Mitchison and Kirschner 1984)]: some MTs depolymerized completely while others elongated excessively, as demonstrated in vitro by Horio and Hotani (1986). This behavior was interpreted as reflecting the presence or absence of GTP caps at MT ends (Mitchison and Kirschner 1984). When its GTP cap is lost, a MT depolymerizes completely. At present the problem of dynamic instability and GTP caps is controversial and it is assumed that the GTP loss is either permanent, and consequently MTs tend to depolymerize completely and are inherently labile (Kirschner and Mitchison 1986), or that after the GTP cap is lost it can be regained (Caplow et al. 1985) and consequently MTs are inherently stable (Farrell et al. 1987).

After any perturbation of MTs in vitro (Farrell et al. 1987) (e.g. shearing) dynamic instability occurs transiently; some MTs "catastrophically disassemble" completely, while others elongate at the same time. Dynamic instability, however, does transform into a new (conventional) steady state which favors the existence of a larger number of MTs with a shorter mean length. MAPs suppress dynamic instability and confer MT stability. In order to determine whether dynamic instability occurs in vivo, we have rexamined the changes in MT arrays and actin filaments that occur in plant endosperm cells as they are prepared for in vivo observation.

MT ORGANIZATION IN ENDOSPERM. Despite known biochemical differences between plant and animal tubulins (Dawson and Llyod 1985, Morejohn and Fosket 1986, Morejohn et al. 1987), the overall course of mitosis is similar in Haemanthus and animal cells. Haemanthus endosperm cells like all angiosperms, however, do not have a centriolar/centrosome complex and therefore offer an unique opportunity to study some intrinsic properties of plant MTs in vivo under conditions where the distal ends of kinetochore MTs are not embedded in

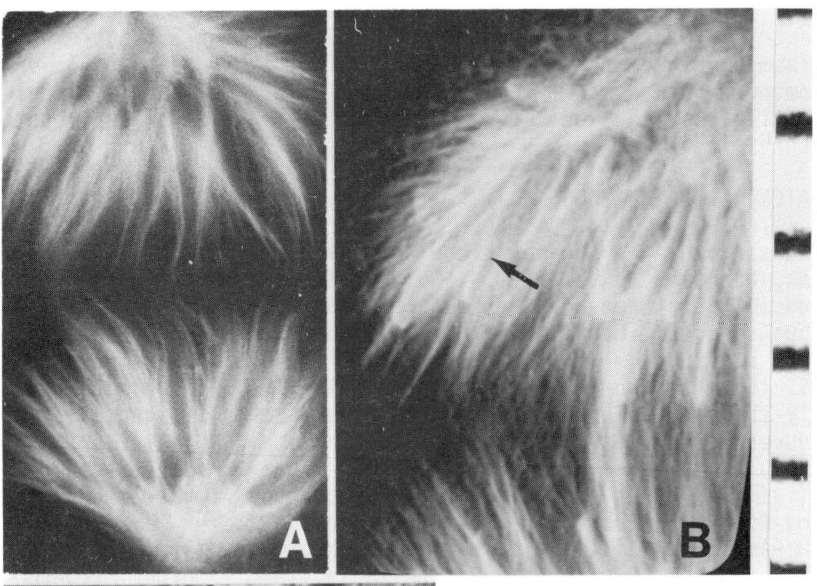

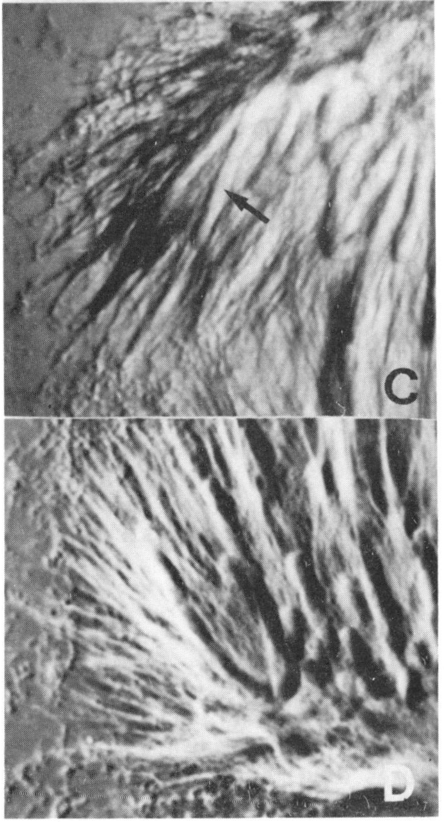

Figure 1. Haemanthus endosperm cells labeled with anti-tubulin. (A) A fluorescently labeled secondary antibody was used with conventional fluorescence microscopy. (B - D) The same cell processed with IGSS/VEM using different methods of image enhancement. Arrows (on B and C) mark the same chromosome arm. (B) The contrast of this brightfield image, taken with crossed polarizing prisms, was digitally increased and sharpened by frame averaging and background subtraction. C and D are of the same optical plane, but the contrast in D is inverted compared to C. In comparison to immunofluorescence, many more MTs are detected and with greater clarity in VEM. Note also that different features of this one cell are best revealed by different types of image enhancement. Scale for B - D; 10 μm intervals. A is printed at approximately 4/5 the magnification of G.

pericentriolar material. In addition, the formation of anuclear, cytoplasmic fragments during the preparation of endosperm, allows us to examine spontaneous MT reorganization, MT nucleation and MT related motility in the absence of two known MT organizers, chromosomes and centrioles.

MATERIAL AND METHODS

Endosperm of the African globe (blood) lily, Haemanthus katherinae Bak. was used as material. The methods of observations in vivo, IGS and IGSS for Haemanthus endosperm, and modification of the microscope are described elsewhere (Bajer and Molè-Bajer 1986 a and Bajer et al. 1986). Cells were stained with IGS (15 nm immuno-gold) or IGSS (5 nm immuno-gold/silver enhanced). The IGSS method is basically the same as used by Goode and Maugel (1987), except that all procedures were done at room temperature and an IGSS Kit (Janssen Pharmaceutica (Beerse, Belgium) was used. The video signals were selectively enhanced by digital image processing (Image-1, Universal Imaging Corp., Falmouth, MA). Video processed images were recorded with a Panasonic Optical Memory Disc Recorder (OMDR), model TQ-2025f, and with a 3/4 inch Sony time-lapse video recorder, TVO-9000. All micrographs are of images recorded on the OMDR and were taken from a nine inch Conrac video monitor. Immuno-fluorescence and cell lysis procedures were described previously (Vantard, 1984). Monoclonal anti-actin, which reacts with higher plant actin as demonstrated by nitrocellulose blots (Marc and Gunning 1986), was N.350 (Amersham, Arlington Heights, IL). Cytochalasin-D (Sigma, St. Louis, MO) was used at concentrations from 10 to 20 uM. Taxol was obtained from DHEW-NCI. Immuno-gold was obtained from Janssen Pharmaceutica (Beerse, Belgium).

RESULTS

SELF REORGANIZATION OF MTS. During preparation of endosperm for observations in vivo some cells break and anucleated, cytoplasmic fragments are formed (Bajer and Molè-Bajer 1986 a). Both in fragments and in nucleated cells, the MT meshwork perturbed during preparation self-reorganizes. Initially distorted mitotic spindles are "reconstructed" and mitosis proceeds (Fig. 1). The role of MTs in these reorganizations is seen more clearly, however, in cell fragments since they lack the complications of nuclear structures.

The fragments arise from shearing of the endosperm syncytium, which presumably also induces shearing of MTs within the fragments. Due to technical problems fragments can not be observed until 1 min after the perturbation; i. e., preparation. Small fragments observed at this time contain short MTs. These do not reorganize. In somewhat larger fragments, irregular MTs arrays are seen a few minutes after preparation (Fig. 2 A-B). In large fragments MTs invariably reorganize into regular structures of a few different types (Fig. 2 E). In both whole cells and in large cell fragments the initially irregular MT meshwork transforms in 1-3 minutes into

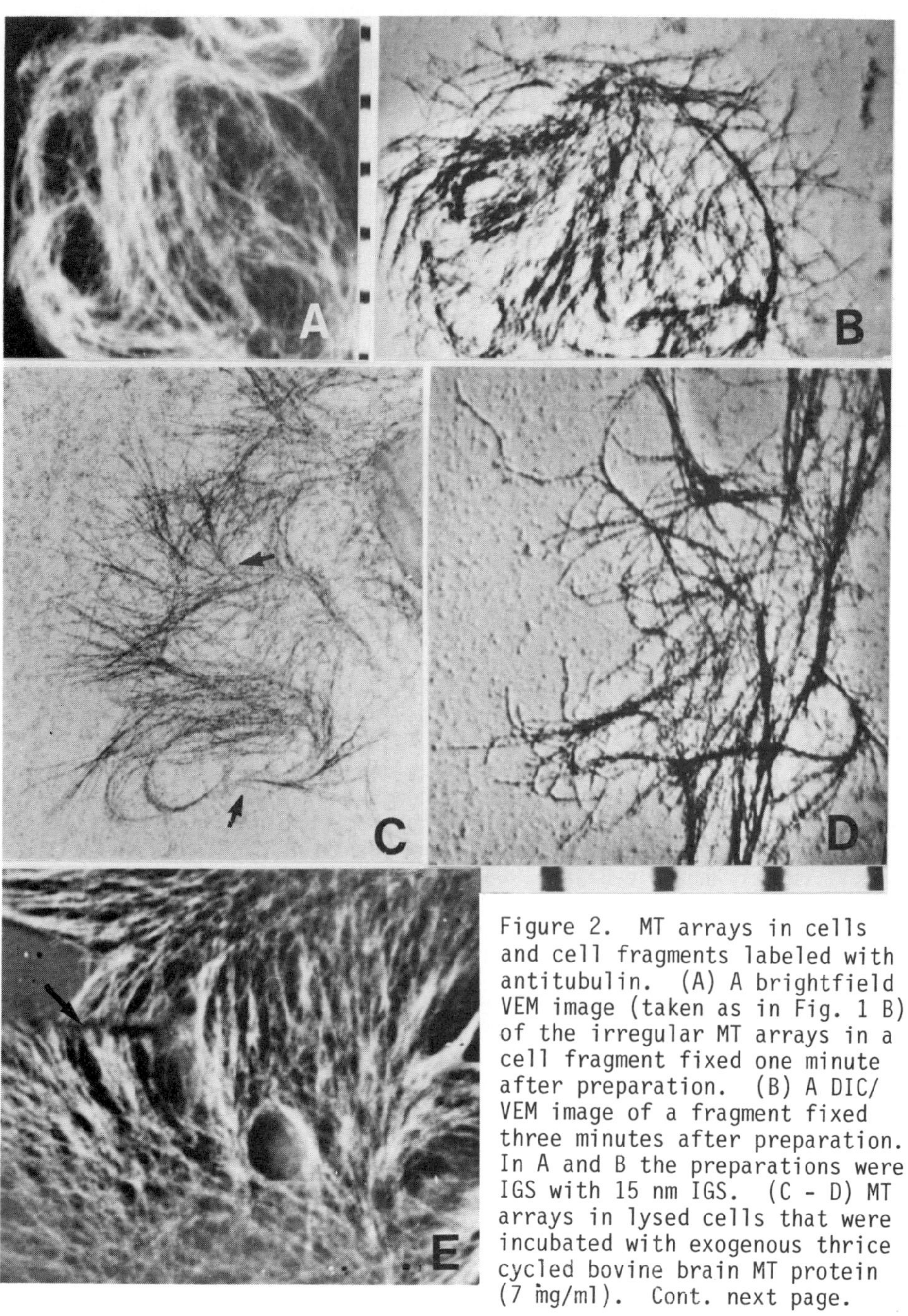

Figure 2. MT arrays in cells and cell fragments labeled with antitubulin. (A) A brightfield VEM image (taken as in Fig. 1 B) of the irregular MT arrays in a cell fragment fixed one minute after preparation. (B) A DIC/VEM image of a fragment fixed three minutes after preparation. In A and B the preparations were IGS with 15 nm IGS. (C - D) MT arrays in lysed cells that were incubated with exogenous thrice cycled bovine brain MT protein (7 mg/ml). Cont. next page.

Cont. from previous page. (C) A bright field IGSS image in conventional microscopy that has not been video enhanced. Compared with control cells not incubated with MT protein, many more MTs arrays are seen with skew MTs forming numerous converging centers (arrows). (D) A DIC/VEM image of an IGSS cell model. The thinnest fibrils are most likely single MTs. (E) A chromosome free spindle/- phragmoplast processed with 15 nm IGS. The contrast of this VEM image has been inverted. Note the thick intertwined arrays of MTs which gradually transform into the phragmoplast with the cell plate (arrow). Scale: 10 μm intervals. Scale for A is next to A. Scale for B, D and E is below D. C is printed at approx. 2/3 the magnification of B.

---

orderly arrays. In large fragments these MTs arrays invariably transform into symmetrical "spindle-like" structures (Fig. 2 E) and finally into multiple and symmetrical phragmoplasts with a cell plate. These chromosome-free spindles and phragmoplasts exhibit the non-kinetochore transport seen in mitotic cells (Bajer et al. 1987). Very large fragments do not form oversize spindles, but split into multiple spindle/phragmoplasts with the same, characteristic average MT length found in mitotic cells.

MT NUCLEATION. In centriolar cells, organizing centers such as centrioles with pericentriolar material affect the dynamics of MT ends (Brinkley 1985). It is not surprising therefore that MT turnover is expressed most clearly in the absence of such a strong organelle organizing/nucleating center. Typical MT nucleating centers in higher plants are MT converging centers (see review in Gunning and Hardham 1982). Such MT configurations have been reported in endosperm cells of *Haemanthus* (Bajer and Molè-Bajer 1982, Molè-Bajer and Bajer 1983), but VEM has provided more details concerning their organization and function. Although the spontaneous nucleation of single MTs also occurs in endosperm, organizing centers are invariably shaped as converging cones which transform into "MT fir trees" (Bajer and Molè-Bajer 1986 a). Following the incorporation of exogenous tubulin into permeabilized cell models with VEM (Fig. 2 C, D) has allowed us to understand several features of nucleation and its role in MT reorganization.

During spontaneous reorganization, MT converging centers appear to act as nucleating sites for MTs. After nucleation, the converging center transforms rapidly into a "MT fir tree" (Bajer and Molè-Bajer 1986 a). During mitosis in whole cells "MT fir trees" usually form at the distal end of the kinetochore fibers. However, the same basic structure is formed spontaneously in fragments, at the "poles" of developing spindle-like structures (Bajer and Molè-Bajer 1986 a). Thus it appears that the development of the "MT fir tree" is a characteristic feature of the self-reorganization of MT arrays that is not yet well understood.

The structural polarity of the MT converging centers and "MT fir trees" seen in cell fragments is unknown. Similar converging centers in animal cells have the plus ends of the MTs distal to the nucleating center (Euteneuer and McIntosh 1980). We assume, therefore, that the polarity of the MTs in these converging centers, is most likely the same and that the minus MT ends are at the top, poleward point, of the "fir trees". It is not known, however, how such polarity is established especially in transitional stages.

MT NUCLEATION AND ORGANIZATION OF THE KINETOCHORE FIBER COMPLEX. At present, the generally accepted conceptual image of spindle organization is based on the assumption that most of the MTs which are essential for spindle function, are parallel to the long axis of the kinetochore fiber. However, methods adequate to visualize and analyze MTs skewed to the spindle axis did not exist until recently. Consequently little is known about these skewed MTs in any spindle.

The demonstration of the birefringent kinetochore fibers in *Haemanthus* spindles (Inoué and Bajer 1961) had important theoretical implications (Inoué and Sato 1967). It is therefore equally important that abundant skewed MTs have been detected in the same material (Inoué et al. 1985). A kinetochore fiber is a complex composed of stable subfibers attached to the kinetochore, and a more loosely organized converging polar region resembling a "fir tree" with branches (Bajer and Molè-Bajer 1986a). The latter are disassembled easily by drugs and temperature, while the four stable subfibers persist even for 10-15 minutes at $0^{\circ}C$, or in $10^{-4}M$ free Ca^{2+} in lysed cells (Vantard and Bajer, 1987).

Treatment of whole cells or cell fragments with the drugs APM or oryzalin causes the disassembly of MTs (Bajer and Molè-Bajer 1986 b, Morejohn et al. 1987). Mitosis is arrested immediately by these agents even though kinetochore MTs are more resistant to the depolymerization than are other MTs. Phragmoplasts are also very resistant to depolymerization by these drugs. Chromosome movements are arrested most rapidly (as fast as their translocation can be monitored) with the drugs oryzalin and APM (Bajer and Mole-Bajer 1986 b, Molè-Bajer and Bajer 1985). The arrest of movement takes place when the branches of the "MT fir trees" start to disassemble. However, even these drugs do not disassemble the subfibers of the kinetochore fiber. This is most likely an universal feature applicable not only to *Haemanthus*.

MT RELATED MOTILITY: NON-KINETOCHORE TRANSPORT AND F-ACTIN DISTRIBUTION. The pattern of the distribution of F-actin in interphase and prophase cells differs negligibly until the formation of the mitotic spindle (Schmit et al. 1985). Thus, the reorganization of actin coincides with stages when MTs are organized into spindles and phragmoplasts. F-actin rearrangement mimics the behavior of other spindle inclusions such as starch grains, mitochondria, or chromo-some fragments. Some filaments caught at the equator in late anaphase, are incorporated into the cell plate ring (Fig. 3). The final distribution of F-actin during mitosis is consistent with the interpretation that its redistribution is due to bidirectional non-kinetochore transport.

APM and oryzalin also cause the collapse of the F-actin network coincident with MT depolymerization. When MTs are stabilized by taxol, however, the actin network seems to be stretched as the MTs elongate. Cytochalasin-D, which disrupts microfilaments, had no effect on either chromosome movement or cell plate formation (data not shown).

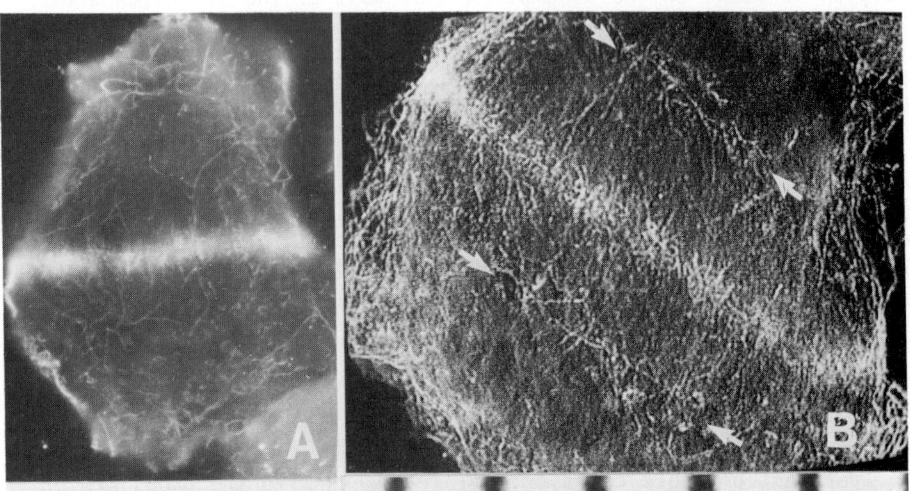

Figure 3. Telophase cells labeled with antiactin. (A) Immunofluorescence microscopy. Most of the actin filaments at the cell plate are visible as a diffuse band, although some long fibrils are seen throughout the cell. (B) An IGSS cell in the same stage as A whose brightfield VEM image (taken as in Fig. 1 B) has been sharpened by convolution. The actin filaments in the cell plate are clearly seen as numerous short fibrils. In addition, long fibrils have accumulated at the edges of the phragmoplast (between the arrows), especially between the phragmoplast and the forming nuclei, where very few MTs are present. Scale for B: 10 µm intervals; **A** is printed at approx. 2/3 the magnification of B.

DISCUSSION

The MT system in endosperm cell fragments has the remarkable ability to self-reorganize spontaneously into highly complex structures, spindles and phragmoplasts. In these anucleate, acentriolar cell fragments certain intrinsic properties of MTs and MT arrays are most clearly revealed. The complex MT behavior observed in these cytoplasmic fragments has not been reported in vitro although the structures may be reminiscent of those reported by Weisenberg et al. (1986). The unsuspected self-reorganizing properties of MTs may be superimposed on several processes in vivo. They may be a general feature of MTs in living cells and be utilized in various MT associated motilities.

MT DYNAMICS IN HAEMANTHUS DURING MITOSIS IN VIVO. The conceptual image of mitosis arising from a comparison of in vivo and in vitro systems is that spindle reorganization is an expression of intrinsic properties of MTs. In the cell these properties lead to states which may be equivalent to steady states of MT organization in vitro. Perturbations of the steady state can drive the system to a new steady state. The dynamic nature of MT assembly and disassembly, and particularly dynamic instability, may be important in vivo to allow progress from one steady state (stage) to the next during the course of mitosis.

The whole course of mitosis can then be discussed theoretically as three ordered steady states, and the relatively disordered transitions between them. These three transitions occur during: the breaking of the nuclear envelope, the onset of anaphase, and phragmoplast formation during cytokinesis. Each transition is triggered by a perturbation which starts an initial rapid elongation of some MTs concurrently with complete disassembly of others, comparable to shearing MTs in vitro (Farrell et al. 1987). Thus each transition leading to long MTs is always the result of a perturbation of a relatively stable long lasting stage, characterized by shorter MTs.

Breaking of the nuclear envelope - metaphase. During prophase numerous MTs are apparently attached to the nuclear envelope. The rapid breakage of the nuclear envelope (within 30 - 60 seconds in Haemanthus) triggers dynamic instability within the cell and some MTs elongate rapidly. This is coincident with the fast penetration of MTs into the nuclear area. The rate of spindle elongation (11 μm/min, measured in vivo in polarized light [Sato et al. unpubl.)] is nearly 10 times faster than the migration of chromosomes during anaphase. The prometaphase spindle is quite long (often 2 times longer than the metaphase spindle), and then it gradually shortens during metaphase until the onset of anaphase.

Onset of anaphase. The onset and progress of anaphase may trigger dynamic instability, comparable to shearing MTs in vitro. Thus MT assembly/disassembly and elongation during anaphase [massive formation of polar (astral) MTs and disassembly of the core of the kinetochore fiber] are an expression of dynamic instability.

Cytokinesis. The new phragmoplast forms at the anaphase-telophase transition. During this period the interzone is quite rapidly totally reorganized. The initial "young" phragmoplast is quite long. It shortens gradually and becomes very resistant to disassembly.

NON-KINETOCHORE TRANSPORT. The segregation of chromosomes is only one aspect of the motile phenomena occurring during mitosis.

Another universal, superimposed, but often overlooked, "non-kinetochore" mitotic transport [Jensen 1982 (referred to as "passive transport", Bajer and Molè-Bajer 1972)] distributes spindle inclusions randomly, but in a predictable manner and direction, depending on the stage of mitosis (Bajer and Molè-Bajer 1972, Bajer et al. 1987). Its molecular mechanism and its relation to chromosome segregation are unknown.

Non-kinetochore transport undergoes two separate 180^0 reversals in the direction of transport during the course of mitosis followed by another reversal during cytokinesis (phragmoplast formation) (Bajer et al. 1987). During prometaphase and metaphase all spindle inclusions are transported poleward. During mid to late anaphase, particles and small chromosome fragments reverse their poleward migration and move predominantly towards the cell's equator. This movement coincides with the formation and elongation of polar MTs. Later, during early telophase, inclusions at but not caught within the forming cell plate are transported poleward. Our data suggest that microfilaments behave like other spindle inclusions and are transported passively, trapped within the cell plate, and are not involved in the mechanism of cell plate formation, as suggested by the immunofluorescence studies on Allium (Clayton and Lloyd 1987). The data suggest the actin is not directly involved in mitosis.

The non-kinetochore transport seen within spindle-like MT arrays in cytoplasmic fragments is the same as that seen in mitotic spindles. During "metaphase and anaphase" of these "pseudo-mitoses", inclusions such as starch grains are distributed randomly and with the same velocity as chromosomes towards the two poles, where they accumulate and form "pseudo-nuclei." The transport appears before the cell plate is seen and continues for several hours after its appearance (Bajer et al. 1987).

CONCLUSIONS. In Haemanthus endosperm there are at least two cyto-skeletal filament systems (MTs and F-actin) associated with the spindle. The dynamic nature of these polymers, observed in vitro, is expressed in mitosis and interphase of endosperm cells in vivo. Both filamentous systems undergo dramatic changes during the course of mitosis. If MTs in the cell are subject to dynamic instability when they proceed from one steady state (stage) to another, they may be following two phase dynamics (Hill and Chen, 1984). However, F-actin appears to be in a steady state throughout mitosis.

There are also at least two transport phenomena associated with mitotic spindles. One of them distributes kinetochores/chromosomes and the other spindle inclusions. The disruption of MTs is reflected in pronounced changes in actin distribution, experimental perturbation of actin filaments does not effect MT distribution. This implies that during mitosis the location of actin filaments is related to MT rearrangements. However, the relationships between the filament systems present in spindles and the observed transport remains to be determined.

ACKNOWLEDGEMENTS. Immunogold and some antibodies were generous gifts from Dr. Jan De Mey, (Janssen Pharmaceutica, Beerse, Belgium). It is a pleasure to thank Dr. Tim Mitchison (Dept. Bioph./Biochem., Univ. of CA Med School, San Francisco, CA) for stimulating comments and constructive criticism. We would like also to thank Ms. Jeanne Thompson for editorial help. Supported by NSF (DCB 850-1264) and NSF-CNRS (INT-8211725).

REFERENCES

Bajer, A. S., and J. Molè-Bajer (1972) Spindle dynamics and chromosome movements. Int. Rev. Cytol. Suppl. 3: 1-271.

Bajer, A. S., and J. Molè-Bajer (1982) Asters, poles and transport properties within spindle-like microtubule arrays. In: Cold Spring Harbor Symp. 46, pp. 263-283.

Bajer, A. S., and J. Molè-Bajer (1986 a) Reorganization of Microtubules in Endosperm Cells and Cell Fragments of the Higher Plant Haemanthus in vivo. J. Cell Biol. 102: 263-281.

Bajer, A. S., and J. Molè-Bajer (1986 b) Drugs with colchicine-like effects that specifically disassemble plant but not animal microtubules. Ann. N.Y. Acad. Sci. 466: 767-784.

Bajer, A. S., J. Molè-Bajer, and H. Sato (1986) Video microscopy of colloidal gold particles in improved rectified DIC and epi-illumination. Cell Struct. Funct. 11: 317-330

Bajer, A. S., M. Vantard, and J. Molè-Bajer (1987) Multiple mitotic transports expressed by chromosome and particle movement. Progress in Zool. 34: (In press).

Brinkley, B. R. (1985) Microtubule organizing centers. Ann. Rev. Cell Biol. 1: 145-72.

Caplow, M., J. Shanks, and B. P. Brylawski (1985) Concerning the anomalous kinetic behavior of microtubules. J. Biol. Chem. 260: 12675-12679.

Clayton, L., and C. W. Lloyd (1985) Actin organization during the cell cycle in meristematic plant cells. Exp. Cell Res. 156: 231-238.

Dawson, P. J., and C. W. Lloyd (1985) Identification of multiple tubulins in taxol microtubules purified from carrot suspension cells. EMBO J. 4: 2451-2455.

Euteneuer, U., and J. R. McIntosh (1980) Polarity of midbody and phragmoplast microtubules. J. Cell Biol. 87: 509-515.

Farrell, K. W., Jordan, M. A. Miller, H. P. and L. Wilson (1987) Phase dynamics at MT ends: the coexistence of dynamic instability and treadmilling. J. Cell Biol. 104: 1035-1046.

Goode, D., and T. K. Maugel (1987) Backscattered electron imaging of immunogold-labeled and silver-enhanced microtubules in cultured mammalian cells J. EM Techn. 5:263-273.

Gunning, B. E. S., and A. R. Hardham (1982) Microtubules. Ann. Rev. Plant Physiol. 33: 651-698.

Hill, T. L., and Y.-D. Chen. 1984. Phase changes at the end of a microtubule with a GTP cap. Proc. Natl. Acad. Sci. U. S. A. 81: 5772-5776.

Horio, T., and H. Hotani (1986) Visualization of the dynamic instability of individual microtubules by dark field microscopy. Nature 321: 605-607.

Inoué, S., and A. Bajer (1961) Birefringence of endosperm. Chromosoma. 12: 43-63.

Inoué, S., and H. Sato (1967) Cell motility by labile association of molecules. J. Gen. Physiol. 50: 259-292.

Inoué, S., J. Molè-Bajer, and A. S. Bajer (1985) Three-dimensional distribution of microtubules in Haemanthus endosperm cells. In: Microtubules and microtubule inhibitors. Eds. M. De Brabander and J. De Mey. Elsevier Sci. Publ. Amsterdam. pp. 269-276.

Jensen, C. G. (1982) Dynamics of spindle microtubule organization: Kinetochore fiber microtubules of plant endosperm. J. Cell Biol. 92: 540-559.

Kristofferson, D., T. Mitchison and M. Kirschner (1986) Direct Observation of steady-state microtubule dynamics. J. Cell Biol. 102: 1007-1019.

Marc, J., and B. E. S. Gunning (1986) Immunofluorescent localization of cytoskeletal tubulin and actin during spermatogenesis in Pteridium aquilinum (L.) Kuhn. Protoplasma 134: 163-177.

Mitchison, T., and M. Kirschner (1984) Dynamic instability of microtubule growth. Nature 312: 237-242.

Molè-Bajer, J., and A. Bajer (1983) The action of taxol on mitosis. Modification of microtubule arrangements of the mitotic spindle. J. Cell Biol. 96: 527-540.

Molè-Bajer J., and A. S. Bajer (1985) Organization of the mitotic spindle in endosperm of a higher plant, Haemanthus, demonstrated by experimental disassembly of microtubules. Cell Motility: mechanism and regulation. Eds. H. Ishikawa, S. Hatano and H. Sato. pp. 435-448, Univ. Tokyo Press, Tokyo, 1985.

Morejohn, L. C., and D. E. Fosket (1986) Tubulins from plants, fungi, and protists. In: Cell and molecular biology of the cytoskeleton. Ed. J. W. Shay. Plenun Press. N. Y. pp. 257-329.

Morejohn, L. C., T. E. Bureau, J. Molè-Bajer, A. S. Bajer and D. E. Fosket (1987) Oryzalin, a dinitroaniline herbicide, binds to plant tubulin and inhibits microtubulepolymerization in vitro. Planta (in press).

Schmit, A-C., M. Vantard and A-M. Lambert (1985) Microtubule and F-actin rearrangement during the initiation of mitosis in acentriolar higher plant cells. In: Cell Motility: Mechanism and regulation (Eds. H. Ishikawa, S. Hatano and H. Sato), pp. 415-433, Univ. Tokyo Press, Tokyo.

Vantard, M. (1984) Role du calcium et de la calmoduline dans la dynamique de la tubuline chez les plantes: etude in vivo et sur modeles cellulaire. Ph. D. These, L'Universite Louis Pasteur ed Strasbourg. France.

Vantard, M., and A. S. Bajer (1987) New features of kinetochore organization revealed by selective disassembly of microtubules by free Ca^{2+} (submitted).

Weisenberg, R. C., R. D. Allen and S. Inoue (1986) ATP-dependent formation and motility of aster-like structures with isolated calf-brain microtubule protein. Proc. Natl. Acad. Sci. 83: 1728-1732.

MICROTUBULE ASSEMBLY AND REGULATION

BIOCHEMICAL ASPECTS OF THE REGULATION OF MICROTUBULE ASSEMBLY.

Ricardo B. Maccioni, Coralia Rivas & Juan C. Vera
University of Colorado Health Sciences Center, Denver CO 80262, U.S.A.

ABSTRACT

The molecular basis underlying the regulatory role of the carboxyl-terminal domain of tubulin have been studied using controlled proteolysis, interaction with microtubule associated proteins and structure-activity analyses.

Controlled proteolysis of tubulin with subtilisin results in the cleavage of both α and βsubunits yielding S-tubulin and two 4kDa fragments containing the carboxyl-terminal domain of tubulin subunits. S-tubulin exhibits an increased propensity to self-associate as compared with tubulin. Treatment with S. aureus V-8 protease resulted in removal of a 4kDa C-terminal fragment (V-tubulin) with a concomitant stimulation of the assembly. The propagation constant for V-8 treated tubulin,where only β -subunit is cleaved, was one-third of that for S-tubulin assembly. Polymers of S-tubulin do not incorporate MAPs, but we observed a partial incorporation of MAPs into V-tubulin polymers relative to microtubule controls. On the other hand cleavage with trypsin, which removes a 16 kDa C-terminal fragment from α -subunit or limited digestion of tubulin with carboxypeptidase Y, which removes only 6-8 aminoacid residues from both subunits, have no regulatory effect while removal of the 4 kDa fragment does. In both cases, treatment with the endo and exopeptidases produces inhibition of the assembly associated with significant changes in tubulin conformation.

The substructure of the tubulin's regulatory domain was further examined by binding of ^{3}H-acetylated peptides from the variable region of the C-terminal domain, α (430-441) and β (422-433) to MAPs. The binding data showed a preferential interaction of β -peptide with MAP-2 and tau as analyzed by Airfuge ultracentrifugation and zone filtration chromatography. The α (430-441) peptide interacted with tau with a higher affinity than MAP-2. These studies support our hypothesis that the C-terminal domain hinders interactions responsible for tubulin assembly and that cleavage at or near the 4kDa end is critical to relieve this hindering effect.

INTRODUCTION

Microtubules, one of the fundamental components of the cytoskeleton are involved in cellular shape, mitosis, as frames for intracellular movements of granules, cell differentiation and modulation of surface receptors (Soifer, 1986; Maccioni, 1986). This functional versatility in addition to their cellular dynamics point to the existence of regulatory signals in microtubule assembly and orientation. Recent studies from our laboratory as well as others (Serrano et al, 1984 a and b; Bhatacharyya et al, 1985; Maccioni et al, 1984 and 1985a) have demonstrated that the carboxyl-terminal moiety of tubulin plays a major role in regulating its assembly into microtubules as well as in the colchicine interaction with tubulin (Avila et al, 1987). One-site subtilisin cleavage produces S-tubulin with a concomitant enhancement of the assembly forming microtubule bundles and polymers containing hooks. The uncontrolled S-tubulin assembly, in the absence of MAPs indicates that the 4kDa segment normally modulates tubulin self-association. Thus, interaction of MAPs with the acidic C-terminal domain appears to promote tubulin assembly. The 4kDa C-terminal moiety which is located in the outer surface of the microtubule lattice (Breitlung and Little, 1986; Serrano et al, 1986b) is directly involved in the interaction of microtubule associated proteins, MAP-1, MAP-2 and tau and the binding of Ca^{2+} ions to tubulin (Serrano et al, 1984b; Serrano et al, 1986b).

Soon after the findings of the regulatory domain of tubulin (Serrano et al, 1984a; Maccioni et al, 1985) a growing interest in tubulin gene families and isotypes became evident and a number of molecular biological studies on this domain have been published (Lewis et al, 1985; Little, 1985; Cleveland, 1987). This studies at a genetic and molecular level along with recent research work with tubulin chimeras incorporated in microtubules (Bond et al, 1986) provide additional support to the regulatory significance of the C-terminal domain.

In this report we have focused our attention on the structure-function relationships of the C-terminal moiety of tubulin and the substructure of the regulatory domain. We have addressed this problem through a comparative study of controlled proteolysis of tubulin with C-terminal directed proteases and analyzed the interaction of synthetic peptides from the C-terminal sequence of both tubulin subunits with MAP-2 and tau.

STRUCTURAL AND FUNCTIONAL DOMAINS OF TUBULIN

Research on tubulin sub-domains displaying specific functions has yielded valuable information on the localization of the major structural sites of tubulin. Structural and protein-chemical studies of microtubules have given us a view of tubulin as a macromolecule containing spatially discrete sequences that constitute different domains implicated in the major tubulin functions, i.e, self-association into microtubules, binding of colchicine, interaction with MAPs and with other modulatory ligands. The available information has been obtained essentially from affinity labeling with reactive probes directed to sites for specific tubulin ligands (Maccioni and Seeds, 1983b; Nath et al, 1985), limited proteolysis combined with binding studies (Maccioni and Seeds, 1983a; Brown and Erickson, 1983; Serrano et al, 1984 a, b and c) and immunochemical analysis of tubulin and peptide fragments antigens (Mandelkow et al, 1985). High resolution X-ray diffraction should be a powerful tool to obtain further information on the tertiary structure of native tubulin, a needed complement for the protein chemistry studies. However, difficulties in obtaining a crystallized tubulin appropriate for these experiments have preclude development of these studies.

Domain mapping studies and research on structure-activity relationships led us to postulate the existence of two major regions in tubulin with functional significance: the carboxyl-terminal regulatory domain and a larger region containing the sites for the interaction of tubulin subunits with each other and with specific ligands involved in tubulin assembly (Maccioni et al, 1984). Calcium ions and MAPs bind to the regulatory domain while the E and N GTP sites and the colchicine site have been found topographically located in the large domain. Figure 1 summarizes the information available on tubulin domains. The C-terminal regulatory domain is shown in both subunits.

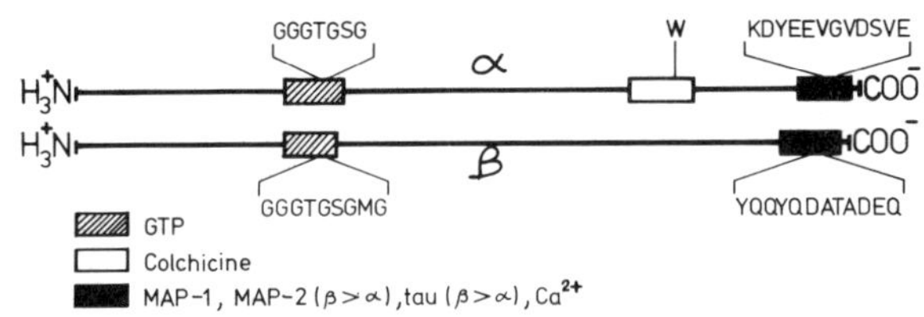

FIGURE 1.. Schematic diagram of α and β tubulin containing the domains for the interaction of GTP (Maccioni and Seeds, 1983b), colchicine (Avila et al, 1987), high affinity Ca^{2+} (Serrano et al, 1986) and microtubule associated proteins (Serrano et al, 1984b) on the basis of the porcine brain tubulin sequence (Ponstingl et al, 1981). The glycine cluster in the GTP site, tryptophan 346 in the colchicine site and the variable sequence within the MAPs site are shown.

ANALYSIS BY COMPARATIVE LIMITED PROTEOLYSIS

Limited proteolysis has proven to be a useful approach for studies on microtubular proteins and their assembly. Early studies on the chymotryptic cleavage of microtubule bound MAP-2 were essential to identify the binding domain interacting with tubulin and the larger projection segment (Valle and Borisy, 1977). Studies on controlled proteolysis of tubulin clearly have provided the main framework for the detailed analysis of tubulin sub-domains (Maccioni, 1986). We have examined the domain structure of the C-terminal region of tubulin using limited digestion with proteases cleaving tubulin at one single site within this region, under limited cleavage conditions (see Table 1). The changes in tubulin structure, assembly and shape of polymers resulting from cleaved tubulin have been analyzed. Proteolysis with subtilisin, cleaving tubulin at α -Glu417-Phe418 and β

Glu407-Phe408 produces an enhancement of tubulin self-association in agreement with previous reports (Serrano et al, 1984a; Maccioni et al, 1985a).

Subtilisin digestion products, S-tubulin and the 4kDa fragment containing the regulatory domain have been purified to homogeneity and characterized (Maccioni et al, 1985a). Cleavage with V-8 _S. aureus_ protease also resulted in removal of a 4kDa terminal fragment (V-tubulin) with a concomitant stimulation of the assembly. The propagation constant for the assembly of S-tubulin, lacking the 4kDa fragment in both subunits was three-fold higher than that of V-tubulin from which β -subunit was cleaved, suggesting that α -subunit may also contribute to the regulatory effect. No incorporation of total MAPs and purified MAP-2 was observed in S-tubulin polymers . A partial incorporation, was observed in V-tubulin polymers as compared with undigested microtubule controls.

Limited action of proteases cleaving tubulin subunits proximally and distally with respect to the subtilisin cleavage site were examined. Carboxypeptidase digestion for 120 min. resulted in removal of 6-8 amino-acid residues from the C-terminal moiety of both α and β subunits as revealed by amino-acid analysis of supernatants obtained after sulfosalycilic acid precipitation of carboxypeptidase Y treated tubulin. The removal of the acidic moiety around the C-terminus of α and β tubulin by carboxypeptidase Y produced a biphasic decay of tubulin assembly. MAPs incorporated into C-tubulin polymers suggesting that the last few C-terminal residues are not directly involved in the selective interaction of MAPs with tubulin. Limited cleavage with trypsin, which removes a 16 kDa C-terminal peptide, resulted in a marked decrease of tubulin self-assembly with pseudo-first-order inactivation kinetics and the assembled polymers also contained MAPs.

TABLE 1

CONTROLED DIGESTION OF TUBULIN BY C-TERMINAL PROTEASES

ENZIME	CLEAVAGE PRODUCTS	CLEAVAGE SITE(S)	EFFECT ON ASSEMBLY	MAP DEPENDENCE	MORPHOLOGY OF POLYMERS
Subtilisin	48 and 4kDa (α and β subunits)	Glu^{417}-Phe^{418} in α; Glu^{407}-Phe^{408} in β subunit	Increase Cr=0.3 mg/ml	No	MT bundles + hooked polymers (constant polarity)
S.aureus V8	48 and 4kDa (β-subunit)		Increase Cr=1 mg/ml	Partial	Open tubules, MT hooks (bidirectional)
Trypsin	36 and 16kDa (α-subunit)	Lys^{338}-Arg^{339} in α-tubulin	Decrease	Yes	Microtubules (discont. in MT array)
Carboxypept.Y	51 and 1 kDa	Gradual removal last 5-8 residues	Decrease	Yes	MT like structures (irregularities, frayed ends, rings)

The structural characteristics of cleavage products from either the trypsin or carboxypeptidase Y treatment of tubulin were examined by fluorescence spectroscopy and circular dichroism. Emission spectra showed that a fluorescence quenching accompanied tubulin proteolysis with 1% w/w trypsin, which increased as digestion progressed. Carboxypeptidase Y also promoted a quenching effect. Interestingly, an increase in the maximum of emission spectra at 340 nm was followed by a gradual decrease over a period of 120 minutes. The quenching effect may result from tryptophan residues previously exposed in native tubulin becoming buried as a consequence of a conformational alteration of tubulin during proteolysis. In addition the circular dichroism spectra showed that both tryptic cleavage as well as carboxypeptidase treatment of tubulin resulted in a significant decrease of its α-helical conformation. The effects of tryptic proteolysis could be explained as the production, resulting from the internal cleavage of α-tubulin at peptide bond Lys338-Arg339, of two fragments of 36 and 16 kDa (Avila et al, 1987) the latter containing the entire C-terminal sequence which can reassociate into the assembled polymer. Thus, it is likely that assembly is inhibited as a result of a significant conformational change. Removal of a few amino-acids from the acidic C-terminal moiety of α and β -tubulin also results in detectable changes in tubulin conformation which appear to be responsible for the alteration of its assembly capacity. These observations support our earlier hypothesis that the 4kDa carboxyl-terminal regulatory domain of tubulin normally hinders the interactions responsible for tubulin self-association and that proteolytic removal of the 4kDa domain or binding of MAPs is critical in order to overcome the normal barrier to self-assembly.

Polymers assembled from tubulin cleaved with the different proteases have different shapes as analyzed by electron microscopy. Subtilsin digested tubulin, in agreement with previous studies (Serrano et al, 1984a) assembles into abundant bundles of microtubules and polymers hooks. V-tubulin polymers consisted mostly of hooks growing bidirectionally with respect to a central tubule, in contrast to S-tubulin polymers which are characterized by a constant polarity in the growth of tubules and hooks in the bundle structure (Figure 2).

STRUCTURAL SUBDOMAINS OF THE C-TERMINAL REGION OF TUBULIN: Interaction with MAPs.

The interaction of MAPs with the C-terminal tubulin domain was probed using synthetic peptides corresponding to subdomain sequences of the regulatory region. Two different peptides from the variable sequence of and 4kDa C-terminal moiety were analyzed for their binding to MAP-2 and tau.

CT-1 (α -430-441): Lys[430]-Asp-Tyr-Glu-Glu-Val-Gly-Val-Asp-Ser-Val-Gly[441]

CT-2 (β -422-433): Tyr[422]-Gln-Gln-Tyr-Gln-Asp-Ala-Thr-Ala-Asp-Glu-Gln[433]

The conformation of both peptides and their acetylated products was studied to examine the structural-functional aspects of these regulatory fragments. Circular dichroism studies of both synthetic peptides indicate an unordered structure in water as solvent. Different CD spectra were recorded at various pH values in the range 3.7-9.2 without any evidence of helix formation. However, the CD analysis in several alcohols showed that the peptide adopted a helical conformation, similar observations as those on the conformation of the 4kDa fragment (Maccioni et al, 1986). Proton nuclear magnetic resonance studies are in agreement with the CD results for the α -430-441 peptide (Sigiura et al, 1987).

In order to examine the binding behavior of both CT-1 and CT-2 peptides we have acetylated the C-terminal peptides with ^{3}H-acetic anhydride and the interaction with MAP-2 and tau analyzed. Zonal filtration chromatography of mixtures of α-430-441 with MAP-2 and tau indicate a relatively weak interaction of the α-peptide with both MAPs. The α-peptide interacted with tau factor with an apparently higher affinity than MAP-2. Filtration experiments using Sephacryl S-300 to analyze the ineteraction of CT-1 with MAP-2 showed that the labeled peptide eluted in a major peak of free peptide while a minor radioactivity peak coeluted as undissociated complex with MAP-2. Further details of the interaction of the peptide with MAP-2 were analyzed by ultracentrifugation in the Airfuge, corroborating the weak interaction found in the filtration experiment (Ka= 1×10^{4} M^{-1}). The relatively low association constant and the observation of a small peak of peptide-MAP-2 complex in the filtration experiment suggest a kinetically controlled process with a slow dissociation rate constant for the complex between MAP-2 and the CT-1 peptide. The interaction of the synthetic peptide CT-2 (β -422-433) was analyzed using the Airfuge procedure (Howlett et al, 1978 ; Maccioni et al, 1986b) and a significantly stronger interaction (Ka= 4×10^{5} M^{-1}) as compared with the α -peptide was observed (Figure 3A). Finally the interaction of CT-2 with tau was analyzed by filtration chromatography on precalibrated Sephacryl S-200 columns of mixtures of the β -peptide with tau factor. In this experiment CT-2 peptide and tau were preincubated (as in the experiments with the α peptide) at 30°C for 16 min. and the samples applied to the Sephacryl column (Fig. 3B). Radioactivity coeluted in a large peak corresponding to the tau- β peptide complex partially retained during the filtration chromatography. No radioactivity peak was observed in the void volume indicating the absence of aggregation.

CONCLUDING REMARKS

Controlled proteolysis combined with analyses of the assembly and shape of the resulting polymers has been used as an integrated approach to define the substructure of the C-terminal regulatory domain of tubulin. removal of the 4kDa fragment by subtilisin proteolysis facilitates tubulin self-assembly and make it MAP independent. Digestion with the V-8 protease provide additional evidence to define the critical size of the modulatory domain within the 4kDa C-terminal moiety. Controlled proteolysis resulting in removal of long C-terminal peptides, as in the

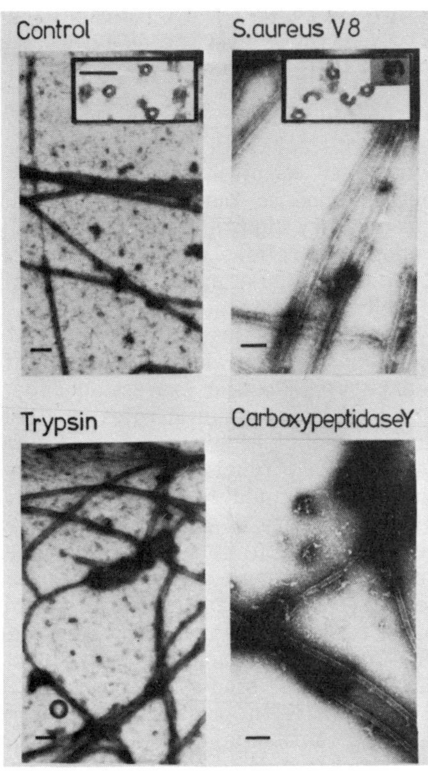

Figure 2. Electron microscopy of polymers obtained by assembly of tubulin cleaved by different C-terminal proteases.

After polymerization, samples were fixed in warm Mes buffer containing 2% glutaraldehyde, stained with 1% uranyl acetate and observed in a Philips 300 Electron microscope. Polymers from tubulin cleaved with S. aureus protease (B), trypsin (C), carboxypeptidase Y (D). Microtubules from undigested tubulin control are shown in (A). The inserts in A and B represent cross sections of the stained polymers. Bars correspond to 0.1 μ.

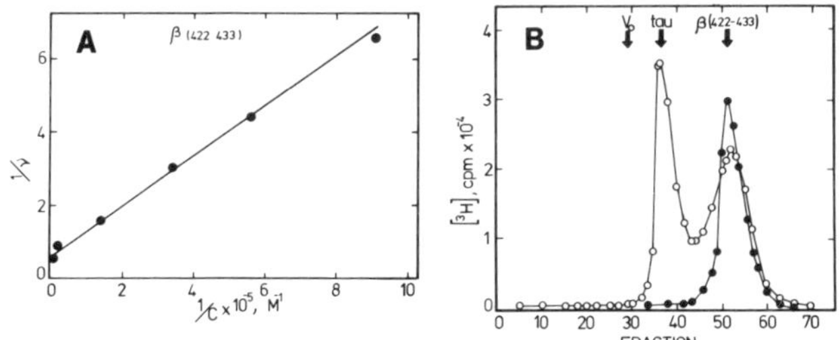

FIGURE 3. A: Binding of CT-2(β -422-433) peptide to MAP-2. Double reciprocal plot of the ratio of moles of peptide bound per moles of MAP-2 against the equilibrium concentration of the β -peptide, C. Data were obtained from the Airfuge ultracentrifugation binding experiment. B: Binding of the CT-2 (β-422-433) peptide to tau protein. The labeled CT-2 peptide was mixed with tau protein in a final volume of 0.45 ml, preincubated for 16 min. at 30 and the mixture eluted in a Sephacryl column (O). Control experiment: 0.1 mg of the ^{3}H-acetylated peptide was chromatographed on a Sephacryl S-200 column (0.8 x 45 cm) equilibrated in buffer 0.05 M Mes, pH 6.8 containing 1 mM MgCl2 (●). radioactivity of 0.5 ml fractions was measured.

tryptic cleavage of α-tubulin, or removal of a very short fragment from the C-terminal moiety have no regulatory effect. The substructure of this region can thus be dissected with regard to function.

Peptides corresponding to the variable sequence within the C-terminal domain, α-430-441 and β-422-433 interact with MAP-2 and tau. The interaction with the β-peptide has stronger affinity for both MAPs. These observations along with our previous findings on the interactions of MAPs (Maccioni et al, 1984) point to the regulatory significance of the binding of MAPs to tubulin. This interaction appear to relieve the hindering effect of the C-terminal region thus promoting microtubule assembly and stabilizing the microtubular structure. The selective interaction could be finely regulated at the molecular level by factors including a differential affinity of the various MAPs with tubulin subunits and possibly with different tubulin isoforms within a determined tissue. Tubulin expression appears to be tissue specific, and certain isoforms may play specific roles in a selected cell type. In addition, tissue specificity of MAPs suggest that different MAPs could be involved in modulating MAPs function depending on their cell distribution. Another regulatory factor could reside in the nature of MAPs phosphorylation controlled by the activity of phosphorylating enzyme systems, i.e. Ca^{2+}/calmodulin dependent protein kinase, cAMP dependent protein kinase and by dephosphorylation by specific phosphatases. Phosphorylation of MAPs decreases its capacity to promote microtubule assembly (Jameson et al, 1980). On the basis of direct binding measurements we have recently found that MAP-2 phsphorylation partially blocks its interaction with the 4kDa peptide. Additional studies in progress on the interaction of discrete C-terminal peptides with purified MAPs combined with immunocytochemical studies using antibodies to these C-terminal peptides, to analyze the control of microtubule organization in cultured cells, may contribute new clues toward the elucidation of the molecular and cellular basis of the regulation of microtubule assembly.

ACKNOWLEDGEMENTS

We gratefully acknowledge the support of the Council for Tobacco Research, U.S.A.(Grant 1913) and the Milheim Foundation (Grant 86-17) to this research program. The critical reading of the manuscript by Dr. E. Balbinder is also acknowledged.

REFERENCES

Avila J, Serrano L, and Maccioni RB (1987) Regulatory aspects of the colchicine interactions with tubulin. Molecular and Cell Biochemistry 73: 29-36.

Bhattacharyya B, Sacket DL, and Wolf J (1985) Tubulin hybrid dimers and tubulin-S. J Biol Chem 260:10208-10216.

Bond J, Fridovich-Kail JL, Pillus L, Mulligan RC, and Solomon F

(1986) A chicken-yeast chimeric β -tubulin protein is incorporated into mouse microtubules *in vivo*. Cell 144:461-468.

Breitlung F and Little M (1986) Carboxy-terminal regions on the surface of tubulin and microtubules. J Mol Biol 189:367-370.

Brown HR and Erickson HD (1983) Assembly of proteolytically cleaved tubulin. Arch Biochem Biophys 220:46-51.

Cleveland D (1987) The multitubulin hypothesis revisited : What we have learned ?. J Cell Biol 104:381-383.

Howlett GJ, Yeh E and Schachman HK (1978) Protein-ligand binding studies with a table-top, air-driven, high-speed centrifuge. Arch Biochem Biophys 190:809-819.

Jameson L, Frey T, Zeeberg B, Dalldorf F and Caplow M (1980) Inhibition of microtubule assembly by phosphorylation of microtubule associated proteins. Biochemistry 19:2472-2479.

Little M (1985) An evaluation of tubulin as a molecular clock. Biosystems 18:241-247.

Lewis SA, Lee MGS and Cowan NJ (1985) Five mice isotypes amd their regulated expression during development. J Cell Biol 101:852-861.

Maccioni RB (1986) Molecular Cytology of Microtubules. Basque Country University Press, Bilbao. 136 pp.

Maccioni RB and Seeds NW (1983a) Limited proteolysis of tubulin and microtubule assembly. Biochemistry 22:1567-1572.

Maccioni RB and Seeds NW (1983b) Affinity labeling of tubulin's GTP exchangeable site. Biochemistry 22:1572-1579.

Maccioni RB, Serrano L and Avila J (1984) Structural and functional domains of tubulin. Bio Essays 2:165-169.

Maccioni RB, Serrano L, Avila J and Cann JR (1985a) Characterization and structural aspects of the enhanced assembly cApacity of tubulin after removal of its carboxyl-terminal domain. Eur J Biochem 156:375-381.

Maccioni RB, Cann JR and Stewart JM (1985b) Interaction of tubulin with substance P. Eur J Biochem 154:423-435.

Mandelkow EM, Herman M and Ruhl U (1985) Tubulin domains probed by limited proteolysis and subunit specific antibodies. J Mol Biol 185: 311-327.

Nath JP, Eagle GR and Himes RH (1985) Direct photoaffinity labeling of tubulin with guanosin-triphosphate. Biochemistry 24:1555-1560.

Ponstingl H, Krauhs E, Little M, Kempf M, Hofer-Warbinek R, Ade W (1982) Aminoacid sequence of α and β -tubulins from pig brains: heterogeneity and regional similarity to muscle proteins. Cold Spring Harbor Symp Quant Biol 46:191-198.

Serrano L, de la Torre J, Maccioni RB and Avila J (1984a) Involvement of the carboxyl-terminal domain of tubulin in microtubule assembly and regulation. Proc Natl Acad Sci USA 81:5989-5993.

Serrano L, Avila J and Maccioni RB (1984b) Controlled proteolysis of tubulin by subtilisin. Localization of the site for MAPs interaction. Biochemistry 23:4675-4681.

Serrano L, Avila J and Maccioni RB (1984c) Limited proteolysis of tubulin and the localization of the binding site for colchicine. J Biol Chem 259:6607-6611.

Serrano L, Valencia A, Caballero R, and Avila J (1986a) Localization of the high affinity colchicine binding site on tubulin molecule. J Biol Chem 261:7076-7081.

Serrano L, Wandosell F, and Avila J (1986b) Location of the regions recognized by five commercial antibodies to the tubulin molecule. Analytical Biochemistry 159:253-259.

Sigiura M, Maccioni RB, Cann J, York E, Stewart JM, and Kotovych G (1987) A proton magnetic resonance and a circular dichroism study of the solvent dependent conformation of the synthetic tubulin fragment Ac-tubulin alpha (430-441) amide and its interaction with substance P J. Biomolecular Structure and Dynamics (In Press).

Soifer D (1986) Dynamic Aspects of Microtubule Biology. v.466, Ann NY Acad Sci USA. 978pp.

Vallee RB and Borisy GG (1977) Removal of the projections from cytoplasmic microtubules in vitro by digestion with trypsin. J Biol Chem 252:377-382.

DYNAMIC INSTABILITY AND TREADMILLING OF BOVINE BRAIN MICROTUBULES IN VITRO: INFLUENCE OF MICROTUBULE-ASSOCIATED PROTEINS

L. Wilson[1], M.A. Jordan[1], H.P. Miller[1], and K.W. Farrell[2]

Department of Biological Sciences[1], and The Neuroscience Program of the Institute of Environmental Stress[2], University of California, Santa Barbara, California 93106, USA

ABSTRACT

We have examined the dynamic instability and treadmilling behaviors of microtubule (MT)-associated protein (MAP)-rich and MAP-depleted preparations of bovine brain MTs at polymer mass steady state under a variety of experimental conditions. With both MAP-rich and MAP-depleted preparations, the MTs exhibited dynamic instability behavior (increases in mean lengths, decreases in numbers) for a short time after attaining polymer mass steady state, or after the MTs were perturbed. With time, if left undisturbed, both the MAP-rich and MAP-depleted populations relaxed to a state in which no dynamic instability behavior could be detected (i.e., length distributions were stable). The extent of dynamic instability with MAP-rich MTs was small (mean-length increases of 1-2 um/h). In contrast, MAP-depleted MTs exhibited extensive dynamic instability behavior (mean-length increases of 40-50 um/h). Under steady state conditions of constant polymer mass and stable MT length distributions (i.e., no detectable dynamic instability), both MAP-rich and MAP-depleted MT preparations exhibited treadmilling behavior (growth at one MT end and equivalent shortening at the opposite ends). The treadmilling rates of MAP-depleted MTs were rapid (approximately 50 um/h), while treadmilling rates of MAP-rich MTs were slow (approximately 1 um/h). Our data indicate that (1) dynamic instability and treadmilling can occur within a single MT population, (2) the degree to which the MTs will exhibit one behavior or the other can be modulated by perturbations which affect the MT ends and (3) MAPs strongly suppress both the dynamic instability and treadmilling behaviors. Our data is discussed in terms of the possible control of MT assembly dynamics in cells by MAPs.

MICROTUBULE ASSEMBLY DYNAMICS

Although it is known that MTs are involved in many cell functions (Roberts and Hyams, 1979; Dustin, 1984), it is generally unclear how they perform these functions or how their activities in cells are regulated. One approach to these questions has been the study of MT polymerization in vitro, while a second approach has been the study of MT dynamics in cells.

MTs initially were viewed as polymers in simple equilibrium with a fixed (critical) concentration of tubulin subunits (Johnson and Borisy, 1977; 1979). Evidence that this view was overly simple was first obtained by Margolis and Wilson (1978), who used radiolabeled GTP as a probe for tubulin addition and loss at the ends of MAP-rich bovine brain MTs in vitro. They found that in the presence of GTP and a GTP-regenerating system, these MTs were not at true equilibrium, but were at steady state. They interpreted their data as indicating that net addition of tubulin was occurring at one MT end and an exactly balanced net loss of subunits was occurring from the opposite end, a behavior called "treadmilling" or "flux". Evidence consistent with treadmilling in vitro was subsequently obtained for other MT preparations (Farrell et al., 1979; Bergen and Borisy, 1980; Cote and Borisy, 1981; Rothwell et al., 1985) and has included direct observation by electron microscopy (Rothwell et al., 1985). The concept of subunit flux was first described by Wegner for actin filaments (1976). Wegner recognized that the hydrolysis of ATP during actin polymerization could result in different critical concentrations for growth at the two ends of the filaments, and thus give rise to net addition of actin at one filament end and loss of actin at the opposite end. For MTs, the energy source is thought to be GTP hydrolysis rather than ATP hydrolysis.

A different kind of dynamic behavior of MTs in vitro was described by Mitchison and Kirschner (1984a,b) and by Kristofferson et al. (1986). They observed that the mean length of MAP-depleted brain MTs increased while the number of MTs in suspension decreased with time, even though the polymer mass remained constant. Thus growing and shortening MTs existed in the same suspension. They called this behavior "dynamic instability", and suggested that it was due to transitions between shrinking and growing phases at MT ends. The mechanistic basis for the putative phase transitions is not known, but the gain and loss of regions of GTP-tubulin at the ends of MTs ("GTP caps", Carlier and Pantaloni, 1981) was proposed as a strong possibility (see Kirschner and Mitchison, 1986, and Caplow, in this monograph).

Recently, the development of appropriate molecular probes has made it possible to examine the dynamic behavior of MTs in cells. The results from several independent studies are in good agreement: many MTs in cells are remarkably dynamic, and their behavior is consistent with the dynamic instability model (discussed in Kirschner and Mitchison, 1986). Kirschner and Mitchison have further proposed that dynamically unstable MTs, which repeatedly shorten and regrow, could be stabilized selectively in cells, as for example, by attachment to kinetochores. Selective stabilization of otherwise unstable MTs could thus play an important role in spindle morphogenesis and in cell shape development.

In contrast, there is little evidence that subunit addition at one MT end and balanced loss from the opposite ends (i.e., treadmilling) might be involved in MT-mediated cell movement (e.g., chromosome movement during mitosis; Margolis et al., 1978). Data consistent with a poleward flux of subunits in kinetochore MTs at metaphase has been obtained (Mitchison et al., 1986). However, most evidence indicates that the participation of treadmilling in MT-mediated movement may be rare, and treadmilling does not appear to occur at rates that are fast enough to account for rapid movement (e.g., see Scherson et al., 1984; Soltys and Borisy, 1985; Wadsworth and Salmon, 1986). The differences in critical concentrations at opposite ends of individual MTs that give rise to treadmilling behavior could be critical in cells for certain other aspects of MT function. For example, the reactions at the opposite MT ends that produce treadmilling in vitro could determine the growth polarity of MTs in cells.

Dynamic instability data were initially perceived (Mitchison and Kirschner, 1984a,b) as being incompatible with data obtained both with MAP-rich MTs and MAP-depleted MTs, which indicate that MTs at polymer mass steady state attain stable length distributions and exhibit treadmilling. Mitchison and Kirschner suggested that the exchange of tubulin subunits that accompanies steady-state MT length redistributions might account for the kinetics of radiolabeled GTP-tubulin exchange with MTs at steady state, which we and others have interpreted as treadmilling (Margolis and Wilson, 1978; Farrell et al., 1979; Cote and Borisy, 1980).

During the past year we have examined the relationship between dynamic instability and treadmilling in MAP-rich and MAP-depleted MT preparations at polymer mass steady state to determine how MTs could exhibit such apparently incompatible behaviors. Our evidence suggests that dynamic instability and treadmilling are not mutually exclusive behaviors, but that one or the other behavior predominates depending upon the experimental conditions. Further, both dynamic instability and treadmilling are strongly suppressed by MAPs, which raises interesting questions regarding the control of MT assembly dynamics in cells. The purpose of this presentation is to describe briefly the data that led us to our conclusions, and to discuss the implications of the data in terms of the control of MT stability and organization in cells.

ASSEMBLY BEHAVIOR OF MAP-RICH MICROTUBULES

<u>Assembly dynamics of unperturbed MT suspensions at polymer mass steady state</u>. Much of the recent research carried out in our laboratory on MT polymerization has utilized a bovine brain MT protein preparation that has a high content of MAPs. The MAPs, which comprise 25-30% of the total protein, consist of the MAP I, MAP II, and the tau proteins, as well as a large number of other proteins present in relatively minor quantities that have not been characterized. The polymerization properties of these MTs have

been studied extensively (e.g., see Margolis and Wilson, 1978; Asnes and Wilson, 1979; Farrell and Jordan, 1982; Jordan and Farrell, 1982; Wilson and Farrell, 1986; Farrell et al., 1987), and most of our studies on treadmilling have been carried out with this preparation. We have studied tubulin exchange at the ends of these MTs at polymer mass steady state using a double isotope procedure that permits us to quantitate the net assembly and disassembly reactions at both ends of these MTs simultaneously in the same suspension. The method and the rationale for its validity has been described elsewhere (Farrell and Jordan, 1982; Jordan and Farrell, 1982; Wilson and Farrell, 1986; Farrell et al., 1987). Briefly, when MTs are polymerized to polymer mass steady state in the presence of [^{14}C]GTP and a GTP regenerating system, they become uniformly labeled with ^{14}C-guanine nucleotide. Upon pulsing with [^{3}H]GTP, the ^{3}H-guanine nucleotide becomes incorporated at the ends that grow at steady state (the operationally-defined A ends) and serves as the marker for the steady state rate of A end net tubulin addition, while the ^{14}C-guanine continues to be added at A ends and lost from D-ends in an unaltered fashion. The rate of steady state tubulin loss at D ends is determined by monitoring ^{14}C-guanine nucleotide loss after adding excess unlabeled GTP. The labeled MTs are collected on glass fiber filters or by sedimentation in a Beckman Airfuge for analysis of isotope exchange. Using this procedure, we have found that MAP-rich MTs treadmill slowly, growing at the A ends and shortening at the D ends at approximately 1 um per hr.

It is important to emphasize that analysis of isotope uptake and loss in terms of treadmilling behavior requires that there be no change in the MT length distributions at polymer mass steady state. Neither the mean lengths of the MTs nor the shapes of the distributions around the means can change. That length distributions are stable with MAP-rich MTs at polymer mass steady state in an unperturbed condition has been amply documented (e.g., see Wilson et al., 1985; Wilson and Farrell, 1986; Farrell et al., 1987). The reason for concern is that dynamic instability and treadmilling behaviors can result in similar patterns of isotope uptake and loss (see Mitchison and Kirschner, 1984a,b). For example, if there were a constant decrease in MT numbers due to complete depolymerization of some MTs and increase in mean lengths due to addition of subunits derived from depolymerizing MTs onto the ends of the remaining MTs, the rate of tritium incorporation during a pulse would be due to addition of tubulin subunits to both ends of the elongating MTs, while the loss of carbon-14 would be due to complete disassembly of those MTs which had become unstable. If both treadmilling and dynamic instability were occurring simultaneously in the same suspension, the uptake and loss rates would be the sum of both behaviors (see below).

Dynamics of MAP-rich MTs shortly after reaching polymer mass steady state, and after perturbation. One difference between the MT preparation used in the Kirschner laboratory and the one used in our laboratory was the content of MAPs. Further, most but not all dynamic instability was observed with MTs that had been sheared or suspensions that had been diluted prior to making the measurements, whereas, the treadmilling experiments were done with MT suspensions that were unperturbed for at least 30 min after reaching polymer mass steady state. Thus, we asked to what extent did the differences in MAP content and differences in the experimental conditions used contribute to different assembly behaviors.

In all of our previous experiments, the dynamics of the MAP-rich MT preparation were analyzed between 30 min and 2-3 hours after reaching polymer mass steady state. Thus in the present work we examined length distributions of unperturbed MT suspensions shortly after they reached polymer mass steady state. In many experiments, no variation in the shapes of the length distributions or the mean lengths could be detected once polymer mass steady state was reached. However in some experiments we detected a small extent of dynamic instability which occasionally continued for as long as 1 hr after the suspension attained polymer mass steady state. Increases in mean length (at a rate of approximately 1 um/hr) occurred. Concomitantly the length distributions broadened and became more normally distributed about the mean. The changes in the length distributions were always small, and the distributions always eventually stabilized.

We also found that we could induce dynamic instability by shearing or by dilution. For example in one experiment, the steady state mean length of a MAP-rich MT population was stable at 5.0 ± 0.1 um prior to shearing. Immediately after shearing, the mean length was reduced to 2.3 um ± 0.1 um but then it increased by 0.9 um during the next 30 min, then remained constant for the next 120 min. Although we repeatedly observed that shearing enhanced steady state length redistributions, the length redistributions and the mean length increases were always small, and were at least an order of magnitude smaller than the changes in sheared MAP-depleted MT preparations seen by Mitchison and Kirschner (1984b). In addition, the length distributions always eventually stabilized.

An example of a dilution experiment exhibiting induced dynamic instability is shown in Fig. 1. Because isotope uptake at MT ends can reflect both treadmilling and dynamic instability behaviors, we also included [^{3}H]GTP in the dilution buffer, so that we might visualize both behaviors as they occurred. The mean length of the MT population immediately before dilution was stable at 3.3 ± 0.1 um. After dilution, the mean length initially decreased to 2.5 ± 0.1 um and then increased to 3.0 ± 0.1 um during the next 60 min, after which no further change in the mean length or length distribution (data not shown) occurred. The mean length of 2.5 um observed immediately after dilution and the 3.0 um mean length observed finally are different from the expected mean of 2.8 um calculated from the distribution before dilution and the dilution

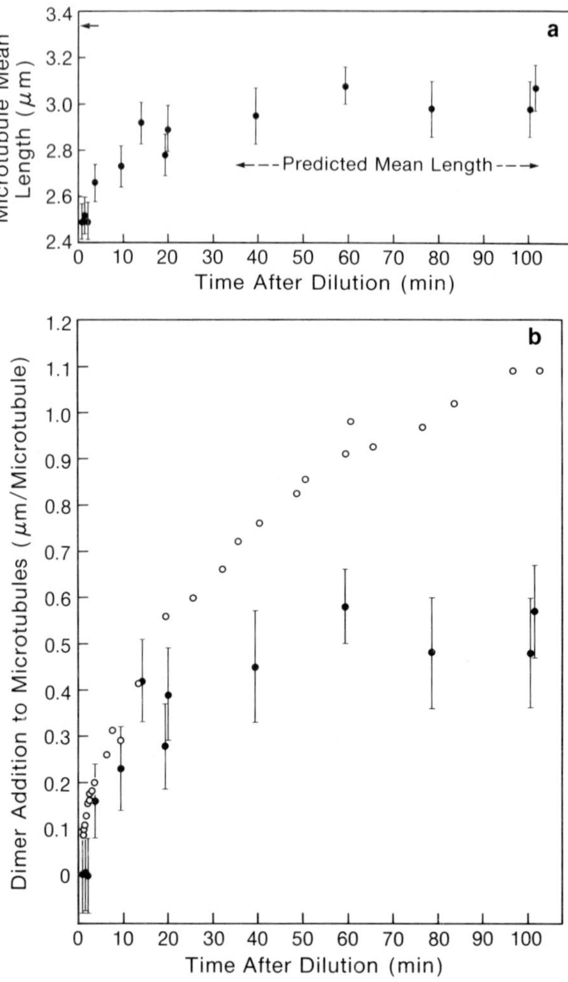

Fig. 1. Effect of dilution on MT length distributions of MAP-rich MTs. MAP-rich MTs at polymer mass steady state were diluted fourfold into reassembly buffer containing [^{3}H]GTP (for complete details of this experiment, see Farrell et al., 1987). (a) MT mean length before dilution (arrow) and for 100 min after dilution. Error bars indicate ±1 SEM. (b) Dimer addition to the MTs was calculated from the increase in the mean length of the MTs shown in a (solid circles) and from the amount of radiolabeled guanine nucleotide incorporated into the MTs (open circles). Reproduced with permission, from Farrell et al., 1987).

factor, assuming a simple equilibrium. These results have been observed repeatedly, but with some variation in the amount of mean length increase. The mean length increases observed were always small, and the length distributions always eventually stabilized.

Rapid uptake of radiolabel into the MTs occurred immediately after dilution concomitantly with the length redistributions. Uptake continued at a slow approximately linear rate after the length distribution stabilized. The net addition of tubulin subunits to MTs during the dilution-induced length redistribution was calculated from the radiolabel incorporated into the MTs, and, independently, from the increase in mean length of the MT population. The two estimates agreed closely during the initial period after dilution, during which time most (~70%) of the mean length increase occurred (first 15 min after dilution; Fig. 1). The data are consistent with the interpretation that some subunits lost from MTs by depolymerization subsequently regrew onto surviving MTs to contribute to the mean length increase. With time, the MT length distribution stabilized, yet label uptake, while slowed, continued at an approximately linear rate. The slow, linear phase of incorporation in the absence of MT length changes is consistent only with the treadmilling reaction. The data indicate that when MT length increases are occurring, tubulin addition due to MT growth can predominate and overwhelm that due to treadmilling. When the length distribution stabilizes, tubulin addition due to treadmilling predominates.

ASSEMBLY BEHAVIOR OF MAP-DEPLETED MICROTUBULES

The rapid and extensive length redistributions reported by Mitchison and Kirschner (1984a,b) were observed using MAP-depleted MTs that had been sheared. We therefore examined whether and to what extent length redistributions occurred in unperturbed suspensions, and whether shearing enhanced length redistributions in MAP-depleted MT populations. We found that the behavior exhibited by MAP-depleted MTs was qualitatively similar to that exhibited by MAP-rich MTs, but that it was strikingly different quantitatively.

Specifically, in unperturbed suspensions, length redistributions were often observed shortly after MT suspensions reached polymer mass steady state with MAP-depleted MTs. As with MAP-rich MTs, MAP-depleted MTs always relaxed to a state in which no further length redistributions could be detected, usually within 60 min of reaching polymer mass steady state. Simliar to earlier results of Cote and Borisy (1980), the MAP-depleted MTs also exhibited behavior consistent with treadmilling, as determined with radiolabel exchange experiments. Also as with MAP-rich MTs, shearing MAP-depleted MTs induced dynamic instability.

The striking difference in the behaviors of the MAP-rich and MAP-depleted MT suspensions was in the extent of the dynamics. Both treadmilling and dynamic instability were strongly suppressed by the MAPs. With MAP-depleted MTs, the population mean length increased after shearing at a rate of approximately 40-50 um/hr; while mean lengths in MAP-rich MT suspensions increased only at a rate of approximately 1 um/hr. Treadmilling rates were similarly affected by the MAPs. The flux rate of MAP-depleted MTs was approximately 50-60 um/hr, while treadmilling rates of MAP-rich MTs were approximately 1 um/hr.

PHASE DYNAMICS AT MICROTOBULE ENDS: THE COEXISTENCE OF TREADMILLING AND DYNAMIC INSTABILITY

The molecular mechanisms underlying the diverse MT behaviors is unknown, but may be due to phase transitions at MT ends (Hill and Chen, 1984). One mechanistic basis for the existence of two phases at MT ends could be the presence (Phase I) or absence (Phase II) of short regions of GTP-tubulin (i.e., GTP caps; Carlier and Pantaloni, 1981; Carlier et al., 1984; see also Caplow in this monograph). If MT ends that lose the GTP cap are readily recapped in an unperturbed MT suspension, the MTs would lose relatively small numbers of subunits at their ends; rarely would a MT disassemble completely. In MT suspensions at polymer mass steady state, the rapid recapping of uncapped MT ends would lead to length distributions that changed so slowly with time as to appear stable. Under these conditions, although both ends of microtubules would continuously shorten and regrow (evidence consistent with this behavior has been obtained by Horio and Hotani, 1986), the time-averaged result would be a net addition of tubulin subunits at one end and an equivalent net loss from the opposite MT end. This would give rise to the flux of subunits through the MTs, inferred from the kinetics of radiolabeled nucleotide exchange with steady state microtubules (discussed in Farrell et al., 1987).

In contrast, shearing MTs or diluting MT suspensions would increase the number of short MTs and/or the proportion of MT ends in Phase II (uncapped), and result in enhanced depolymerization (Farrell et al., 1983; Carlier et al., 1984). If a significant number of MTs were sufficiently short that they depolymerized completely before Phase I could be reestablished (cap regain), an increase in the mean length of the MT population would occur due to regrowth of the tubulin subunits lost from these MTs onto the Phase I ends of the surviving MTs. With time, however, the system would relax to a steady state in which the microtubule ends continually shorten and regrow without causing a net change in the length distribution, and, simultaneously, there would be net growth at one end and equivalent net shortening at the other end.

Another way to account for the stable length distributions is for GTP caps to be lost only very rarely at the ends of MTs in unperturbed suspensions. Conceivably, perturbation of MT ends by shearing or dilution may temporarily increase the duration of time that ends remain uncapped, perhaps by disrupting the normal steady state structure of ends. This may account in part for the relatively long delay between the time of perturbation, and the reestablishment of stable length distributions.

THE ROLE OF MAPS IN CONTROLLING MICROTUBULE STABILITY AND ORGANIZATION

It is becoming clear that MAPs play a major role in the structural organization and stability of MTs in cells. They are undoubtedly critical in the determination and control of MT function as well. A good example of the involvement of MAPs in determining the structural organization and stability of MTs exists in the case of sperm tail outer doublet MTs, which are extremely stable doublet MT structures. Interestingly, outer doublet tubulin separated from the many MAPs that associate with the outer doublet MTs are highly dynamic and exhibit many of the properties of MTs from cytoplasmic sources (Farrell and Wilson, 1978; Binder and Rosenbaum, 1978; Farrell et al., 1979). Thus, the MAPs must be responsible for the doublet structure and the marked stability of the MTs. Similar conclusions regarding the role of MAPs in microtubule stability, organization, and function are being drawn using other approaches (e.g., see Cowan, in this monograph).

Like the outer doublet microtubules of cilia and flagella, most cytoplasmic MTs appear to have an extensive array of MT-associated proteins on their surfaces (e.g., Vallee, 1982). However, our knowledge regarding the functions of MAPs is still in its infancy. It is conceivable that specific MAPs may have multiple functions, as for example, contributing to MT stability or organization but also functioning in the interaction of MTs with other cytoskeletal elements. Only the most prevalent MAPs from brain have been studied in any detail. It is not known how many other MAPs exist in brain, and our knowledge regarding the occurrence and roles of MAPs from non-neural cells and tissues is even more limited.

The effects of the mixture of MAPs in our MAP-rich MT preparation on assembly dynamics in relation to possible phase changes at MT ends are shown diagrammatically in Fig. 2. It is interesting that the MAPs strongly suppressed both the extent of dynamic instability and treadmilling rates. These data add additional support to the hypothesis that the MAPs, not the tubulin backbone itself, are important for controlling MT behavior and function in cells.

Kirschner and Mitchison (1986) have emphasized the potential importance of dynamic instability in the control of MT organization and stability in cells. The dynamic instability model is attractive, as it can account for much of the behavior exhibited by MTs in cells. However, according to this model, MTs are inherently instable; i.e., MT ends frequently transition from a growing to a shortening phase and the MTs disappear completely.

Our data, however, indicate that MT suspensions are inherently stable: if unperturbed, they relax to a state in which dynamic instability is undetectable. Either transitions to a shortening phase are very rare, or the transitions between shortening and growing phases are so rapid that in effect little net disassembly can occur. If our data are representative of most MTs in cells, and if dynamic instability is important in the control of MT dynamics, then one must conclude that there is a mechanism in cells for actively depolymerizing MTs, perhaps by removing, or suppressing reformation, of GTP caps. MAPs would be obvious candidates for possessing such regulatory capability.

However, the data raise a paradox. The MAP-rich MT protein preparation we have used in our studies contains a complex mixture of MAPs. If one assumes that the suppression of MT assembly dynamics by the MAPs present in the MAP-rich MT preparation is representative of most MAPs that associate with MTs in cells, then it is surprising that MTs in cells, which appear to have an extensive array of MAPs associated with them, are so dynamic. The mixture of MAPs present in the MAP-rich bovine brain MT preparation suppressed assembly dynamics rather than augmented them.

One possibility is that the results we have obtained with the mixture of MAPs in the MAP-rich bovine brain MT preparation are not representative of MAPs that associate with dynamic cytoplasmic MTs in cells but rather, the stabilization of assembly dynamics that we have observed is a specific property of brain MAPs that reflects the functions of these MAPs in stabilizing axonal and dendritic MTs. It is also quite possible that the stabilizing behavior of the MAP mixture is conferred by a small number of the many MAPs present (e.g., MAP II and/or the tau proteins), and that other MAPs might individually affect assembly dynamics quite differently than that observed with the mixture of MAPs in our preparation. It is also possible that tubulin diversity plays a significant role in the regulation of assembly dynamics by individual MAPs. For example, MTs composed of neuronal tubulin may have unique binding sites on their surfaces that recognize and respond specifically to neuronal cell-specific MAPs. We suggest that the analysis of the effects of specific MAPs on MT assembly dynamics with tubulins from functionally-defined sources could prove to be a fruitful avenue of research, as it could lead to an understanding of whether and how MT organization, stability, and perhaps function are determined and controlled by specific MT-associated proteins.

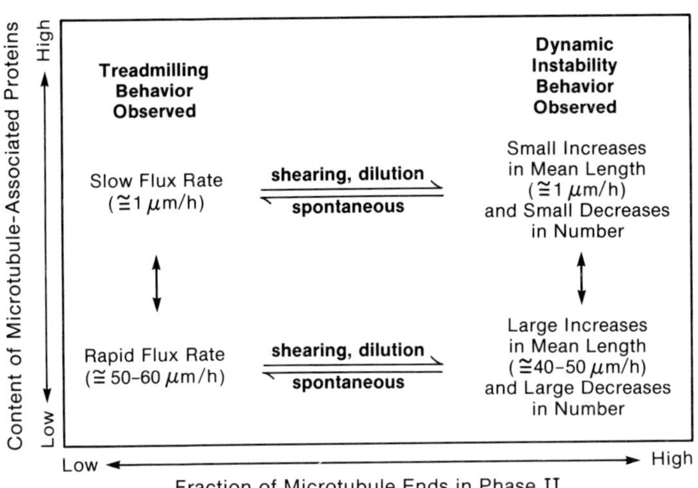

Fig. 2. The phase dynamics model: the effect of MAPs on population dynamics. The observed macroscopic behaviors of microtubule populations are shown in relation to the MAP content and the fraction of microtubule ends in Phase II (see text).

ACKNOWLEDGEMENTS

This work was supported by United States Publich Health Service grant NS13560 (L. Wilson) and American Cancer Society grant CD 316 (K.W. Farrell).

REFERENCES

Asnes, C.F. and Wilson, L. (1979). Isolation of bovine brain microtubule protein without glycerol: polymerization kinetics change during purification cycles. Analyt. Biochem. 98, 64-73.

Bergen, L.G. and Borisy, G.G. (1980). Head-to-tail polymerization of microtubules in vitro. Electron microscope analysis of seeded assembly. J. Cell Biol. 84, 141-150.

Binder, L.I. and Rosenbaum, J. (1978). The in vitro assembly of flagellar outer doublet tubulin. J. Cell Biol. 79, 500-515.

Caplow, M. (1987). Microtubule dynamics: the role of GTP hydrolysis. In The Cytoskeleton in Cell Differentiation and Development (ed. R.B. Maccioni & J. Arechage). ICSU Press.

Carlier, M.-F. and Pantaloni, D. (1981). Kinetic analysis of guanosine 5'-triphosphate hydrolysis associated with tubulin polymerization. Biochemistry 20, 1918-1924.

Carlier, M.-F., Hill, T.L. and Chen, Y.-D. (1984). Interference of GTP hydrolysis in the mechanism of microtubule assembly: an experimental study. Proc. Natl. Acad. Sci. U.S.A. 82, 771-775.

Cote, R.H. and Borisy, G.G. (1981). Head to tail polymerization of microtubules in vitro. J. Mol. Biol. 150, 577-602.

Cowan, N.J., Lewis, S.A., Sarkar, S. and Gu, W. (1987). Free intermingling of tubulin isotypes in the assembly of functionally distinct microtubules. In The Cytoskeleton in Cell Differentiation and Development (ed. R.B. Maccioni & J. Arechaga). ICSU Press.

Dustin, P. (1984). Microtubules. Springer-Verlag.

Farrell, K.W. and Wilson, L. (1978). Microtubule reassembly in vitro of S. purpuratus sperm tail outer doublet tubulin. J. Mol. Biol. 121, 393-410.

Farrell, K.W., Kassis, J.A. and Wilson, L. (1979). Outer doublet tubulin reassembly: evidence for opposite end assembly-disassembly at steady state and a disassembly end equilibrium. Biochemistry 12, 2642-2647.

Farrell, K.W. and Jordan, M.A. (1982). A kinetic analysis of assembly-disassembly at opposite microtubule ends. J. Biol. Chem. 257, 3131-3138.

Farrell, K.W., Himes, R.H., Jordan, M.A. and Wilson, L. (1983). On the nonlinear relationship between the initial rates of dilution-induced microtubule disassembly and the initial free subunit concentration. J. Biol. Chem. 258, 14148-14156.

Farrell, K.W., Jordan, M.A., Miller, H.P. and Wilson, L. (1987). Phase dynamics at microtubule ends: the coexistence of microtubule length changes and treadmilling. J. Cell Biol. 104, 1035-1046.

Hill, T.L. and Chen, Y.-D. (1984). Phase changes at the end of a microtubule with a GTP cap. Proc. Natl. Acad. Sci. U.S.A. 81, 5772-5776.

Horio, T. and Hotani, H. (1986). Visualization of the dynamic instability of individual microtubules by dark-field microscopy. Nature (London) 321, 605-607.

Johnson, K.A. and Borisy, G.G. (1977). Kinetic analysis of microtubule self-assembly in vitro. J. Mol. Biol. 117, 1-31.

Johnson, K.A. and Borisy, G.G. (1979). Thermodynamic analysis of microtubule self-assembly in vitro. J. Mol. Biol. 133, 199-216.

Jordan, M.A. and Farrell, K.W. (1982). Differential radiolabelling of opposite microtubule ends: methodology, equilibrium exchange-flux analysis, and drug poisoning. Analyt. Biochem. 130, 41-53.

Kirschner, M. and Mitchison, T. (1986). Beyond self-assembly: from microtobules to morphogenesis. Cell 45, 329-342.

Kristoffersen, D., Mitchison, T. and Kirschner, M. (1986). Direct visualization of steady state microtubule dynamics. J. Cell Biol. 102, 1007-1019.

Margolis, R.L. and Wilson, L. (1978). Opposite end assembly and disassembly of microtubules at steady state in vitro. Cell 13, 1-8.

Margolis, R.L., Wilson, L. and Kiefer, B.I. (1978). Mitotic mechanism based on intrinsic microtubule behaviour. Nature (London) 272, 450-452.

Mitchison, T. and Kirschner, M. (1984a). Microtubule assembly nucleated by isolated centrosomes. Nature (London) 312, 232-237.

Mitchison, T. and Kirschner, M. (1984b). Dynamic instability of microtubule growth. Nature (London) 312, 237-242.

Mitchison, T., Evans, L., Schulze, E. and Kirschner, M. (1986). Sites of microtubule assembly and disassembly in the mitotic spindle. Cell 45, 515-527.

Oosawa, F. and Asakura, S. (1975). Thermodynamics of the polymerization of protein. Academic Press, London.

Roberts, K. and J.S. Hyams. (1979). Microtubules. Academic Press, Inc., New York.

Rothwell, S.W., Grasser, W.A. and Murphy, D.B. (1985). Direct observation of microtubule treadmilling by electron microscopy. J. Cell Biol. 101, 1637-1642.

Scherson, T., Kreis, T.E., Schlesinger, J., Littauer, U.Z., Borisy, G.G. and Geiger, B. (1984). Dynamic interactions of fluorescently labeled microtubule-associated proteins in living cells. J. Cell Biol. 99, 425-434.

Soltys, B. and Borisy, G. (1985). Polymerization of tubulin in vivo: direct evidence for assembly onto microtubule ends and from centrosomes. J. Cell Biol. 100, 1682-1689.

Vallee, R.B. (1982). A Taxol-dependent procedure for the isolation of microtubules and microtubule-associated proteins (MAPs). J. Cell Biol. 92, 435-442.

Wadsworth, P. and Salmon, E.D. (1986). Analysis of the treadmilling model during metaphase of mitosis using fluorescence redistribution after photobleaching. J. Cell Biol. 102, 1032-1038.

Wegner, A. (1976). Head-to-tail polymerization of actin. J. Mol. Biol. 108, 139-150.

Wilson, L., Miller, H.P., Farrell, K.W., Snyder, K.B., Thompson, W.C. and Purich, D.L. (1985). Taxol stabilization of microtubules in vitro: dynamics of tubulin addition and loss at opposite microtubule ends. Biochemistry 24, 5254-5262.

Wilson, L. and Farrell, K.W. (1986). Kinetics and steady state dynamics of tubulin addition and loss at opposite microtubule ends: the mechanism of action of colchicine. Ann. N.Y. Acad. Sci. 466, 690-708.

GTP REQUIREMENT FOR *IN VITRO* AND *IN VIVO* MICROTUBULE ASSEMBLY AND STABILITY

Michael Caplow and John Shanks

Department of Biochemistry, University of North Carolina at Chapel Hill, Chapel Hill, NC 27514-7231 USA

ABSTRACT

The nucleotide triphosphate dependence of microtubule assembly and stability have been found to be identical with PTK cells and with pure tubulin dimer in the presence of glycerol. In both cases nucleotide triphosphate is required for assembly, but not for stabilizing microtubules. These conclusions are derived from the following observations: (a) The *in vitro* elongation of microtubule seeds or flagellar axonemes with tubulin subunits requires GTP; there is no elongation with tubulin-GDP subunits, even at high concentrations; (b) There is no assembly of microtubules in nocodazole-treated cells when the nocodazole is washed out in the presence of dinitrophenol and deoxyglucose, or in the presence of azide; (c) Microtubules formed *in vitro* in a glycerol-promoted reaction are stable, when the GTP in a reaction assembled to steady state is enzymically hydrolyzed to GDP; (d) Microtubules in PTK cells are stable in the presence of a mixture of dinitrophenol and deoxyglucose. Results (a) and (c), and (b) and (d) present a paradox: assembled microtubules are stable under conditions where microtubule elongation does not occur. We have attempted to account for this behavior and for nucleotide effects on nocodazole-induced disassembly, by a scheme in which microtubule assembly and disassembly is influenced by nonproductive subunit addition, and by a scheme in which microtubule-associated protein (MAP) phosphorylation decreases the stability of microtubules.

INTRODUCTION

We are concerned with the *in vitro* and *in vivo* requirements for guanine nucleotides in assembling and stabilizing microtubules. *In vitro* microtubule assembly requires GTP or a nonhydrolyzable GTP analog (Arai and Kaziro, 1976; Weisenberg and Deery, 1976; Penningroth and Kirschner, 1977; Karr, *et al.*, 1979). There are, however, exceptions, such as assembly with taxol (Schiff and Horwitz, 1981) or pyrophosphate (Bayley and Manser, 1985), where tubulin-GDP subunits will assemble. The effect of guanine nucleotides on stabilizing microtubules is intriguing. Despite the fact that tubulin-GDP subunits are inactive (Jameson and Caplow, 1980; Engelborghs and Van Houte, 1981; Carlier and Pantaloni, 1982), or only weakly active (Zackroff, *et al.*, 1980) in promoting *in vitro* microtubule elongation, when microtubules are assembled to a steady state with GTP, and the GTP is then enzymically converted to GDP, the microtubules do not rapidly disassemble (Zeeberg and Caplow, 1981; Margolis, 1981; Lee, *et al.*, 1982). Under some conditions rapid disassembly follows the GTP depletion (Hamel, *et al.*, 1984). This stabilization of microtubules in the absence of GTP has been accounted for by a mechanism in which tubulin-GDP subunits add nonproductively to microtubule ends (Zeeberg and Caplow, 1981; Caplow and Reid, 1985). Nonproductively bound subunits serve to cap the ends, so as to prevent subunit dissociation. The aberrant binding of these terminally-located subunits prevents subsequent subunit addition for elongation.

The nucleotide requirement for microtubule assembly in PTK cells is apparently markedly different from that observed *in vitro*. There appears to be no GTP requirement for assembly, since when nocodazole is used to induce disassembly and the nocodazole is subsequently washed out in the presence of the metabolic inhibitor azide, very extensive microtubule assembly occurs (DeBrabander, *et al.*, 1981; DeBrabander, *et al.*, 1982). Recent work, where dinitrophenol and deoxyglucose were used to deplete the nucleotide triphosphate pool, has demonstrated that microtubule assembly is reduced, but still significant when nocodazole is removed in the presence of these agents (Spurk, *et al.*, 1986). With regard to the stabilization of cellular microtubules, it appears that nucleotide triphosphates are not required, since metabolic inhibitors do not cause disassembly of the microtubule cytoskeleton (DeBrabander, *et al.*, 1981; DeBrabander, *et al.*, 1982; Bershadsky and Gelfand, 1981; Hepler and Palevitz, 1985). This parallels the *in vitro* results, where, as described above, the polymers formed with GTP are stable in the absence of GTP.

We find that when nocodazole-treated PTK cells are exposed to dinitrophenol and deoxyglucose for two hours, microtubule assembly does not occur on removal of the nocadazole, in the presence of these metabolic poisons. This result suggests that GTP is required for microtubule assembly in PTK cells. We have confirmed earlier studies (DeBrabander, *et al.*, 1981; DeBrabander, 1982; Bershasky and Gelfand, 1981; Hepler and Palevitz, 1985) showing that the microtubule cytoskeleton is stable under conditions where there is presumably no GTP.

Our study of the nucleotide requirements for forming and stabilizing MAP-free microtubules *in vitro* has shown that the glycerol-promoted reaction parallels the *in vivo* process. That is, tubulin-GDP does not elongate microtubules, and microtubules are stable after enzymic depletion of GTP from tubulin that had been assembled to a steady state in the presence of GTP. The *in vitro* behavior of microtubules formed from pure tubulin dimer in the absence of glycerol is different. Microtubule elongation requires GTP, but microtubules are unstable in the presence of tubulin-GDP.

EXPERIMENTAL

Materials: Microtubular protein was prepared from beef brain, as described previously (Zeeberg, *et al.*, 1980). Pure tubulin was obtained by chromatography using phosphocellulose (Williams and Detrich, 1979). Hexokinase and nucleotide diphosphokinase were purchased from Sigma, and *Strongylocentrotus purpuratus* sperm axonemes were a gift from Dr. Mary Porter (University of Colorado).

Methods: In studies of the effect of GTP depletion on microtubule stability tubulin was polymerized at 37° with ATP (100 µM), 38 U/ml nucleoside diphosphokinase, and 2.5 U/ml hexokinase. 2 mM glucose was added after the attainment of a steady state. Proof that the GTP-depleting system was effective was obtained by including a trace amount of [^{3}H]GTP in the assembly reaction, as a marker for HPLC analysis. The GTP-depleting system was also used to generate tubulin-GDP subunits for assembly studies. In this case the enzymes and glucose were added to unassembled tubulin at 0° and the reaction mixture was incubated for 10 min. at 37°. It was established that GTP is more than 95% converted to GDP under these conditions. Assembly of these tubulin-GDP subunits was initiated by adding either a 1% volume of freshly sheared microtubules, that had been assembled in a reaction promoted with 3.4 M glycerol and 100 µM GTP, or an aliquot of a concentrated solution of purified axonemes. Assembly was monitored by electron microscopic analysis of negatively-stained samples and by a radiolabel assay. In this case [^{3}H]GTP was used in the initial GTP depletion reaction and at varying times after nucleating the assembly reaction aliquots were centrifuged and the amount of pelletable radioactivity was measured. Centrifugation was done in a Beckman Airfuge at 30 PSI (3 min), maintaining a constant temperature by doing all manipulations in a 37° thermostated room. The temperature of reaction mixtures was 38.5° after the centrifugation. Rigorous temperature control is required since a temperature change from 37° to 35° results in more than 10% disassembly in the reaction without dimethyl sulfoxide or glycerol (unpublished results); there is no significant effect of the centrifuge temperature (23-37°) in the glycerol or dimethyl sulfoxide-promoted reactions.

Rat kangaroo cells (PTK) were grown on cover slips as described previously (Zeive, *et al.*, 1980). Cells were treated with a saturated nocodazole solution, prepared by filtering a mixture of 100 µl of 0.1M nocodazole (in dimethyl sulfoxide) and 25 ml of culture medium. An hour incubation in this solution was followed by two hours in PBS saturated with nocodazole and 10 mM 2-deoxyglucose, 0.5 mM dinitrophenol. This was followed by a 40 min. incubation in either PBS/dinitrophenol/deoxyglucose, or culture media. The microtubule cytoskeleton was visualized by immunofluorescence microscopy, as described elsewhere (Saxton, *et al.*, 1984).

RESULTS

GTP Requirement for *In vitro* Microtubule Assembly: Tubulin subunits were treated with an active GTPase system to convert E-site-bound and unbound GTP to GDP. Prior to the GTP depletion the protein was equilibrated with [^{3}H]GTP, and more than 95% of this radioactive nucleotide was in the form of [^{3}H]GDP at the point (10 min) when microtubule assembly was nucleated by addition of sheared microtubules or axonemes. Assembly was quantitated by measuring the amount of radioactive nucleotide that could be pelleted by high speed centrifugation (Table 1), and by electron microscopy. When GTP had been hydrolyzed there was no pelletable radioactivity in the reactions with or without glycerol. In accord with this we found that axonemes were not elongated under these conditions (results not shown). In contrast, in the presence of dimethyl sulfoxide a small amount of radioactive nucleotide was incorporated into polymer (Table 1) and both free and axoneme-associated tubulin polymers were observed.

TABLE 1. Effect of GTP Depletion on Pelletable Radioactivity in Microtubules Assembled with [^{3}H]Guanine Nucleotide

Assembly Promoter	Radioactivity Pelletted with GTP (% total)		GTP-depleted (% total)	
	20 min.	60 min.	20 min.	60 min.
None	19	13	1	0
Glycerol	26	36	0	0
Dimethyl Sulfoxide	34	30	3.4	2.8

[a]Assembly was carried out with tubulin (14.6 μM for glycerol- and dimethyl sulfoxide-promoted reactions, 36 μM in reaction without these agents), a trace amount of [^{3}H]GTP (approximately 600,000 CPM), 333 μM ATP, hexokinase and nucleoside diphosphokinase. The assembly was preceded by a 10 min incubation at 37°; 1.9 mM glucose was included in this incubation for the GTP-depletion reactions.

The assembly observed with dimethyl sulfoxide does not correspond to a reaction of residual tubulin-GTP, since at the time when the assembly was initiated only about 2% of the nucleotide remained as GTP; this percentage would be expected to decrease on further incubation. With only 2% of the nucleotide as GTP the tubulin-GTP concentration would be only .02 x 14.6 μM = 0.3 μM, which is less than the previously measured (Caplow, *et al.*, 1985) 1.3 μM critical concentration for tubulin-GTP under these conditions. Thus, we conclude that tubulin-GDP is assembling, with an apparent critical concentration that is less than 14.6 μM.[1]

GTP Requirement for Stabilizing *In Vitro* Microtubules: When microtubules were assembled to a steady state with GTP, and a hexokinase-nucleotide diphosphate kinase GTPase system was then activated by addition of glucose, microtubule disassembly resulted (Fig. 1). The disassembly went to completion when glycerol or dimethyl sulfoxide were not present in the reaction mixture. This disassembly was associated with a 75% decrease in solution turbidity, with the residual turbidity stable to chilling to 2°. The residual turbidity is apparently due to nonspecific protein aggregates, since electron microscopy of the chilled reaction mixtures revealed a total absence of microtubules. When axonemes were included in the reaction mixture these were found to be extensively elongated at both ends before the addition of glucose. There were, no microtubules at axoneme ends 30 min. after addition of glucose. In contrast to the just-described results, when glycerol or dimethyl sulfoxide were present when GTP was depleted the decrease in turbidity did not go to completion, since a significant additional decrease was observed when the reaction was subsequently chilled (Fig. 1). HPLC analysis of the nucleotide composition of the reaction with glycerol present revealed that GTP represented 4% of the guanine nucleotide 2.5 min. after glucose addition and 0.4% after 25 min. Electron microscopic analysis of the reaction mixtures 25 min after glucose addition revealed that numerous microtubules were present. Also, when axonemes were included in the glycerol-induced assembly reaction and samples analyzed when the reaction was at steady state, 100% of these were elongated at both ends, with an average of about 7 microtubules/end. Analysis of this reaction 30 min after glucose addition showed that 96% of the axoneme ends were elongated, with about 4 microtubules/end.

---

[1]A theoretical analysis (Hill and Chen, 1984) of microtubule assembly indicates that in the presence of stable microtubule nucleating centers microtubules can be formed at subunit concentrations that are below the critical concentration. This factor does not, however, account for the polymerization observed in the presence of dimethyl sulfoxide, since we observe both free and axoneme-associated microtubules in this reaction. Also, we found pelletable radioactive tubulin polymer in a reaction which did not contain stable microtubule nucleating centers, such as axonemes (Table 1).

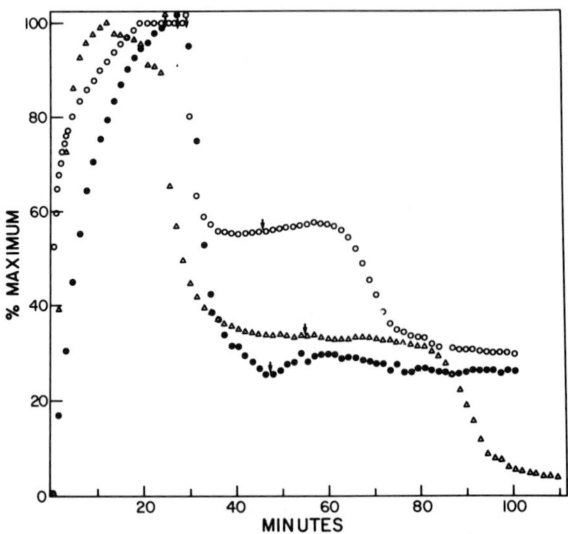

Fig. 1. Effect of GTP depletion on assembled microtubule. Tubulin was assembled with either 7.5 mM $MgCl_2$ (●), 12% dimethyl sulfoxide and 7.5 mM $MgCl_2$ (O), or 4.0 M glycerol and 7.5 mM $MgCl_2$ (Δ). The tubulin concentrations were 32 μM for the reaction without added organic solvent and 15 μM in the glycerol and dimethyl sulfoxide-promoted raction. Nucleoside diphosphokinase and hexokinase were present from the start of the reaction, which was initiated by addition of a 1% volume of sheared microtubules. The turbidity at steady state was about 3.5-fold higher in the dimethyl sulfoxide- and glycerol-promoted reaction; in the latter reaction the small decrease in turbidity prior to addition of glucose is due to instrument drift, since it was not observed in other reactions. Time of glucose (2 mM) addition and decrease in bath temp. are indicated by 1st and 2nd arrow; the cuvette temp. fell 12 min. later.

GTP Requirement for *In vivo* Microtubule Assembly: Microtubules were disassembled in PTK cells by treatment with culture media saturated with nocodazole. As reported previously (DeBrabander, *et al.*, 1981; DeBrabander, *et al.*, 1982; Spurck, *et al.*, 1986), disassembly with nocodazole is extensive, but not complete. When these cells were subsequently treated for 2 hours with dinitrophenol and deoxyglucose and the nocodazole then washed out, there was no microtubule assembly if the cells were kept energy-depleted by dinitrophenol/deoxyglucose. In contrast, if nocodazole was removed from cells in glucose-rich culture media microtubules were formed. Our observation of a GTP requirement for microtubule assembly in PTK cells agrees with results from similar studies with macrophages (Frankel, 1976); we are unable to account for the previous results with PTK cells (DeBrabander, *et al.*, 1981; DeBrabander, *et al.*, 1982; Spurk, *et al.*, 1986).

GTP Requirement for Stabilizing *In vivo* Microtubules: We find, in agreement with earlier studies (DeBrabander, *et al.*, 1981; DeBrabander, *et al.*, 1982; Hepler and Palevity, 1985; Spurk, *et al.*, 1986) that microtubules are stable in energy depleted cells. We are not able to analyze the composition of tubulin-E-site nucleotide in cells. However, since microtubules do not form from free tubulin subunits in the presence of the same dinitrophenol-deoxyglucose mixture (see above), it seems reasonable to assume that this treatment results in GDP occupancy of the E-site in free tubulin subunits. Thus, it appears that cellular microtubules are stable in the presence of tubulin-GDP subunits that are unable to participate in net microtubule elongation.

DISCUSSION

We attempt to account for the effect of ATP on *in vivo* and *in vitro* microtubule assembly and disassembly, in the presence and absence of nocodazole, by two mechanisms. Neither mechanism is able to satisfactorily account for the full range of results:

I. Microtubules react with tubulin subunits both productively and nonproductively, according to the scheme shown here:

$$\begin{array}{cccccc} (MT)_n & \rightleftharpoons & (MT)_{n+1} & \rightleftharpoons & (MT)_{n+2} & \rightleftharpoons \\ \downarrow\uparrow & & \downarrow\uparrow & & \downarrow\uparrow & \\ (MT')_{n+1} & & (MT')_{n+2} & & (MT')_{n+3} & \end{array} \qquad \text{(Eq. 1)}$$

Microtubules having nonproductively bound subunits at an end (MT') cannot be elongated, until these subunits are lost. It is proposed that nonproductive binding predominates with tubulin-GDP, while productive addition is the principal reaction path for tubulin-GTP subunits. This would account for the facile assembly of tubulin-GTP and the observed stabilization of *in vitro* and *in vivo* microtubules in the presence of tubulin-GDP, under conditions where tubulin-GDP subunits do not assemble. The instability of glycerol-free *in vitro* microtubules when GTP is hydrolyzed (see Results) probably results from the fact that the tubulin-GDP subunit dissociation rate from microtubule ends is about 80-times faster in the absence of glycerol (unpublished results). If subunit dissociation from microtubule ends is extremely rapid, then the metastable state produced by nonproductive subunit addition will be short lived.

Although the nonproductive binding mechanism would account for the *in vivo* and *in vitro* behavior of microtubules in the presence of GDP and GTP, additional factors must account for the effects of nucleotides on microtubule behavior in the presence of nocodazole. To account for the lack of disassembly in the presence of GDP within the framework of the nonproductive binding hypothesis, it must be assumed that tubulin-GDP-nocodazole subunits add nonproductively to microtubule ends, and tubulin-GTP-nocodazole subunits do not react with microtubule ends. Effective nonproductive binding by tubulin-GDP-nocodazole would prevent nocodazole induced disassembly. The presumed nonreactivity of tubulin-GTP-nocodazole is problematic, since we have found that tubulin-GTP-podophyllotoxin subunits, which might be expected to behave similarly, are able to generate a metastable steady state under conditions where such subunits do not assemble into microtubules (Caplow and Zeeberg, 1981). This behavior was taken to indicate that tubulin-GTP-podopyllotoxin subunits are able to add nonproductively to microtubule ends. Finally, the GTP requirement for nocodazole-induced microtubule disassembly could be accounted for if tubulin-GTP but not tubulin-GDP subunits were able to bind nocodazole. We have found, however, that this is not true, since the relative affinities of tubulin for GDP and GTP are unchanged by the presence of nocodazole (unpublished results).

II. It is presumed that the dephosphorylation of MAPs that is likely to occur in the absence of ATP increases the MAP-microtubule interaction energy (Jameson, *et al.*, 1980). This increased stabilization would serve to prevent microtubule disassembly by nocodazole, under energy-depleting conditions. On the other hand, with ATP present the destabilization of microtubules that would be induced by MAP phosphorylation is apparently insufficient to prevent assembly of tubulin-GTP subunits. These two hypotheses are not mutually incompatible. According to the proposed mechanism, the observed reaction of tubulin-GDP subunits do, however, present a paradox. Under conditions where the MAPs will be dephosphorylated (i.e., in the presence of metabolic inhibitors) the MAPs are able to stabilize microtubules; however, under these conditions MAP stabilization of microtubules is insufficient to allow nucleated assembly of tubulin-GDP subunits.[2] This paradox suggests that MAP phosphorylation does not account for the nucleotide triphosphate requirements of microtubule assembly and disassembly.

ACKNOWLEDGEMENTS

This work was supported by NIH grant DE03246. We are grateful for Dr. J. Richard McIntosh for his hospitality and advice when some of this work was done in his laboratory and to Michael Levy, for his critical comments.

---

[2]In the presence of metabolic inhibitors, nocodazole treated cells were seen to have short microtubules eminating from a centrosome. Consequently, the lack of assembly is not related to there not being a nucleation center.

REFERENCES

Arai, T. and Kaziro, Y. (1976). Effect of guanine nucleotides on the assembly of brain microtubules: Ability of 5'-guanylyl imidodiphosphate to replace GTP in promoting the polymerization of microtubules *in vitro*. *Biochem. Biophys. Res. Commun.* 69, 369-376.

Bayley, P.M. and Manser, E.J. (1985). Assembly of microtubules from nucleotide-depleted tubulin. *Nature* 318, 683-685.

Bershadsky, A.D. and Gelfand, V.I. (1981). ATP-dependent regulation of cytoplasmic microtubule disassembly. *Proc. Natl. Acad. Sci. USA* 78, 3610-3613.

Caplow, M. and Reid, R. (1985). Directed elongation model for microtubule GTP hydrolysis. *Proc. Natl. Acad. Sci. USA* 82, 3267-3271.

Caplow, M., Shanks, J. and Brylawski, B.P. (1985). Concerning the anomalous kinetic behavior of microtubules. *J. Biol. Chem.* 260, 12675-12679.

Caplow, M. and Zeeberg, B. (1981). Incorporation of radioactive tubulin into microtubules at steady state: Experimental and theoretical analysis of the effect of podophyllotoxin. *J. Biol. Chem.* 256, 5608-5611.

Carlier, M.F. and Pantaloni, D. (1982). Assembly of microtubule protein: Role of guanosine di- and triphosphate nucleotides. *Biochemistry* 21, 1215-1224.

DeBrabander, M., Geuens, G., Nuydens, R., Willebrords, R. and De Mey, J. (1981). Microtubule assembly in living cells after release from nocodazole block: The effects of metabolic inhibitors, taxol and pH. *Cell Biol. Inter. Reports* 5, 913-920.

DeBrabander, M., Geuens, G., Nuydens, R., Willebrords, R., and De Mey, Jr. (1982). Microtubule stability and assembly in living cells: The influence of metabolic inhibitors, taxol and pH. *Cold Spring Harbor Symp. Quant. Biol.* 46, 227-240.

Engelborghs, Y. and Van Houte, A. (1981). Temperature jump relaxation study of microtubule elongation in the presence of GTP/GDP mixtures. *Biophys. Chem.* 14, 195-202.

Frankel, F.R. (1976). Organization and energy-dependent growth of microtubules in cells. *Proc. Natl. Acad. Sci. USA*, 73, 2798-2802.

Hamel, E., del Campo, A.A. and Lin, C.M. (1984). Stability of tubulin polymers formed with dideoxyguanosine nucleotides in the presence and absence of microtubule-associated proteins. *J. Biol. Chem.* 259, 2501-2508.

Hepler, P.K. and Palevitz, B.A. (1985). Metabolic inhibitors block anaphase A *in vivo*. *J. Cell Biol.* 102, 1995-2005.

Hill, T.L. and Chen, Y. (1984). Phase changes at the end of a microtubule with a GTP cap. *Proc. Natl. Acad. Sci. USA* 81, 5772-5776.

Jameson, L. and Caplow, M. (1980). Effect of guanosine diphosphate on microtubule assembly and stability. *J. Biol. Chem.* 255, 2284-2292.

Jameson, L., Fray, T., Zeeberg, B., Dalldorf, F. and Caplow, M. (1980). Inhibition of microtubule assembly by phosphorylation of microtubule-associated proteins. *Biochemistry* 19, 2472-2479.

Karr, T.L., Podrasky, A.E. and Purich, D.L. (1979). Participation of guanine nucleotides in nucleation and elongation steps of microtubule assembly. *Proc. Natl. Acad. Sci. USA* 76, 5475-5479.

Lee, S.H., Kristofferson, D. and Purich, D.L. (1982). Microtubule interactions with GDP provide evidence that assembly-disassembly properties depend on the method of brain microtubule isolation. *Biochem. Biophys. Res. Commun.* 105, 1605-1610.

Margolis, R.L. (1981). Role of GTP hydrolysis in microtubule treadmilling and assembly. *Proc. Natl. Acad. Sci. USA* 78, 1586-1590.

Oosawa, F. and Asakura, S. (1975). Thermodynamics of the Polymerization of Proteins, Academic Press, Orlando, FL.

Penningroth, S.M. and Kirschner, M.W. (1977). Nucleotide binding and phosphorylation in microtubule assembly *in vitro*. *J. Mol. Biol.* 115, 643-673.

Saxton, W.M., Stemple, D.L., Leslie, R.J., Salmon, E.D., Zasortink, M. and McIntosh, J.R. (1984). Tubulin dynamics in cultured cells. *J. Cell Biol.* 99, 2175-2186.

Schiff, P.B. and Horwitz, S.B. (1981). Taxol assembles tubulin in the absence of exogenous guanosine 5'-triphosphate or microtubule-associate proteins. *Biochemistry* 20, 3247-3252.

Spurck, T.P., Pickett-Heaps, J.D. and Klymkowsky, M.W. (1986). Metabolic inhibitors and mitosis. *Protoplasma* 131, 60-74, 1986.

Weisenberg, R.C., Deery, W.J. (1976). Role of nucleotide hydrolysis in microtubule assembly. *Nature (London)* 263, 792-793.

Williams, R.C., Jr., Detrich, H.W., III (1979). Separation of tubulin from microtubule-associated proteins on phosphocellulose. Accompanying alterations in concentrations of buffer components. *Biochemistry* 18, 2499-2503.

Zackroff, R.V., Weisenberg, R.C. and Deery, W.J. (1980). Equilibrium and kinetic analysis of microtubule assembly in the presence of guanosine diphosphate. *J. Mol. Biol.* 139, 641-677.

Zeeberg, B. and Caplow, M. (1981). An isoenergetic exchange mechanism which accounts for tubulin-GDP stabilization of microtubules. *J. Biol. Chem.* 256, 12051-12057.

Zeeberg, B., Cheek, J. and Caplow, M. (1980). Preparation and characterization of [^{3}H]ethyl-tubulin. *Anal. Biochem.* 104, 321-327.

Zieve, G.W., Turnbull, D., Mullins, J.M. and McIntosh, J.R. (1980). Production of large numbers of mitotic mammalian cells by use of the reversible microtubule inhibitor nocodazole. *Exp. Cell Res.* 126, 397-405.

MICROTUBULES ENRICHED IN DETYROSINATED TUBULIN *IN VIVO* ARE LESS DYNAMIC THAN THOSE ENRICHED IN TYROSINATED TUBULIN

G.G. Gundersen, S. Khawaja, and J.C. Bulinski
Department of Biology, UCLA, Los Angeles, CA 90024

ABSTRACT

Microtubules (MTs) in proliferating cells in culture contain varying levels of tyrosinated (Tyr) and detyrosinated (Glu) tubulin. By analyzing the ends of MTs *in vivo*, we found that whereas MTs enriched in Tyr tubulin added monomeric tubulin to their growing ends, those enriched in Glu tubulin did not. Coupled with other recent studies that have demonstrated the decreased turnover and enhanced stability of Glu MTs *in vivo*, we propose a model for the role of detyrosination of MTs (and perhaps other post-translational modifications of tubulin) in MT function.

INTRODUCTION

Tubulin undergoes a reversible post-translational tyrosination-detyrosination in which a tyrosine residue is added to or removed from the C-terminus of the alpha subunit. By localizing the Tyr and Glu forms with specific antibodies, we have previously shown that a small proportion of the MTs in proliferating cultured cells are enriched in Glu tubulin (Glu MTs), while the majority contain predominantly Tyr tubulin (Tyr MTs; Gundersen *et al.*, 1984; Gundersen and Bulinski, 1986; Geuens *et al.*, 1986). These distinct populations of MTs are maintained *in vivo* by a cycle (see Fig. 1) in which MTs polymerize from a pool of Tyr subunits and are then

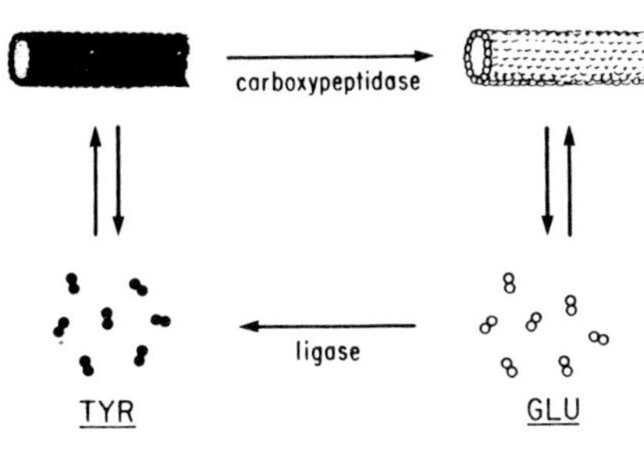

Fig. 1. Cyclic model for the tyrosination-detyrosination of tubulin. The cyclic nature of this model results from the restriction of the two enzyme activities to separate pools of tubulin: tubulin tyrosine ligase maintains the monomeric pool in the Tyr form, while the tubulin carboxypeptidase acts on the polymeric pool to generate MTs with increasing levels of Glu subunits as they age. Reproduced from Gundersen et al., 1987, with permission of J. Cell Biol.

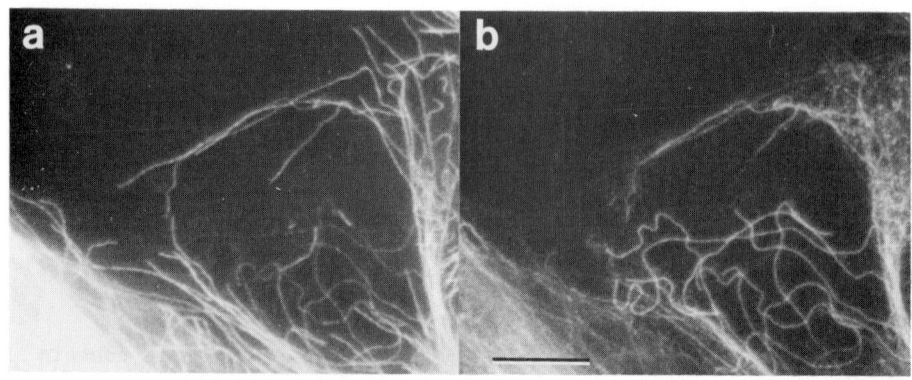

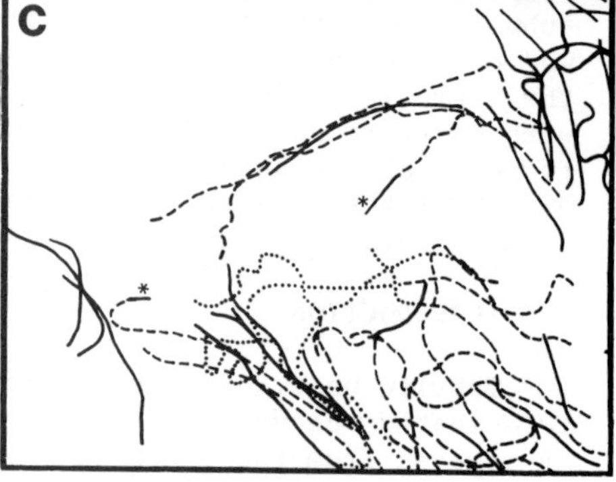

Fig. 2. Distribution of Tyr and Glu tubulin at the periphery of a TC-7 cell. TC-7 cells were fixed and prepared for IF as previously described (Gundersen et al., 1984). Fixed cells were stained with a 1/50 dilution of rabbit Glu antiserum (Gundersen et al., 1984) and a 1/250 dilution of rat monoclonal Tyr antibody, designated YL 1/2 (Kilmartin et al., 1982; Wehland et al., 1983). YL 1/2 was the generous gift of Dr. J.V. Kilmartin (MRC). Immunoreactivity was then detected with appropriate second antibodies conjugated with fluorescein and rhodamine. a,b) fluorescent images showing distribution of Tyr and Glu tubulin, respectively. c) interpretive diagram produced by tracing patterns in a and b onto a sheet of acetate (———, Tyr MTs; ------, Glu-Tyr MTs; ······· Glu MTs). The * denotes two examples of distinct Tyr segments at the ends of Glu-Tyr MTs. Bar, 10 μM.

detyrosinated by the tubulin carboxypeptidase to yield Glu MTs (Gundersen _et al_., 1987). Retyrosination of subunits by the tubulin tyrosine ligase occurs after the Glu MTs depolymerize. For simplicity, the model shows the complete conversion of a Tyr to a Glu MT; however, because all MTs in proliferating cells in culture contain at least some level of each form (Geuens _et al_., 1986), it is likely that MT breakdown occurs before a MT becomes completely detyrosinated. Our evidence for this cycle can be summarized as follows: Glu MTs reappear later than Tyr MTs during recovery from cold or drug treatments; increasing levels of Glu MTs are observed following taxol, or other stabilizing treatments; and Glu tubulin composes only a small percentage (<2%) of the pool of monomeric tubulin (Gundersen _et al_., 1987). In a separate study, Webster _et al_. (1987) found that microinjected Glu tubulin is rapidly retyrosinated, providing further evidence for the last step in the cycle, the conversion of Glu to Tyr tubulin. Because of the time-dependent, post-polymerization conversion of Tyr to Glu tubulin, the extent of detyrosination of a MT is an indication of its age; those MTs enriched in Glu tubulin would have persisted longer and hence would be expected to be more stable. Here we present direct evidence for the enhanced stability of Glu MTs.

ANALYSIS OF MT ENDS

In the first type of experiment, we determined whether Glu MTs are growing *in vivo*. Since the end of the MT distal to the centrosome is the site of subunit addition *in vivo* (Soltys and Borisy, 1985; Schulze and Kirschner, 1986) and > 98% of the monomer pool is Tyr tubulin (Gundersen *et al*., 1987), we expected that growing MTs would have only Tyr immunofluorescence (IF) at their distal ends. For MTs with Glu IF at their distal ends, we can conclude that the carboxypeptidase has had enough time to generate Glu tubulin and that additional (Tyr) subunits have not been added recently (i.e., the MT is not growing). Tyr IF should be visible along a considerable length at the distal end of a growing MT, because the conversion of Tyr to Glu tubulin in MTs (~25 min to form a Glu MT; Gundersen *et al*., 1987) is significantly slower than the rate of subunit addition *in vivo* (3-4 μm/min; Schulze and Kirschner, 1986). Therefore, we compared the IF staining of the distal end with the staining along the rest of the MT. For this analysis, we have defined the end of a MT as the most distal 3μm, which represents ~1 min of growth. Fig. 2 shows typical Tyr and Glu IF at the edge of a double stained monkey kidney cell (TC-7)

Table 1. Analysis of MT Ends

MT Type	End Type		
	Glu	Glu-Tyr	Tyr
Glu	123	1	0
Glu-Tyr	6	157	99
Tyr	4	1	664

(N = 1055)

TC-7 cells were double-stained with Glu and Tyr antibodies (as in legend to Fig. 1) and then photographed. MTs over 10 μm long and with identifiable ends in the photographic prints of Glu and Tyr fluorescent images were traced onto acetate transparencies. The individual transparencies were then aligned on the print of the other fluorescence image (e.g., Glu transparency on Tyr print) and the type of staining at the distal end (3 μm) and the remainder of the MT (3 μm to ≥ 10 μm) was noted.

and Table 1 summarizes results from 30 cells. Although most MTs did show only Tyr IF at their distal ends, a significant population of MTs exhibited Glu IF at their ends (see Fig. 2). The presence of Glu subunits on the ends of MTs, when the pool of monomeric tubulin is almost exclusively Tyr, strongly suggests that these MTs have not recently added subunits, i.e., they are not growing. Nongrowing MTs (those with either Glu or Glu-Tyr ends) represented ~26% of the MTs examined, however, the true percentage of nongrowing MTs is lower since we were able to detect only ~1/3 of the Tyr MT (growing MT) ends. Almost all MTs with proximal Glu IF (staining from the centrosome to within 3μm of the distal end) and ~60% of the MTs with proximal Glu-Tyr IF were nongrowing by this assay. Although it is clear from this analysis that nongrowing MTs have not recently added subunits, it is not evident whether they persist in this nongrowing phase (i.e., they are stabilized MTs) or whether they are actually losing subunits and depolymerizing (i.e., they are shrinking MTs). Experiments described below suggest that many are in the former category. The lack of subunit addition is important since it suggests that the ends of Glu MTs may be modified to prevent further growth. It would appear from Table 1 that virtually all MTs with proximal Tyr IF are growing; however, other interpretations, for example, that these MTs are shrinking rather than growing, or have stopped growing too recently to display Glu IF, are also consistent with these results. Microinjection studies have shown that most MTs in cultured cells are growing (Soltys and Borisy, 1985; Schulze and Kirschner, 1986), suggesting that most of the Tyr MTs are, in fact, growing.

It is worth noting that our analysis of MT growth was obtained without any perturbation of the cells. The usual method for determining MT dynamics in cultured cells involves the microinjection of derivatized brain tubulin (e.g., Soltys and Borisy, 1985; Schulze and Kirschner, 1986) and this alters the tubulin pools in two ways: 1) the steady-state level of tubulin subunits is increased, which might alter the characteristics of polymerization, and 2) the steady state level of Glu and Tyr tubulin is changed since brain tubulin is ~50% Glu, while the tubulin in cultured cells is only ~5% Glu (Gundersen _et al_., 1987). Thus the microinjection of brain tubulin in an amount equivalent to 10% of the cellular tubulin (usually at least this much tubulin is microinjected) would increase the cellular Glu tubulin level ~2-fold and this could potentially alter the normal dynamics of the MTs. Until the effect of detyrosination of MTs in cultured cells has been determined, we recommend a cautious interpretation of turnover data obtained from studies involving the microinjection of brain tubulin.

The analysis of MT ends described above clearly shows that Glu and Tyr MTs exhibit different behaviors with respect to growth _in vivo_. Nonetheless, only by microinjection of fluorescein- or

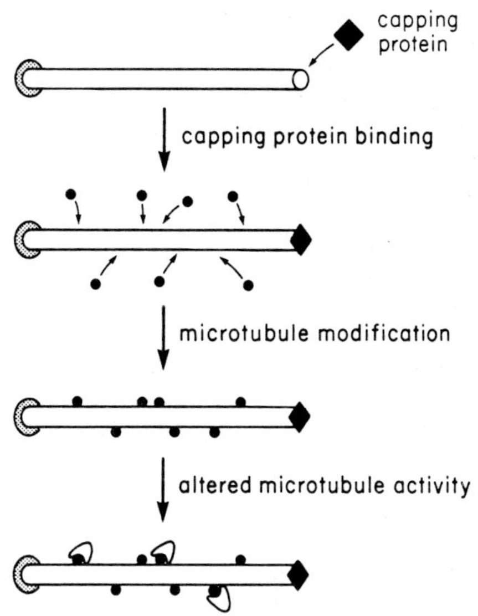

Fig. 3. Induction Model for the generation of specialized MT function by post-translational modification. A typical growing MT with one end attached to the centrosome (on the left) is shown binding a capping protein at its free end. This prevents further polymerization and stabilizes the MT. Subsequent covalent modification identifies the MT as a stable one and permits the interaction of a specific effector molecule. See text for details.

biotin-derivatized tubulin can the longevity of the nongrowing phase of Glu MTs be determined. By such an analysis, Webster *et al.* (submitted for publication) have found that Glu MTs persist for hours (>20% of the Glu MTs did not contain derivatized subunits even after 16 h). This shows that many of the Glu MTs are not part of the pool of dynamic (and presumably Tyr) MTs, which turnover with an estimated half-life of ~10 min (Schulze and Kirschner, 1986).

STABILITY OF GLU AND TYR MTS

In a second type of experiment, reported in detail elsewhere (Khawaja *et al.*, submitted), we have found that Glu MTs *in vivo* exhibited enhanced stability in comparison with Tyr MTs toward nocodazole treatment and toward dilution following detergent extraction of the cells. Little difference was observed in the stability of Tyr and Glu MTs toward cold treatment. In extracted cells, Tyr MTs converted to Glu MTs by treatment with exogenous carboxypeptidase did not show enhanced stability, suggesting that the stability of Glu MTs *in vivo* is not solely a function of detyrosination level. The differential stability of Glu MTs toward nocodazole and dilution, but not toward cold, further supports the idea that cellular Glu MTs are modified at their ends, since depolymerization by nocodazole and by dilution are thought to be end-mediated processes, while cold treatment may fragment MTs and allow depolymerization at sites other than at the ends (Mandelkow and Mandelkow, 1985).

MODEL: POST-TRANSLATIONAL INDUCTION

A hypothetical model for the role of tyrosination-detyrosination in MT function is shown in Fig. 3. In this model we propose that a MT is removed from the dynamic pool by some factor that interacts with the MT to stabilize it. We have shown this as the binding of a capping protein to the growing end, since this would be consistent with the lack of subunit addition to Glu MTs and the enhanced stability of Glu MTs to end-mediated depolymerization. Once the MT has been stabilized by the capping protein, it is subjected to covalent modification (in this case, detyrosination). This is consistent with the fact that Tyr MTs are converted to Glu MTs and that agents which stabilize MTs lead to extensive detyrosination of MTs (Gundersen et al., 1987). Because the level of Glu tubulin in a MT does not appear to directly affect stability (see above), we propose that detyrosination serves as a signal to some other effector molecule in the cell. When the level of Glu tubulin in the MT reaches a certain level, this factor would associate with the MT and induce a novel function, e.g., cross-linking of MTs to other filaments or transporting specific organelles. According to this model, then, the detyrosination of stabilized MTs provides a mechanism by which the cell "recognizes" stable MTs so that they can be utilized for specific functions. A particularly attractive feature of this model is that action at the MT end (e.g., capping) is transmitted to the entire length of the MT (by post-translational modification). Other post-translational modifications of tubulin (e.g., phosphorylation [Eipper, 1974] and acetylation [L'Hernault and Rosenbaum, 1985]) may be signals for other specialized functions. Since dynamic MTs turn over before they become extensively detyrosinated, this model provides a molecular mechanism by which specialized functions are restrained from inadvertantly becoming associated with MTs that have a high probability of transiting to a depolymerization phase. Conversely, dynamic MTs provide a reservoir from which individual MTs can be recruited for specialized functions as the need develops. The importance of this latter aspect has been discussed in detail by Kirschner and Mitchison (1986).

ACKNOWLEDGEMENTS

This research was done during the tenure of a Postdoctoral Fellowship from the Muscular Dystrophy Association to G.G. Gundersen, and was supported by grants from the National Institutes of Health (USPS CA 39755), the Muscular Dystrophy Association, and a National Sciences Foundation Presidential Young Investigator Award to J.C. Bulinski.

REFERENCES

Eipper, B.A. (1974). Properties of rat brain tubulin. J. Biol. Chem. 249, 1407-1416.

Geuens, G., Gundersen, G.G., Nuydens, R., Cornelissen, F., Bulinski, J.C. and DeBrabander, M. (1986). Ultrastructural colocalization of tyrosinated and detyrosinated alpha tubulin in interphase and mitotic cells. J. Cell Biol. 103, 1883-1893.

Gundersen, G.G., Kalnoski, M.H. and Bulinski, J.C. (1984). Distinct populations of microtubules: Tyrosinated and nontyrosinated alpha tubulin are distributed differently *in vivo*. Cell 38, 779-789.

Gundersen, G.G. and Bulinski, J.C. (1986). Microtubule arrays in differentiated cells contain elevated levels of a post-translationally modified form of tubulin. Eur. J. Cell Biol. 42, 288-294.

Gundersen, G.G., Khawaja, S. and Bulinski, J.C. (1987). Post-polymerization detyrosination of alpha tubulin: A mechanism for the subcellular differentiation of microtubules. J. Cell Biol. (in press).

Kilmartin, J.V., Wright, B. and Milstein, C. (1982). Rat monoclonal antitubulin antibodies derived by using a new nonsecreting rat cell line. J. Cell Biol. 93, 576-582.

Kirschner, M. and Mitchison, T. (1986). Beyond self-assembly: From microtubules to morphogenesis. Cell 45, 329-342.

L'Hernault, S.W. and Rosenbaum, J.L. (1985). Chlamydomonas α-tubulin is posttranslationally modified by acetylation on the ε-amino group of a lysine. Biochemistry 24, 473-478.

Mandelkow, E.-M. and Mandelkow, E.R. (1985). Unstained microtubules studied by cryo-electron microscopy: Substructure, supertwist, and disassembly. J. Mol. Biol. 181, 123-135.

Schulze, E. and Kirschner, M. (1986). Microtubule dynamics in interphase cells. J. Cell Biol. 102, 1020-1031.

Soltys, B.J. and Borisy, G.G. (1985). Polymerization of tubulin *in vivo*: Direct evidence for assembly onto microtubule ends and from centrosomes. J. Cell Biol. 100, 1682-1689.

Webster, D.R., Gundersen, G.G., Bulinski, J.C. and Borisy, G.G. (1987). Assembly and turnover of detyrosinated tubulin *in vivo*. J. Cell Biol. (in press).

Wehland, J., Willingham, M.C. and Sandoval, I.V. (1983). A rat monoclonal antibody reacting specifically with the tyrosylated form of α-tubulin.I. Biochemical characterization, effects on microtubule polymerization *in vitro*, and microtubule polymerization and organization *in vivo*. J. Cell Biol. 97, 1467-1475.

TENTATIVE MODEL FOR THE GTP BINDING DOMAIN OF TUBULIN

L. Serrano. Centro de Biología Molecular (CSIC-UAM) Canto Blanco. 28049-Madrid. Spain.

Tubulin, is a heterodimer composed of two subunits that have a 42% homology between them and bind one mol of GTP-Mg^{2+} per mol of subunit. By limited proteolysis of tubulin we have found four possible structural and functional domains comprising aa (1- 130, 131-283, 321-417 and 418- C-terminal) (1-5). Both subunits have the conserved ß-loop-α-ß structure (aa 135-183) involved in the interaction with the phosphate and the ribose or Magnesium, in nucleotide binding proteins (6). Mandelkow, et al. (4) proposed a model in which the sequence VVE (aa 181-183), referred to us as ß-strand-ß, is implicated in the interaction with the ribose of GTP and the sequence LRXG (aa 241-245) is equivalent to the ß-strand referred to as in nucleotide binding proteins. We have analyzed the existance of homologies between tubulin and other GTP-binding proteins or dehydrogenases. Our analysis indicate that the sequence VVE was correctly assigned by Mandelkow et al. (4), although we suggest that this sequence is involved in the binding of Magnesium and in the hydrolisis of GTP (Fig. 1). Also the sequence LRXG was correctly assigned as a ß-strand, but we think it does not correspond to ß-strand$_D$, but to ß-strand$_E$. These

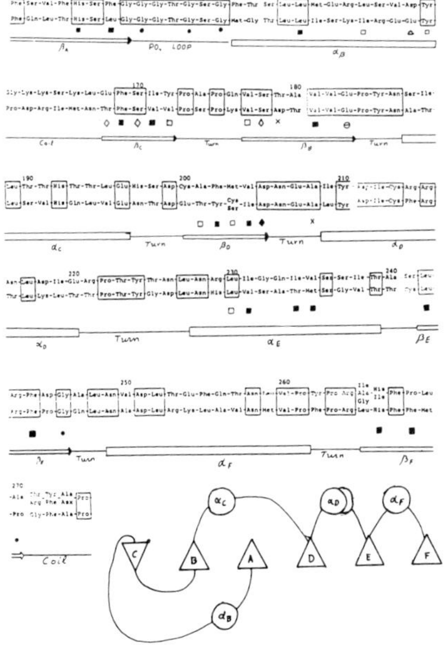

Fig. 1. Secondary and tertiary structure prediction for both tubulin subunits. Secondary structure prediction for tubulin subunits based on homologies with other nucleotide binding proteins and in the algorithm for secondary structure prediction of Cohen et al (8). α-helix (□), ß-strand (---›), Coil, turn (-). Those residues conserved in nucleotide binding proteins are indicated by closed symbols. Those only partially conserved by open circles (o), G,P,A, ( ), A,L,V,I,M,C,F,Y,W,T,S, ( ), T,S,N,Q,K,R,E,D,(θ),D,E. Inset shows a tentative three dimensional model of the GTP-binding domain of tubulin. (Δ) ß-strand. (o) α-helix, (-) connection.

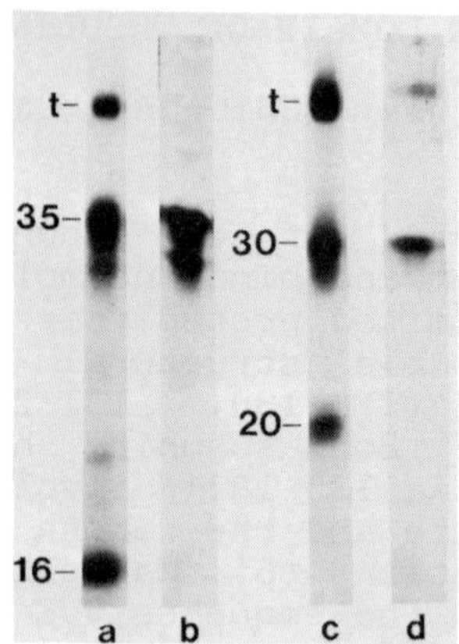

Fig. 2. Partial purification of the tryptic and chymotryptic fragments of α and ß-subunits. Tubulin digested with trypsin (a) or chymotrypsin (c) was incubated with urea 6M, and purified by to a Sepharose 6B chromatography and a ion-exchange colum (to be described elsewhere). The purified amino-fragments were renatured and a sample of each one subjected to electrophoresis. A) Trypsin digested tubulin. B) 35KDa fragment. C) Chymotrypsin digested tubulin. E) 30KDa fragment.

results are based on the homology of transduccin α(Tα) with tubulin, fundamentally the sequences IETKFSV (183-189) and CIIFCAALSAYDMV (220-233) of Tα with the α tubulin sequences SKLEFSI (163-169) and CAFMVDMEAIYDICR. These sequences correspond to ß-strand B and D. However since in tubulin there is not the sequence NXXD (7) present in all GTP-binding proteins, the model presented in Figure 1 must be taken only as an approximation. In order to test the validity of these model we have purifiy the tryptic and chymotryptic fragments, comprising aa 1-319, 331-451 of α-subunit and aa 1-283 and 284-455 of ß-subunit. We are currently analyzing their GTP-binding properties.

REFERENCES

(1) Serrano, L., Avila, J., Maccioni, R.B. (1984). J. Biol. Chem. 259, 6607-6611.

(2) Maccioni, R.B. and Seeds,. N.W. (1983). Biochemistry 22, 1567-1572.

(3) Serrano, L., Avila, J. (1985). Biochem. J. 230,551-556.

(4) Mandelkow, E.M., Kirchner, K. and Mandelkow, E. (1985). In Microtubule and Microtubule Tuhilips. pp 31-50.

(5) Serrano, L., Wandosell, F., De La Torre, J., Avila, J. (1986) Methods inEnzymology Vol. pp.

(6) Wierenga, R.K., Marc, C.H., Maeyer, De Hol. W.G.J. (1985) Biochemistry. 24, 1346-1357.

(7) Masters, B.S., Strond, M.R. and Bowne., R.H. (1986). Port. Engineering. Vol. 1, 47-54.

(8) Cohen, F.E., Marbanel, M.R., Kuntz, I.D. Fletterick, R.J. 81983). Biochemistry 22, 4899-4904.

EFFECTS OF CALPAIN ON MICROTUBULES AND TUBULIN

M. Billger, M. Wallin and J.-O. Karlsson*

Dept. of Zophysiology, Univ. of Göteborg, Box 25059, S-400 31 Göteborg, Sweden and *Inst. of Neurobiology, Univ. of Göteborg, Box 33031, S-400 33 Göteborg, Sweden.

Calpains are a group of Ca^{2+}-activated proteases present in almost all mammalian cells. Their proteolytic effect on several proteins, e.g. intermediate filaments, neurofilaments and axonally transported proteins, has been demonstrated. In spite of this, the biological role(s) of calpains remains unknown (for a recent review see (1)). In this study, we have investigated the effects of calpain I on purified microtubule proteins.

Microtubule proteins were isolated from bovine brain as desribed in (2). Calpain I was isolated from rabbit brain as in (3). After separation, proteins were transferred from their isolation buffers to 100 mM Mes, 0.5 mM Mg^{2+} and 1 mM DTT at pH 6.8 by gel filtration. Proteolysis of tubulin (2 mg/ml) by calpain was initiated by the addition of 4 mM Ca^{2+} at +30°C. After different time intervals, proteolysis was measured fluorometrically as the increase of free primary amino-groups. Tubulin was stimulated to assemble into microtubules by the addition of 20 µM taxol and the assembly was monitored as the increase in absorbance at 350 nm. Electrophoretic analysis were performed on 8% polyacrylamide and 8M urea or 12% polyacrylamide slab gels in the presence of 0.1% sodium dodecylsulphate. Gels were stained in 0.27% Coomassie Brilliant Blue in 45% methanol and 9% acetic acid.

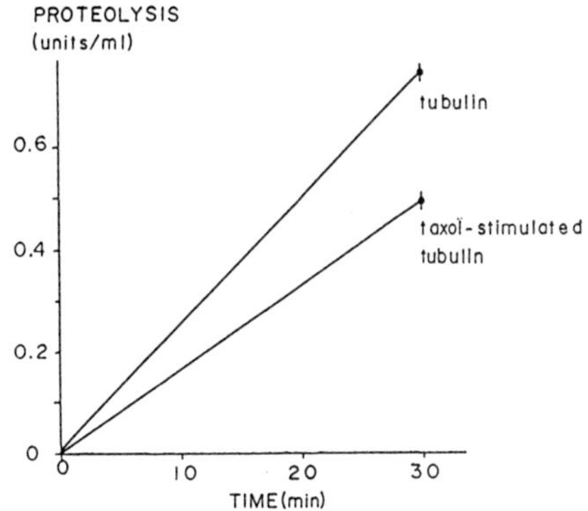

Fig. 1. Proteolysis of tubulin. Tubulin was incubated in the presence or absence of taxol and calpain. After 15 min proteolysis was initiated by the addition of Ca^{2+}. Proteolysis was measured as the increase in free amino-groups.

Calpain inhibited the GTP-induced assembly of microtubule proteins. This was accompanied by the loss of MAP1 and MAP2. When taxol was added to purified tubulin prior to the onset of proteolysis, the proteolytic activity measured after 30 minutes was reduced compared to non-assembled tubulin (Fig. 1). Neither taxol nor dimethyl sulphoxide, which was used to solubilize taxol, exerted any effect on the caseinolytic activity of calpain in the concentrations used. Electrophoretical analysis showed no formation of tubulin fragments on 8% polyacrylamide 8 M urea gels after treatment with calpain. However, on 12% polyacrylamide gels in the presence of 0.1% SDS fragments with molecular weights of 48,000, 43,000, 40,000 and 29,000 Da were found (Fig. 2). When tubulin was pre-assembled by taxol before proteolysis, the peptide mapping of tubulin was changed and only the 48,000 Da peptide was seen in addition to tubulin. The taxol-induced assembly of tubulin was

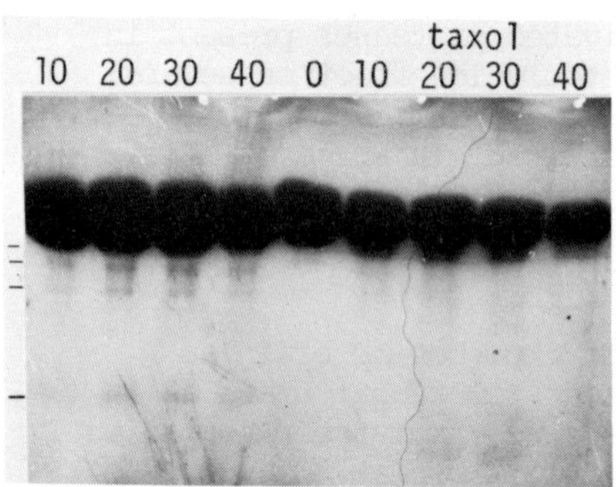

Fig. 2. Peptide mapping of tubulin. Tubulin was preincubated as in Fig. 1. At the indicated time intervals, samples were boiled in electrophoresis buffer and analyzed on 12% polyacrylamide gels. Molecular weights were calculated from standard proteins. Bars indicate proteolytic fragments.

reduced when pre-treated with calpain. Embedding of the assembly product showed the formation of aberrant polymeric structures of tubulin, i.g. short, twisted and partly opened microtubules (Fig. 3).

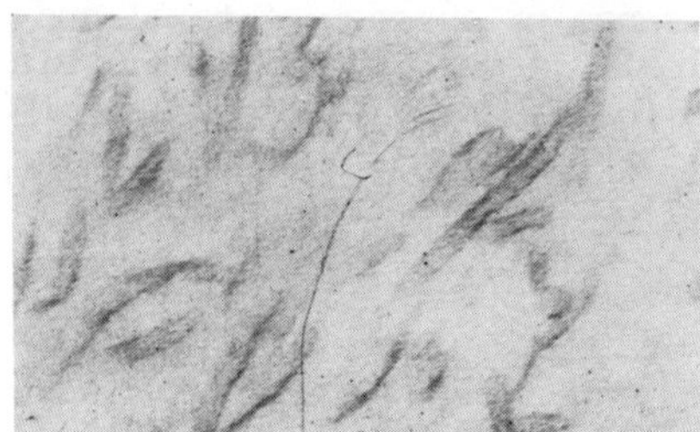

Fig. 3. Structure of calpain-treated tubulin. Tubulin was incubated with calpain. After 12 min taxol was added and the assembly product was centrifuged at 10^5 x g. The pellet was fixed in tannic acid and embedded in Epon. Magnification 90,000X.

In summary, calpain degrades MAP1 and MAP2 leading to inhibition of GTP-induced assembly of microtubule proteins. It has also a proteolytic effect towards tubulin, resulting in an increase in free aminogroups and formation of proteolytic fragments. Proteolysed tubulin is assembled by taxol to aberrant polymers. However, taxol-induced assembly of tublin leads to protection of proteolytic site(s) on the tubulin molecule.

REFERENCES

(1) Murachi, T. (1983) Trends Biochem. Sci. 8, 167-169.
(2) Wallin, M., Nordh, J. and Deinum, J. (1986) Biochim. Biophys. Acta 880, 189-196.
(3) Karlsson, J.-O., Gustavsson, S., Hall, C. and Nilsson, E. (1986) Biochem. J. 231, 201-204.

TUBULIN SYNTHESIS IN ARTEMIA

T.H. MacRae[*], P. Rafiee[*], and J.C. Bagshaw[**]
[*]Department of Biology, Dalhousie University, Halifax, N.S., B3H 4J1, Canada and [**]Department of Biology and Biotechnology, Worcester Polytechnic Institute, Worcester, Ma., 01609, U.S.A.

Brine shrimp possess three α- and two β-tubulins which appear not to change during development from the gastrula to early larval stages (1). For most organisms, however, development is accompanied by isotubulin modulation (2,3). To more clearly understand the interrelationships of the *Artemia* isotubulins we have analyzed the composition of brine shrimp tubulin genes and mRNA, using as molecular probes, cloned tubulin genes from other sources. The results indicate that brine shrimp have only one α- but two or more β-tubulin genes and that some of the *Artemia* isotubulins arise by post-transcriptional mechanisms.

ARTEMIA TUBULIN GENES. Cloned tubulin genes from *Drosophila* hybridized efficiently to Southern blots of restriction digested *Artemia* DNA. The tubulin genes from *Chlamydomonas* (α-253 and β-37, described in Cell, 24, 81-88, 1981) and chicken (pTl and pT2, described in Cell, 20, 95-105, 1980) either did not hybridize to *Artemia* DNA, or hybridized very poorly. The *Chlamydomonas* gene, β-37, hybridized weakly to small fragments not recognized by the *Drosophila* tubulin genes. *Artemia* tubulin genes have apparently diverged significantly from those within the non-protostomes examined.

Hybridization of the *Drosophila* α-tubulin gene, pDmTα1 (described in cell, 24, 97-106, 1981), to *Artemia* DNA digested with various restriction enzymes yielded a maximum of three bands on any one gel lane. The 5'-end of pDmTα1 gave only one hybridization signal per lane, whereas the 3'- end, with one exception, failed to react with *Artemia* DNA. *Artemia* may thus possess only one α-tubulin gene and the gene has diverged from pDmTα1 at the 3'-end. The sequence of pDmTα1 is similar to other tubulin sequences (4), suggesting the 3'-end of the *Artemia* α-tubulin gene may differ from that found in most other α-tubulin genes. As the carboxyl-terminal of tubulin binds microtubule-associated proteins (3), the lack of 3'-end cross-hybridization may reflect sequence differences responsible for the assembly characteristics of *Artemia* tubulin (1,3). The indication of only one α-tubulin gene in *Artemia* is noteworthy, since all other metazoan animals studied have multiple α-tubulin genes.

The *Artemia* β-tubulin gene family was analysed by using the cloned *Drosophila* β-tubulin gene, DTB2 (described in Develop. Biol., 104, 187-198, 1984) . Digestion of *Artemia* DNA with Sal I yielded eight bands upon hybridization to DTB2, at least six of which reacted with the 3'-end of the *Drosophila* gene. The 5'-end of DTB2 hybridized poorly to *Artemia* DNA, probably due to a region in DTB2 with weak interspecies homology. Our results indicate a minimum of two β-tubulin genes in *Artemia*. In contrast to α-tubulin, the 3'-ends of *Artemia* β-tubulin genes and the *Drosophila* β-tubulin gene used herein are similar.

ARTEMIA TUBULIN mRNA. Hybridization of pDmTα1 and DTB2 to brine shrimp RNA demonstrated that the proportion of poly (A^+) mRNA composed of α- and β-tubulin message and the amount of tubulin mRNA remained constant during development. Northern blots revealed two α- and β-tubulin mRNAs. One of the α-tubulin mRNAs formed a diffuse band upon electrophoresis but we do not believe the diffuse character resulted from degradation because : (a) there was no tubulin mRNA in the poly (A^-) fraction; (b) one of the α-tubulin bands remained compact with no hint of breakdown; (c) the β-tubulin mRNA gave two distinct bands on blots; and (d) the same hybridization pattern was obtained with two gel systems.

In summary, our results suggest that two *Artemia* α-tubulins arise as a result of post-transcriptional mechanisms whereas the β-tubulins may be products of different genes. To test these possibilities we are characterizing brine shrimp DNA fragments that cross-hybridize with *Drosophila* tubulin genes.

ACKNOWLEDGEMENTS

We thank Drs. D. Cleveland, J. Natzle, C. Silflow and P. Wensink for cloned tubulin genes. This work was supported by NSERC Grant A7661 to THM and NSF Grant DCB8510471 to JCB.

REFERENCES

(1) Rafiee, P., MacKinlay, S.A. and MacRae, T.H. (1986) Biochem. Cell Biol. 64, 238-249.
(2) Cleveland, D.W. and Sullivan, K.F. (1985) Ann. Rev. Biochem. 54, 331-365.
(3) MacRae, T.H. (1987) BioEssays 6, 128-132.
(4) Theurkauf, W.E., Baum, H., Bo, J. and Wensink, P. (1986) Proc. Natl. Acad. Sci. U.S.A. 83, 8477-8481.

BINDING OF BISCATIONIC FLUOROCHROMES TO PIG BRAIN TUBULIN.

M. Joniau*, E. De Clercq# and M. De Cuyper*
Kath. Univ. Leuven, *Interdisc. Res. Center, B 8500 Kortrijk, and #Rega Inst., B 3000 Leuven, Belgium.

Recently, PANTALONI et al. (1) have reported on the binding to tubulin of the biscationic DNA-binding fluorochrome 4',6-diamidino-2-phenyl indole (DAPI). Tubulin dimer presents a single binding site with K_d = 43 µM (37°C). In an effort to find probes to report on the binding of cationic regulators such as Ca^{2+}, MAPs and polyamines (2,3) to tubulin, we have studied several cationic fluorochromes, some of them analogous to DAPI.

1. Binding of DAPI and structural analogues. The fluorescence enhancement of these compounds (Table 1) was studied by addition of increasing amounts of tubulin to 10^{-5}M fluorochrome.

Table 1 : DAPI and analogues*

N°	R_1	R_2	Spacer
1	Am(5)	Am(3')	-
2	Am(5)	Am(4')	-
3	Am(6)	Am(3')	-
4	AM(6)	Am(4')	-
6	Im(6)	Im(4')	-
54	Am(6)	Am(4')	-CH=CH-
55	Am(6)	Am(4')	-C6H4-O-

R_1 and R_2 are the substituents on resp. the indole and the phenyl rings. The spacer is located between both rings. Am : amidino- Im : imidazolinyl- Compound n° 4 is DAPI

* Obtained through courtesy of Prof O. DANN (Inst.Pharm., Univ. Erlangen, F.R.G.).

Upon binding, DAPI fluorescence increases dramatically, accompanied by a blue shift of up to 20 nm. Also, fluorescence energy transfer occurs from tryptophan (λ ex = 295 nm) to bound DAPI. The orientation of both amidino substituents has an important effect on the degree of fluorescence enhancement, which is in the order : 2>4>3>6. Exchanging the amidino groups for imidazolinyl eliminates the fluorescence increase. Introducing internal spacers severely reduces the fluorescence increase capacity : 4>54>56. In direct competition assays compounds 1 and 3 apparently bind resp. 8 and 15 times less than DAPI itself. The fluorescence of bound DAPI is strongly reduced by several cations :

Mg^{2+}, Ca^{2+}, spermine, in agreement with earlier observations (2,3). Subtilisin digestion strongly affects the binding site. This observation further substantiates the involvement of the anionic C-terminus in DAPI binding.

2. Bisbenzimidazoles and oligopyrroles. The strongly fluorescent, DNA intercalating HOECHST dyes 33258 and 33342 also show a pronounced fluorescence enhancement of the same order as for DAPI, somewhat in contrast to the strict orientational requirements observed above. The non-fluorescent antibiotics netropsin and distamycin could be used as inhibitors of tubulin-enhanced DAPI fluorescence (ID_{50} resp. 5.4 and 2.8×10^{-4}M).

3. Other DNA-intercalating fluorochromes. Propidium iodide shows a more pronounced fluorescence increase upon tubulin binding than ethidium bromide. Both dyes possess the same aromatic nucleus, but the former has a biscationic rather than a monocationic side chain. This again indicates the involvement of a (poly) anionic component in the binding site.

4. Quinines and acridines. The strongly fluorescent quinine and the related, biscationic antimalarial quinacrine do not show any fluorescence increase in the presence of tubulin; neither does the bis-acridinium dye lucigenine. The non-fluorescent chloroquine was unable to inhibit DAPI fluorescence increase ($ID_{50} > 10^{-3}$M).

CONCLUSIONS

The DAPI binding site seems to involve a polyanionic component, possibly the C-terminus of one or both of the tubulin subunits, and to preferentially bind biscationic substances. These may turn out to be useful reporters of the interaction of cationic regulators such as Ca^{2+}, MAPs, spermine e.a. with the anionic C-termini of tubulin.

ACKNOWLEDGEMENTS

We highly appreciate the technical assistance of Mrs. Linda De Sender. Financial support came from the Belgian FGWO and Nationale Loterij.

REFERENCES

(1) Bonne D., Heusele C., Simon C. and Pantaloni D. (1985) J. Biol. Chem. 260, 2819-2825.
(2) Joniau M. and De Cuyper M. (1984) Colloids and Surfaces 10, 233-238.
(3) Joniau M. and De Cuyper M. (1985) in Microtubules and Microtubule Inhibitors 1985 (De Brabander M. and De Mey J., eds.) pp. 287-298, Elsevier , Amsterdam.

THE SPONTANEOUS ORGANIZATION OF MICROTUBULES IN SOLUTION

M. Somers and Y. Engelborghs
Katholieke Universiteit Leuven
Laboratory of Chemical and Biological Dynamics
Celestijnenlaan 200 D, B-3030 Leuven Belgium

INTRODUCTION

Long linear molecules are known to organize themselves into parallel arrays excluding the smaller molecules. Eventually phase separations can occur, as have been observed in the case of Tabac Mozaic Virus (1). The driving force for such a spontaneous organization is the decrease of the excluded volume, when polymers diffuse from the perpendicular configuration into the parallel one (2). Upon the polymerization of tubulin, macroscopic anisotropic zones have been observed through crossed polarizers (3). Here we studied the conditions necessary for spontaneous organization by comparing the appearance of turbidity and birefringence upon polymerization.

MATERIALS AND METHODS

Microtubule protein (MTP) was purified from pig brains according to the method of Shelanski et al. (4), modified as previously described (5). Glycerol was used only in the first cycle to increase the yield . This preparation contains about 15 % of microtubule associated proteins. Pure tubulin (PCT) was prepared by phosphocellulose chromatography (6). The polymerization buffer (at pH 6.4) consists of 50 mM MES, 70 mM KCl, 1 mM $MgCl_2$, 1 mM EGTA, 1 mM NaN_3. 1 mM GTP is (re)generated from 1 mM GDP and 10 mM acetylphosphate with 1 unit acetate kinase. PCT assembly was stimulated using glycerol or Me_2SO at the concentrations indicated. Birefringence is measured at 550 nm using crossed polarizers.

RESULTS

Microtubule protein showed at all concentrations a normalised birefringence curve which was very similar to the turbidity one, except for a short delay.
The appearance of birefringence was independent of how the reaction was initiated: either by temperature jump or by the generation of GTP from GDP with the regenerating system.

When pure tubulin was polymerized in the presence of 25 % glycerol or 8 % Me_2SO anisotropic phases were only very slowly formed. A fast appearance of birefringence could, however, be induced at any time by gently shaking the solution.

Shaking the solutions, after the establishment of maximal birefringence, caused a decrease, followed by a transient recovery.

DISCUSSION

Although the theory predicts a critical concentration for the appearance of anisotropic phases, whenever polymerization was observed with turbidity, birefringence could be measured, provided the sensitivity was sufficiently adjusted. For MTP the speed of the organization is remarkably high, and probably due to a combination of reorientation of shorter microtubules followed by their polymerization. The orientation process is not dependent on the temperature gradient which temporarely exists after a T-jump. The fact that organization is observed in the absence of MAPS, proves that they are not necessary. However, the kinetics are remarkably different. This cannot be attributed to the twofold higher viscosity in 25 % glycerol. It is possible that this deviation is due to differences in average length and length distribution and therefore to nucleation conditions.

In the oriented phases, annealing is expected to occur with high efficiency. The existence of oriented phases may explain the first order concentration dependence observed for annealing of microtubules.

ACKNOWLEDGEMENT

M.S. is research assistent and Y.E. is senior research associate of the Belgian National Fund for Scientific Research. This research is supported by a grant of the University and the E.E.C.

REFERENCES

(1) Oster, G. (1950) J. Gen. Physiol. 33, 445-473.
(2) Onsager, L. (1949) Ann. N. Y. Acad. Sci. 51, 627-659.
(3) Detrich, H.W. and Jordan, M.A. (1986) Annals N.Y. Acad. Sci. USA 466, 529-542.
(4) Shelanski, M.L., Gaskin, F., and Cantor, C.R. (1973) Proc. Natl. Acad. Sci. U.S.A. 70, 765-768.
(5) Engelborghs, Y., De Maeyer, L.C.M. and Overbergh, N. (1977) FEBS Lett. 80, 81-85.
(6) Weingarten, M.D., Lockwood, A.H., Hwo, S.-Y., and Kirschner, M.W. (1975) Proc. Natl. Acad. Sci. U.S.A. 72, 1858-1862.
(7) Herzfeld, J. and Briehl, R.W. (1981) Macromolecules 14, 1209-1214.
(8) Nordth, J., Deinum, J. and Norden, B. (1986) Eur. Biophys. J. 14, 113-122.

MICROTUBULE ASSOCIATED PROTEINS IN CELL DIFFERENTIATION

REGULATION OF THE POLYMERIZATION OF MICROTUBULE PROTEIN

J. Avila, L. Serrano, M.A. Hernández, F. Wandosell, E. Montejo, J. Díaz and A. Hargreaves.
Centro de Biología Molecular (C.S.I.C.-U.A.M.) Madrid, Spain.

ABSTRACT

Limited proteolysis of microtubule proteins,(tubulin and associated proteins, MAPs) has been performed to analyze the regions involved in protein-protein interactions and those which may regulate protein assembly.

INTRODUCTION

Microtubules play important roles in cell morphology, intracellular transport and motility based on two features, their ability to polymerize-depolymerize and to interact with other structures. These two characteristics must be modulated during cell differentiation and development by different ways. Functional differences between microtubules may be due to structural differences in the main component of microtubules, tubulin, or to the presence of different microtubule associated proteins (MAPs). Different tubulin isoforms have been described and it has been proposed (Fulton and Simpson, 1979) that specific isoforms may be involved in specific microtubule roles (see also Cleveland, 1987). Tubulin isoforms may be different gene products or could arise from posttranslational modifications. Several tubulin genes or their products, from different sources, have been already sequenced (Cleveland, 1987; Villasante et al., 1986,

Ponstingl et al., 1982) and in some cases it has been found that structural differences may, correspond to functional differences (Murphy et al., 1984.) Also the presence of different MAPs at different developmental stages or in different cell types has been widely discussed (Nuñez, 1986, Riederer, and Matus, 1985). However, less is known about the involvement of post-translational modifications on the regulation of protein-protein interaction and, in general, in the modulation of microtubule assembly.

The localization of the regions that are involved in protein- protein interaction or that contains modified residues can be done after cleavage of the different microtubule proteins by limited proteolysis.

Thus, limited proteolysis analyses of MAPs (MAP_2 and tau) and tubulin have been performed to determine the binding sites for a MAP present in tubulin and viceversa. Also these analyses were used to localize the modified residues after posttranslational modifications (phosphorylation).

LIMITED PROTEOLYSIS OF MAP_2

Limited proteolysis was first performed on microtubule proteins by Vallee and Borisy (1979). They cleaved microtubule associated protein, MAP_2, with trypsin. Cleaved protein yielded two fragments, the smaller one containing the tubulin binding site and the larger one containing, as indicated by Vallee group (Vallee et al., 1981), the binding site for a cAMP dependent kinase associated with microtubules. Similar results can be obtained when pepsin was used (Figure 1). A more detailed map of the molecule can be obtained by further enzymatic or chemical cleavages of the protein.

LIMITED PROTEOLYSIS OF TAU FACTOR

Tau factor is composed of at least four related peptides with a relative molecular weight ranging from 52000 to 65000 (Cleveland et al., 1977).

The different tau peptides can be isolated by preparative electrophoresis, under denaturing conditions followed by renaturation after removing of the detergent. Limited proteolysis of the three larger peptides with chymotrypsin resulted, in the appearance of a peptide of Mr. 50000, while no cleavage was found with the smaller tau peptide (Figure 2). Thus, a model for tau peptides is suggested. This model implies that tau proteins are composed of two regions which can be obtained after by chymotrypsin treatment. The larger region may be similar to the fastest moving tau peptide. Since this peptide binds to tubulin, that region should contain the tubulin binding site.

LIMITED PROTEOLYSIS OF TUBULIN

By limited proteolysis with nine different proteases, which differ in their residue specificity, three main cleavage sites were found on each tubulin subunit (Fig. 3). These exposed cleavage sites can define four regions (domains?) which we have named as regions I, II, II and IV, starting from the amino to the carboxy terminal end. Since an additional cleavage site can appear in region IV, it can be subdivided in region IVa and region IVb. The main cleavage sites are present around the residues 130, 340 and 410 in α subunit and around the residues 130, 280 and 410 in β subunit (as numbered by Ponstingl et al., 1982).

By sequence analogy with other nucleotide binding proteins, and also taking into account the previous data of other laboratories (Maccioni and Seeds, 1983; Mandelkow et al., 1985), it has been suggested that region I and/or II from each subunit may contain a nucleotide (ATP and/or GTP) binding site. Preliminary data suggest that the GTP binding site may be in region II (L. Serrano, unpublished results) of each subunit although in the β subunit (exchangeable site; Carlier, 1982) it would be more accesible than in the α subunit. Region I from α subunit contains a binding site for tau protein (Littauer et al., 1986). Region II from α subunit and region III of β subunit seem to be involved in the dimer interaction (Serrano and Avila, 1985). Region III from α subunit contains a colchicine binding site (Serrano et al., 1984a). Finally region IV, (the most variable one, when the sequences from different sources are compared) (see for example, Cleveland, 1987); appears to have a regulatory role in microtubule assembly, although it is not directly involved in the interaction between subunits (Serrano et al., 1984b).

A model for the different binding sites on tubulin subunits is indicated in Figure 4.

TUBULIN REGION IV

Region IV of α and β subunits corresponds to the carboxy terminal end in each tubulin subunit. The deletion of this region by proteolytic cleavage results in an increase in tubulin assembly (Serrano et al., 1984b), suggesting its regulatory role on microtubule polymerization. Also, a high affinity calcium binding site is present in each subunit, located at their region IV (Serrano et al, 1986a). Region IV from β subunit can be partially (Chymotrypsin), or totally (V8 S. aureus protease) removed by limited proteolysis. The polymers obtained with chymotrypsin-digested tubulin have an aberrant sheet-like structure. In these structures the interaction between tubulin subunits appear to be different from that found in microtubules (Serrano et al., 1986b) (Figure 5). On the other hand, the polymers obtained with V8 S. aureus protease cleaved tubulin appears to be structures without a defined direction (Figure 5). Removal of the region IV from both subunits restores the directionality of the obtained polymers and increases the lateral interactions between tubulin protofilaments (Figure 5).

Deletion of region IV from both subunits did not affect to the assembly of other microtubule structures such as zinc-induced sheets or vinblastine induced aggregates (Figure 4). This suggests that the zinc and vinblastine binding sites on tubulin molecule are not present in region IV.

TUBULIN-MAP INTERACTION

It has been described that some MAP_s (MAP_2 and tau factor) bind to tubulin through region IV (Serrano et al. 1984c; Serrano et al. 1985; Littauer et al. 1986), although tau factor may bind also through region I of α subunit (Littauer et al., 1986). Preliminary results suggest that other MAPs from brain microtubules as for example MAP_1 B also bind to tubulin through region IV. A protein immunologically related to MAP_1 B has been found in nonneural cells. This protein (p280) has been located in the cell nucleus and it should not be considered as a MAP since it is not present in the microtubule network of interphase cells, but as an spindle associated protein (SAP) since it associate wich the mitotic spindlec. Also preliminary data suggest that this protein binds to tubulin also through region IV.

PHOSPHORYLATION OF TUBULIN

The above results, based on proteolytic analysis, suggest a regulatory role for the C-terminus (region IV) of tubulin subunits. Although such class of irreversible posttranslational modifications would not take place in vivo, other reversible modifications, such as phosphorylations (which could affect to the funcionality of tubulin) may happen. Several protein kinases have been located associated with preparations of microtubule assembled in vitro. One of them is a cAMP dependent kinase which does not phosphorylate tubulin but phosphorylates MAP_s (Sloboda et al., 1975; Rappaport et al., 1976). Recently it has been indicated the association of a calcium/calmodulin (CaM) dependent kinase with microtubules (Vallano et al., 1986). This kinase phosphorylates tubulin only when region IV is not previously removed by proteolysis. Furthermore, by localization of the aminoacids phosphorylated in the undigested protein, it has been found that a CaM-dependent kinase phosphorylates aminoacids present in the region IV of both tubulin subunits. As a consequence of such phosphorylation the ability of tubulin to self assemble is decreased whereas its ability to associate with lipid vesicles is increased. Among the cAMP independent kinases associated with microtubules, an enzyme the activity of which decreases in the presence of heparin, was found in microtubule preparations. This kinase resembles to a casein kinase II and phosphorylates β subunit in its region IV, mainly when the microtobule protein was first treated with phosphatase. In contrast with the CaM-dependent kinase, a higher proportion of tubulin was found phosphorylated by this kinase in the polymerized fraction as compared wich the unpolymerized fraction.

DECREASE IN THE MAP_2-TUBULIN BINDING UPON PHOSPHORYLATION OF TUBULIN

Phosphorylation of tubulin with CaM dependent kinase prevents its interaction with MAP_2, as determined by gel-filtration experiments. Preliminary results suggest that phosphorylation of tubulin with casein kinase also affects to the interaction of the phosphorylated protein with MAP_2.

In summary the regulation of microtubule assembly, during differentiation or development, could depend not only on changes in the amount or nature of specific MAP_s, but also on the presence of other proteins, the activity of which may modify the main component of microtubules, tubulin. Our "in vitro" results suggest that such modifications mainly take place at the region IV, located at the carboxy terminus of tubulin subunits, which appears to have a regulatory role in microtubule assembly.

ACKNOWLEDGEMENTS

This work was supported by grants from CAICYT and U.S. Spain Committee for Scientific and Technological Cooperation.

REFERENCES

1. Carlier, M.F. (1982). Guanosine 5′ triphosphate hydrolysis and tubulin polymerization. Mol. Cell Biochem. 47, 97-107, 1982.
2. Cleveland, D.W., Hwo, S.Y. and Kirschner, M.W. (1977). Purification of tau a microtubule associated protein that induces assembly of microtubules from purified tubulin. J. Mol. Biol. 116, 207-228.
3. Cleveland, D.W. (1987). The multitubulin hypothesis revisited what we learned?. J. Cell Biol. 104, 381-383.
4. Littauer, U.Z., Giveon, D., Thierauf, M., Grinzburg, I and Ponstingl, H. (1986). Tubulin binding sites for microtubule associated proteins. In Microtubules and Microtubule inhibitors (eds. M. De Brabander and J. De Mey) pp. 171-176, Elsevier, Amsterdam.

5. Maccioni, R.B. and Seeds, N.W. (1983). Affinity labelling of tubulin's GTP exchangeable site. Biochemistry 22, 1572-1575.
6. Mandelkow, E.M., Hermann, M. and Ruhl, H. (1985). Tubulin domains probed by limited proteolysis and subunit specific antibodies. J. Mol. Biol. 185, 311-328.
7. Murphy, D.B., Wallis, K.T. and Grasser, W.A. (1984). Expression of a unique β tubulin variant in chicken red cell development. In Molecular Biology of the Cytoskeleton (ed., D.W. Cleveland and D.B. Murphy) pp 59-70. Cold Spring Harbor Publications, NY.
8. Nuñez, J. (1986). Differential expression of microtubule components during brain development. Dev. Neurosci. 8, 125-141.
9. Ponstingl, H., Krauhs, E., Little, M., Kempf, M., Hofer-Warbineck, R. and Ade, W. (1982). Aminoacid squence of α and β tubulins from pig brain: heterogenity and regional similarity to muscle proteins. Cold Spring Harbor Symp. Quant. Biol. 46, 191-198.
10. Purich, D.L. and Scaife, R.M.C. (1985). Microtubule cytoskeletal proteins as targets for covalent inter converting enzymes. Current Topics in Cell. Regul. 27, 107-116.
11. Rappaport, L., Leterrier, J.F., Virion, A. and Nuñez, J. (1976). Phosphorylation of microtubule associated protein, Eur. J. Biochem. 62, 539-549.
12. Riederer, B. and Matus, A.(1981). Differential expresion of distinct microtubule associated proteins during brain development. Proc. Natl. Acad. Sci. USA 82, 6006-6009.
13. Serrano, L., Avila, J. and Maccioni, R.B. (1984a) Limited proteolysis of tubulin and the localization of the binding site for colchicine. J. Biol. Chem. 259, 6607-6611.
14. Serrano, L., de la Torre, J.C., Macconi, R.B. and Avila, J. (1984b). Involvement of the carboxy terminal domain of tubulin in the regulation of its assembly Proc. Natl. Acad. Sci. USA 81, 5989-5993.
15. Serrano, L., Avila, J., and Maccioni, R.B. (1984c). Controlled proteolysis of tubulin by subtilisin: localization of the site for MAP_2 interaction. Biochemistry 23, 4675-4681.
16. Serrano, L. and Avila, J. (1985a). The interaction between subunits in the tubulin dimer. Biochem. J. 230, 551-556.
17. Serrano, L., Montejo, E., Hernández, M.A. and Avila, J. (1985b). Localization of the tubulin binding site for tau protein. Eur. J. Biochem. 153, 595-600.

18. Serrano, L., Valencia, A., Caballero, R. and Avila, J. (1986a). Localization of the high affinity. Calcium binding site on tubulin molecule. J. Biol. Chem. 261, 7076-7081.
19. Serrano, L., Wandosell, F., de la Torre, J. and Avila, J. (1986b). Proteolytic modification of tubulin Methods in Enzymology. 134, 179-190.
20. Sloboda, R.D., Rudolph, H., Rosenbaum, J.L. and Greengard, P. (1975). Cyclic AMP dependent endogenous phophorylation of a microtubule associated protein. Proc. Natl. Acad. Sci. USA 72, 177-181.
21. Vallano, M.C., Goldenring, J., Lasher, R.S. and De Lorenzo, R. (1986). Calcium/Calmodulin dependent kinase II and cytoskeletal function. In Microtubules and Microtubule Inhibitors (ed., M. De Brabander and J. de Mey). pp. 161-169, Elservier, Amsterdam.
22. Vallee, R.B. and Borisy, G.G. (1979). Removal of the projection from cytoplasmic microtubules in vitro by digestion with trypsin. J. Biol. Chem. 252, 377-382.
23. Vallee, R.B., Di Bartolomeis, M.J. and Theurkauf, W.E. (1981). A protein kinase associated with the projection portion of MAP_2 J. Cell. Biol. 90, 568-576.
24. Villasante, A., Wang D, Dobner P., Dolph P., Lewis, S.A. and Cowan N.J. (1986). Six mouse α-tubulin mRNAs encode five distinct tubulin isotypes: testis-specific expression of two sister genes. Mol. Cell. Biol. 6, 2409-2415.

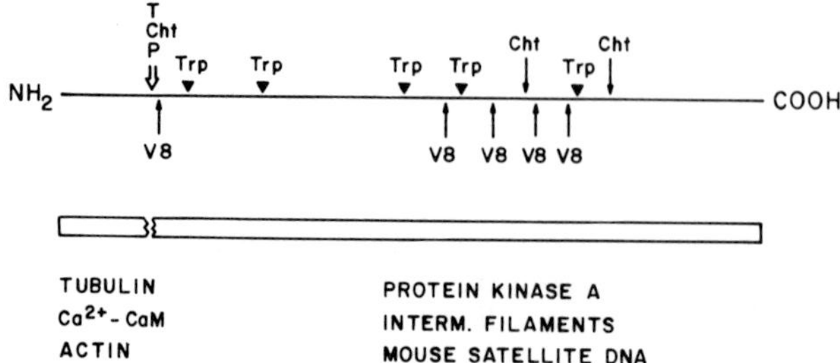

Figure 1. Model of MAP_2 molecule obtained from limited proteolyis data using trypsin (T) (Vallee and Borisy, 1979); chymotrypsin (Cht) and pepsin (P). The fragments obtained by cleaving MAP_2 with N-Chlorosuccinimide (Trp) and V8 S. aureus protease (V8) are indicated. In the lower part is shown the putative binding sites for the different molecules.

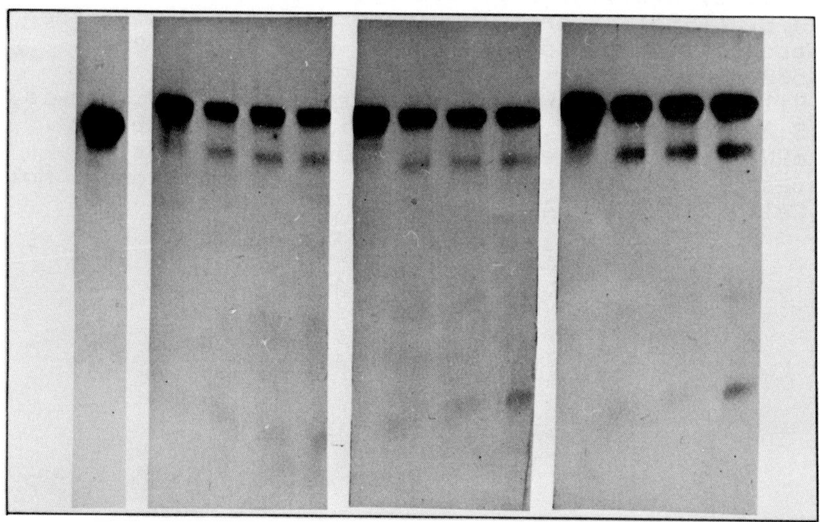

Figure 2. Limited proteolysis of tau peptides. Tau peptides were isolated by gel electrophoresis and cleaved with chymotrypsin (1% w/w) for 0,10,20 and 30 min at 30°C. The cleaved proteins were fractionated by polyacrylamide gel electrophoresis.

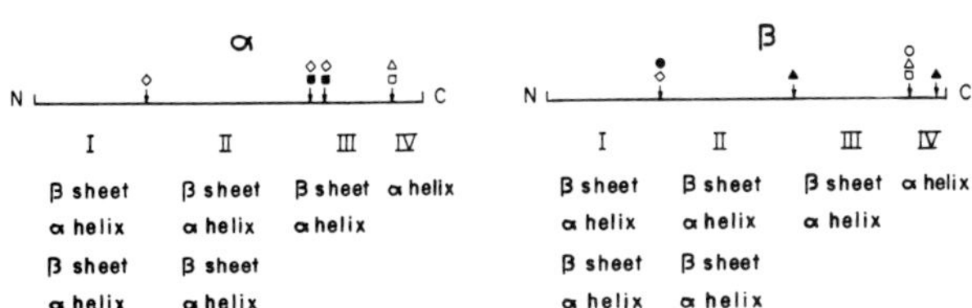

Figure 3. Limited proteolysis of tubulin. Cleavage sites for pronase (◇), trypsin (■), subtilisin (Δ), proteinase K (□), chymotrypsin (▲) and S. aureus V8 protease (o) are shown. The predicted secondary structure for each region is also indicated.

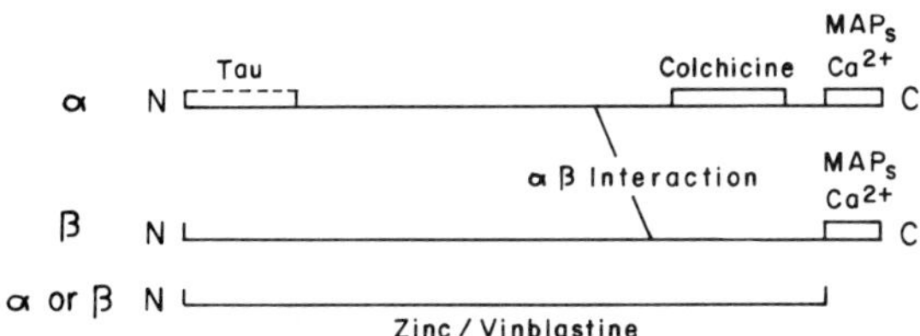

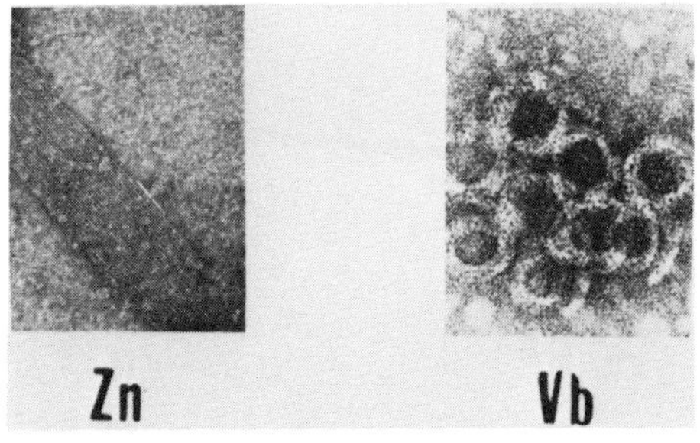

Figure 4. Model for putative binding sites on tubulin for the indicated molecules, based on our work and that of Littauer et al. (1986). Lower panel indicates the structures obtained for subtilisin-cleaved tubulin in the presence of 0.1 mM Zinc (Zn) or 10 μM vinblastine (Vb).

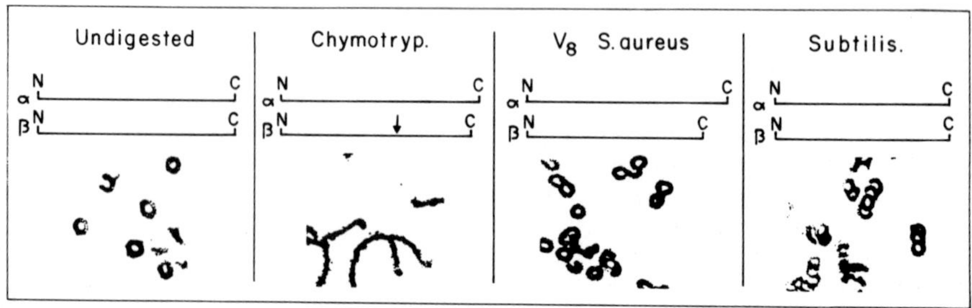

Figure 5. Cleaved tubulin structures-Undigested or Chymotrypsin, V8 S. aureus, or subtilisin digested tubulin samples were polymerized and thin sections, roughly perpendicular to the assembled polymers, were obtained and visualized.

PLECTIN AND HIGH-M_r MAPs: VERSATILE CONNECTING LINKS OF THE CYTOSKELETON

G. Wiche, R. Foisner, H. Herrmann, H. Hirt and G. Weitzer

Institute of Biochemistry, University of Vienna, 1090 Vienna, Austria

ABSTRACT

The first part of this paper describes various aspects of plectin including its conserved primary structure, general molecular properties, ultrastructure of single and aggregated molecules, molecular interaction partners and cellular functions. Subject of the second part are the high M_r microtubule associated proteins MAP-1 and MAP-2, in particular their proposed role as tissue specific mediators of microtubular interactions within the cytoplasm.

INTRODUCTION

Plectin (Wiche et al., 1982) which is homologous (Herrmann and Wiche, 1987) to IFAP 300K (Lieska et al., 1985) is an ubiquitous high M_r (300,000) polypeptide of higher eucaryotic cells. Recent evidence indicates that this protein is a multifunctional crosslinking element of the cytoskeleton. A similar function is ascribed to widespread polypeptides (Wiche et al., 1984) that are structurally related to the major microtubule associated proteins of brain tissue, MAP-1 and MAP-2 (Murphy and Borisy, 1975).

PLECTIN

Structural homology of plectin (soluble and insoluble) from various cultured cell lines. Extraction of cultured rat glioma C6 or BHK-21 cell monolayers with buffers containing 0.77 M NaCl/1% Triton X-100 yielded insoluble cell residues that consisted predominantly of a M_r=300,000 protein and the intermediate filament (IF) subunit proteins vimentin and vimentin plus desmin, respectively (Fig. 1, lanes 2, 4). Corresponding soluble fractions contained multiple bands in the high molecular weight range (Fig. 1, lanes 1, 3). The insoluble high M_r proteins from both cell lines could be immunoprecipitated to a similar extent using antibodies to C6 plectin (Fig. 1, lanes 6, 8). Among the corresponding soluble high M_r proteins only those comigrating with the plectin bands were immunoprecipitated (Fig. 1, lanes 5, 7). Rough estimates from gel scans indicated ratios of 7:3 for soluble versus insoluble plectin from C6, BHK-21 and CHO cells. Thus, in fibroblast cells major parts of plectin were soluble in high salt/Triton X-100, in marked contrast to the IF subunit proteins. Plectin from IFs conventionally prepared

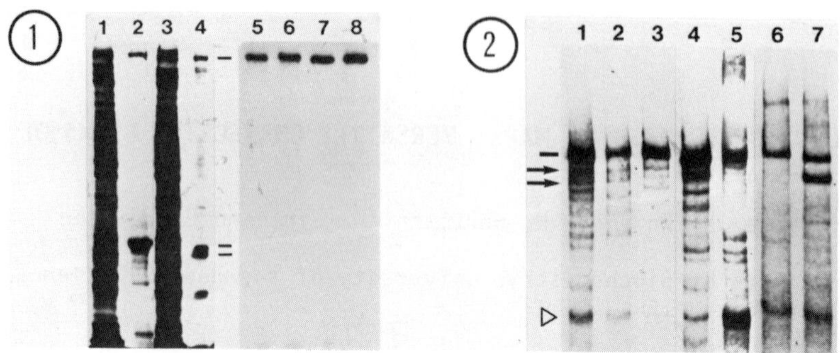

Fig. 1 Identification of plectin in high salt/Triton X-100 soluble and insoluble cell fractions. Rat glioma C6 (lanes 1,2,5,6) and BHK-21 (lanes 3,4,7,8) cell monolayers were labeled with ^{3}H-leucine. Soluble fractions (odd numbers) and insoluble residues (even numbers) were prepared by procedure B (Herrmann and Wiche, 1987) and boiled after the addition of electrophoresis sample buffer. Immunoprecipitation from boiled samples was done with antibodies to C6 plectin as described (Herrmann and Wiche, 1983). Fluorographs of the protein patterns (lanes 1-4) and corresponding immunoprecipitations (lanes 5-8) are shown. The position of plectin, vimentin and desmin is indicated (top to bottom).

Fig. 2 High resolution gel electrophoresis (fluorography) of various plectins. IF residues prepared by various methods were labeled by reductive methylation and components separated on 5% polyacrylamide gels. Lane 1, 2: IFs from two independent preparations using scraped-off C6 cells (procedure A, Herrmann and Wiche, 1987). Lanes 3-5, IFs from CHO, C6 and BHK-21 cells, respectively, prepared by direct extraction of chilled monolayers (procedure B). Lanes 6, 7: IFs isolated from CHO and C6 cells, respectively, as follows: Cells were lysed as in procedure B except that the lysis buffer contained only 0.17M instead of 0.77M NaCl. After 3 min of incubation, the concentration of salt was increased to 0.77M by stepwise addition of 5M NaCl. Only the upper parts of the gel between start and the position of myosin (open triangle) is shown. The position of plectin is marked by lines. Bands used for peptide mapping in Fig. 3A are indicated by arrows.

by high salt/Triton X-100 extraction of suspended cells separated on high resolution gels into one major band comigrating with the microtubule associated protein MAP-2A of M_r=300,000 (Herrmann et al., 1985) and several minor bands of higher electrophoretic mobility (Fig. 2, lanes 1, 2). When fast extractions on ice without prior dislodging of the cells were performed, in some cases, e.g. BHK-21 cells, only one major band was observed, but in other cases, such as C6 or CHO cells, several bands were still seen (Fig. 2, lanes 3-5). However, also the patterns of C6 and CHO plectins became less complex when monolayer cultures were incubated first with solutions containing 1% Triton X-100 and then brought to 0.77 M NaCl by stepwise addition. There was only a single band in the case of CHO cells, and two bands of roughly equivalent quantity in the case of C6 cells (Fig. 2, lanes 6, 7, respectively).

When the three largest anti-plectin immunoreactive polypeptide bands of ^{3}H-radiolabeled IF preparations from C6 cells (Fig. 2, lane 1) were subjected to limited proteolysis with Staphylococcus aureus V8 protease on SDS-polyacrylamide gels, identical peptide maps were obtained (Fig. 3A, lanes 1-3). Likewise, tryptic fingerprinting after direct iodination with ^{125}I of the two uppermost gel bands revealed identical patterns (Herrmann and Wiche, 1987). This clearly demonstrated that the anti-plectin immunoreactive bands observed after high salt/Triton X-100 extraction of C6 cells were degradation products of plectin. Virtually indistinguishable V8 fragmentation patterns were also observed in a direct comparison of ^{3}H-labeled soluble and insoluble C6 plectin species (Fig. 3B, lanes 1, 2).

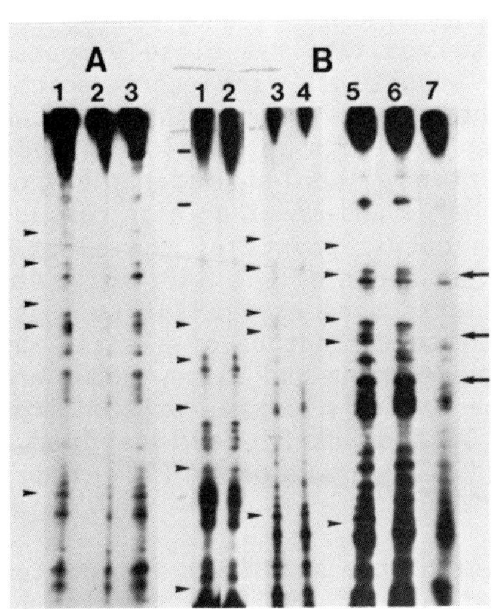

Fig. 3 Peptide mapping analysis of various plectins. Protein bands from fractions shown in Fig. 2 were excised after autoradiographic detection, digested during electrophoresis with 0.5 µg V8 protease and fragments separated on high percentage gels. A, analysis (12.5% gel) of the uppermost 3 bands from C6 IF residues shown in Fig. 2, lane 1. B, lanes 1, 2, analysis (10% gel) of plectins from a C6 IF residue (uppermost band) and corresponding soluble fraction, respectively. Lanes 3, 4, and 5-7, separate analyses (12.5% gel) of plectin bands from IF residues of C6 (3) and CHO(4) and of C6 (5), BHK-21 (6) and HeLa cells (7), respectively. Full length gels are shown. Arrowheads in A and B indicate M_r standards of 55,000, 40,000, 36,000, 31,000, and 13,700; lines 205,000 and 116,000. Arrows, see text.

Peptide maps generated from insoluble C6 and CHO plectin by V8 digestion of ^{3}H-labeled proteins were virtually identical with groups of major fragments around 14 kDa and 24 kDa. In addition, an identical set of fragments in the range of 30 to 35 kDa was generated from both proteins (Fig. 3B, lanes 3, 4). A high degree of similarity was observed also in a direct comparison of C6 cell plectin with plectins from BHK-21 and HeLa cells (Fig. 3B, lanes 5-7). The general fragmentation pattern after V8 digestion was very much alike but the intensities of some bands varied (arrows) indicating different susceptibility of some cleavage sites towards proteolysis. Nearly identical peptide maps were also obtained in a comparison of plectin from C6 and BHK-21 cells using chymotrypsin or subtilisin. Together, these data showed that the plectins expressed in four different cell lines, including three of fibroblastic and one of epitheloid type, possess conserved primary structure.

Molecular properties of plectin. The electrophoretic mobility of plectin on SDS-polyacrylamide gels was dependent on the pH of the electrophoresis running buffer. Titration of the running buffer to pH 8.9, instead of commonly used pH 7.8, resulted in a lower mobility of plectin relative to MAPs shifting its position from that of MAP-2A to that of MAP-1C. Plectin's mobility was also affected by the composition of the sample buffer employed. In the presence of 2-mercaptoethanol plectin's position on the gel remained unaltered independent of the amount of protein loaded on the gel. However, when increasing amounts of plectin were run in the presence of DTT and EDTA, mobility increased.

In 2-d gel electrophoresis ^{35}S-methionine labeled C6 cell plectin focused around pH 4.8, and was resolved into two closely spaced spots of slightly different molecular weights and isoelectric points (Fig. 4a, arrowheads). Presumably, these spots corresponded to the closely spaced double band often seen in 1-d high resolution gels (Fig. 2). Coelectrophoresis of radioactively labeled preparations of C6 cell IFs with unlabeled hog brain MAP-1 and MAP-2 on high resolution 2-d gels revealed that the isoelectric points of the plectin species were slightly more acidic than those of the MAPs (Fig. 4b, c) and that the apparent molecular weights were equivalent to those of MAP-2A (300K) and MAP-2B (280K). The acidic nature of plectin was also demonstrated by chromatofocusing (Herrmann and Wiche, 1987) and amino acid analysis. According to the latter plectin isolated from three sources, bovine eye lenses, C6 and BHK-21 cells showed a predominance of potentially acidic (24%) versus basic (13%) residues.

For purification of plectin, most of the actin and associated proteins contained in crude preparations of IFs were removed by sequential washes with PBS supplemented with nucleases and 0.6 M KI. Upon gel permeation chromatography of the remaining residues in 9 M urea at room temperature plectin fractions of high purity were obtained. Plectin eluted as a broad peak starting after the void volume and trailing into the vimentin peak. This indicated that even in 9 M urea plectin occured as aggregates of varying monomer number. Circular dichroism spectra of purified plectin samples from C6 cells indicated 30-35% α-helix, close to 10% ß-structure and the rest aperiodic structure. CD-measurements in the presence of various concentration of urea showed partial destruction of the plectin molecule in 3.5 M urea and total unfolding in 7 M urea. Heating of the plectin sample to 70°C resulted in loss of higher order conformation.

As a phosphoprotein, plectin is the target of cAMP-independent, cAMP-dependent and Ca/calmodulin-dependent kinases, with serin being

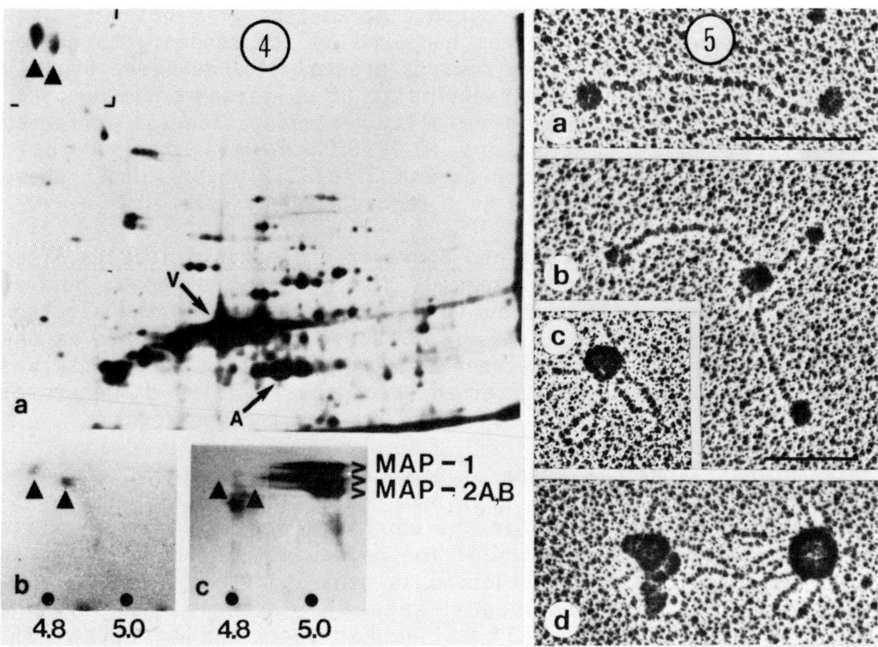

Fig. 4 2-d gel electrophoresis of C6 cell plectin. Proteins were labeled _in vivo_ by incubation of cell cultures with 100µCi/ml of ^{35}S-methionine for 24 hours. a, fluorography of IF preparation; b and c, fluorography and Coommassie Brilliant Blue staining, respectively, of one gel after the analysis of a mixture of the radiolabeled sample shown in a and phosphocellulose-purified hog brain microtubule associated proteins (not radiolabeled). b and c correspond to the upper left hand gel region indicated in a. A, actin; V, Vimentin. Isoelectric focusing was from right (basic) to left (acidic) and SDS-polyacrylamide gel electrophoresis from top to bottom. Numbers in b and c indicate pH at the positions of the full circles. Arrowheads, plectin spots.

Fig. 5 Electron microscopy of rotary shadowed C6 cell plectin. Samples of purified plectin were transfered into 2mM Tris/HCl, pH 8.0 by ultrafiltration and either immediately processed for electron microscopy (a,b,c) or first incubated in the presence of 100mM Tris/HCl, pH 7.0, for two hours at 37°C (d). Protein concentration was 0.05-0.10 mg/ml. a, x 200,000, bar 100nm; b,c,d, x 150,000, bar 100nm.

the major phosphate acceptor (Herrmann and Wiche, 1987). A preliminary domain structure analysis of the molecule showed that plectin's phosphorylation sites, except for the Ca/calmodulin-dependent ones, were located within a 18 kDa domain at the end(s) of the polypeptide chain. This might be important for mechanisms regulating the interaction of plectin with other cellular components.

The determination of hydrodynamic parameters of plectin by analytical ultracentrifugation was hampered by its tendency to aggregate and its high sensitivity towards proteolysis. However, broadly sedimentation boundaries, characteristic of polydispers systems, can be avoided when sedimentation velocity centrifugation was performed at low protein concentrations (0.01-0.05 mg/ml) in low ionic strength buffer solution, such as 2mM Tris/HCl, pH 8.0. Under these conditions plectin sedimented as a major boundary near 10S.

Ultrastructure of single and aggregated plectin molecules. After rotary shadowing of plectin samples in low ionic strength buffer, numerous dumb-bell like structures were observed in the electron microscope (Fig. 5a). The average contour lenght, measured as the center to center distance between the two globular end domains, was close to 200nm. Negatively stained specimens revealed diameters of 9nm for both globular end domains and 2nm for the rod segment.

Assuming a sherical shape of the end domains their M_r could be calculated as 320,000, the M_r of the rod segment as 480,000, yielding a value of 1 120,000 for the whole molecule. Based on these data and the striking symmetry of the molecule we favour a model in which the dumb-bell like structure is generated from 4 equal polypeptide chains. Two chains, each, laterally associate to form one globular domain and one half of the rod segment. The tetrameric molecule is assembled from two halfs interacting at the tips of their rod domains. In fact, we have observed occasionally dumb-bell like structures that were kinked or broken at half distance between the globular domains.

Frequently aggregation of plectin molecules into dimeric or higher order structures was observed (Fig. 5b, c). These structures were formed through intra- and intermolecular associations of plectin's globular head domains. Complex aggregates of plectin were observed particularly at high concentrations of protein and salt. These aggregates consisted of large globular center pieces with interconnecting filaments and filaments radiating and looping out all around them (Fig. 5d). Thus, it seems that plectin molecules have a strong tendency to interact with themselves, preferentially via their globular head domains.

Molecular interaction partners of plectin. In order to detect interaction partners of plectin on the molecular level solid phase binding assays under varied conditions were performed. Using plectin samples radiolabeled with ^{32}P *in vivo* or ^{125}I *in vitro*, a number of prominent binding partners have been identified so far: vimentin, the microtubule associated proteins, MAP-1 and MAP-2, α-spectrin and its brain counterpart, the 240 kDa polypeptide chain of fodrin (Herrmann and Wiche, 1987). The binding of plectin to most of these proteins was also demonstrated by similar overlays using unlabeled plectin whose binding was dedected by incubation with antibodies to plectin followed by secondary antibodies coupled to alkaline phosphatase.

Previous experiments have shown that plectin interacts *in vitro* with microtubules polymerized by temperature- or taxol-dependent procedures (Koszka et al., 1985). Furthermore several lines of evidence indicated that plectin interacted with vimentin filaments. These included the *in vitro* reconstitution of plectin/IF networks from purified components, codistribution of plectin with vimentin in repeated rounds of *in vitro* filament assembly and disassembly, partial codistribution with vimentin filaments demonstrated by immunofluorescence microscopy, and immunoelectron microscopy of whole mount or deep etched replicas of cytoskeletons (Wiche and Baker, 1982; Wiche et al., 1983; and unpublished data). Hence it is possible that plectin is a multifunctional cytoskeletal crosslinking protein. According to the proposed scheme shown in Fig. 6, plectin might interlink IFs and via spectrin type polypeptides anchor these filaments into the plasma membrane. On the other hand plectin might be engaged in the long recognized interactions of IFs and microtubules (Ishikawa et al., 1968; Goldman, 1971; Singer et al., 1982;), likely in concert with MAPs. Furthermore, plectin/MAP interactions might play some role in the interplay of microfilaments with IFs, as MAPs have previously been shown to interact also with actin filaments (Sattilaro et al.,1981, Griffith and Pollard, 1982, Arakawa and Frieden, 1984). Provided this scheme comes close to reality, plectin, in fact, would be the most versatile crosslinking protein of the cytomatrix reported to date.

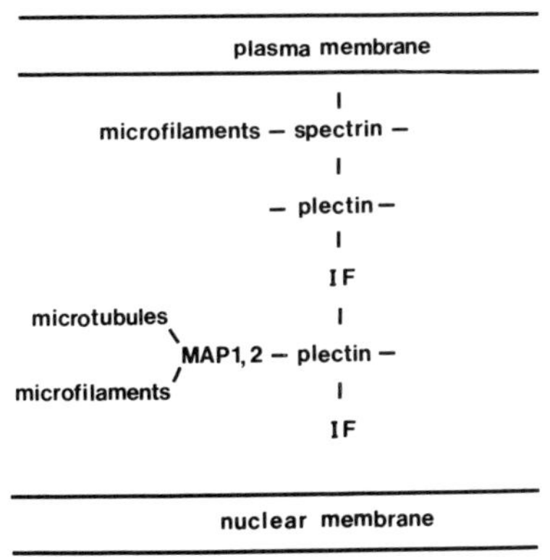

Fig. 6 Schematic model of plectin's engagement in the construction of a three-dimensional lattice between plasma membrane and nucleus. (Reproduced by permission of Amer. Soc. Biol. Chem., Inc.)

HIGH M_r MICROTUBULE ASSOCIATED PROTEINS (MAPs)

The major microtubule associated proteins in brain tissue, MAP-1 and MAP-2, and their counterparts in non-neuronal cells and tissues have been another important area of interest of our laboratory over the last years. Recent results of relevance for the cellular function(s) of these proteins can be summarized under the following five points:

1. Widespread occurrence. Immunofluorescence microscopy of cultured cells and tissue sections as well as immunoblotting of microtubules polymerized *in vitro* with taxol showed that polypeptides immunologically related to neuronal MAP-1 and MAP-2 are widespread also in cells and tissues of non-neuronal origin (reviewed by Wiche, 1985). It is possible, however, that the structural relationship between neuronal and non-neuronal MAPs, especially of MAP-2, is confined to only one or a few common epitopes, as suggested by the work of Lewis et al., 1986.

2. Microheterogeneity. There is evidence for microheterogeneity not only of neuronal but also of non-neuronal MAPs. For instance, microtubules, polymerized *in vitro* from rat glioma C6 cells, like those from mamalian brain, displayed three bands in the MAP-1 region and a number of bands at the MAP-2 position. In immunoblotting experiments the entire MAP-1 region stained with antibodies to brain MAP-1, different antisera to MAP-2 stained bands at the position of MAP-2A (Wiche et al., 1986). Multiple bands with the electrophoretic mobility of MAP-1 and MAP-2 were observed also in microtubule preparations from various non-neuronal tissues and red blood cells (Fig. 7).

3. Tissue-specific primary structure. The immunoreactivities of non-neuronal MAP-antigens with antibodies to neuronal MAP-1 and MAP-2 varied and in some cases were quite limited, indicating differences in the primary structure of MAPs. Examining a whole series of different cell lines and tissues, using immunofluorescence microscopy and immunoblotting, we came to the conclusion that tissue-specificity prevails over species specificity. Thus, particularly when using monoclonal antibodies to non-neural MAPs chances are high that no cross-reactivities will be observed with non-neuronal MAPs.

4. Differential distribution. High M_r MAPs are differentially distributed among different microtubule populations and occasionally they are found at locations where there are no microtubules. The most dramatic case repoted to date probably is the distinct distribution of MAP-1 and MAP-2 to axonal and dendritic microtubules, as we have repoted using polyclonal antibodies, and others using monoclonal antibodies (reviewed by Wiche et al., 1985). Of interest in this context are also the reports by Sato et al., 1984, and Briones and Wiche, 1985, showing the occurrence of a MAP-1 analogue in the nucleus, and the presence of a MAP-related antigen in the extracellular matrix compartment of fibroblast cells, respectively.

5. Periodic protrusions. We have shown by immunoelectron microscopy that the molecular arrangement of non-neuronal MAPs on the surface of *in vitro* polymerized non-neuronal microtubules is very similar to that of neuronal MAPs (Koszka et al., 1985). This suggests that non-neuronal MAPs, like those of neuronal origin, are capable of forming periodically arranged protrusions on the microtubular surface.

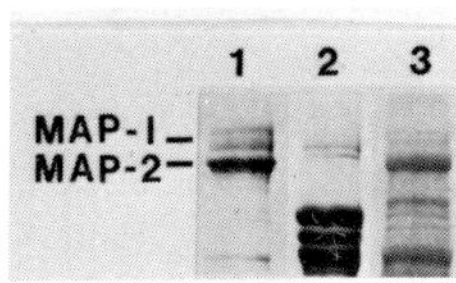

Fig. 7 Microheterogeneity of non-neuronal MAPs. Microtubules assembled in vitro from soluble cell extracts of rat liver (lane 3) and chicken erythrocytes (lane 2) were analysed by high resolution gel electrophoresis. Lane 1, control (hog brain MAPs).

On the basis of these five major conclusions we derived the following working hypothesis regarding the function of MAPs. We believe that the important cellular function of high M_r MAPs is not the promotion of microtubule assembly, as has been infered from early in vitro assembly experiments, but the physical interlinking of microtubules with different cell organelles including other cytoskeletal filaments. For the promotion of microtubule assembly MAPs would probably not need to be structurally diversified, rather one would expect that they have structural features in common. However, for the crosslinking of microtubules to a variety of different organelles and other cell components, variations in the structure of MAPs may be very important. In fact, the microheterogeneity of MAPs, together with their structural tissue specificity, allows for a broad variability in their primary structure. As outlined in the scheme shown in Fig. 8, we postulate that each MAP has at least two binding domains. One enables it to bind to tubulin; the other, to another cellular component. The structure of the tubulin binding domain may be similar for all MAPs; however, the structure of the other binding domain could be unique, and thus specific, for each MAP. Because of this specific binding domain, the interaction partner of each MAP-species could be specific. Therefore, microtubules could specifically interact with different partners, depending on their associated MAP. In this way MAPs could have a dramatic effect on the structure of the cytoplasm. Moreover, the differential expression of MAPs during cell development and differentiation could lead to changes in the cytoarchitecture, which are prerequisites for dynamic alterations in cell morphology and function. The tissue-specificity of MAP structure may also be of importance for specific interactions of MAPs with tubulin isotypes.

At least 2 Binding Domains of MAPs: Tubulin
Other(s) (specific)

↓

Different Binding Partners of MAPs and Microtubules

↓

Effect on Structure of Cytoplasm

↓

Differential Expression of MAPs During Cell Development and Differentiation Could Lead to Changes in Cytoarchitecture (Cell Morphology, Function)

Fig. 8 MAP structure and function - a hypothesis.

ACKNOWLEDGEMENTS

This work was supported by grants from the Austrian Research Fund, the Austrian National Bank and the Verlassenschaft Josefine Hirtl.

REFERENCES

Arakawa, T. and Frieden, C. (1984). Interaction of microtubule-associated proteins with actin filaments. J. Biol. Chem. 259, 11730-11734.

Briones, E. and Wiche, G. (1985). M_r 205,000 sulfoglycoprotein in extracellular matrix of mouse fibroblast cells is immunologically related to high molecular weight microtubule-associated proteins. Proc. Natl. Acad. Sci. 82, 5776-5780.

Goldman, R.D. (1971). The role of three cytoplasmic fibers in BHK-21 cell motility. I. Microtubules in the effects of colchicine. J. Cell Biol. 51, 752-762.

Griffith, L.M. and Pollard, T.D. (1982). The interaction of Actin filaments with microtubules and microtubule-associated proteins. J. Biol. Chem. 257, 9143-9151.

Herrmann, H. and Wiche, G. (1983). Specific in situ phosphorylation of plectin in detergent-resistant cytoskeletons from cultured chinese hamster ovary cells. J. Cell Biol. 258, 14610-14618.

Herrmann, H., Dalton, J.M. and Wiche G. (1985). Microheterogeneity of microtubule-associated proteins, MAP-1 and MAP-2, and differential phosphorylation of individual subcomponents. J. Biol. Chem. 260, 5797-5803.

Herrmann, H. and Wiche, G. (1987). Plectin and IFAP-300K are homologous proteins binding to microtubule associated protein 1 and 2 and to the 240-kilodalton subunit of spectrin. J. Biol. Chem. 262, 1320-1325.

Ishikawa, H., Bischoff, R. and Holtzer, H. (1968). Mitosis and intermediate-sized filaments in developing skeletal muscle. J. Cell Biol. 38, 538-555.

Koszka, C., Leichtfried,F.E. and Wiche, G. (1985). Identification and spatial arrangement of high molecular weight proteins (M_r 300,000-330,000) co-assembling with microtubules from a cultured cell line (rat glioma C6). Eur. J. Cell Biol. 38, 149-156.

Lewis, S.A., Villasante, A., Sherline, P. and Cowan, N.J. (1986). Brain specific expression of MAP 2 detected using a cloned cDNA probe. J. Cell Biol. 102, 2098-2105.

Lieska, M., Yang, H-Y. and Goldman, R.D. (1985). Purification of the 300K intermediate filament-associated protein and its in vitro recombination with intermediate filaments. J. Cell Biol. 101, 802-813.

Murphy, D.B. and Borisy, G.G. (1975). Association of high molecular weight proteins with microtubules and their role in microtubule assembly in vitro. Proc. Natl. Acad. Sci. 72, 2696-2700.

Sattilaro, R.F., Dentler, W. and Le Cluyse, E.L. (1981) Microtubule associated proteins (MAPs) and the organization of actin filaments in vitro. J. Cell Biol. 90, 467-473.

Sato, C., Nishizawa, K., Nakamura, H. and Veda, R. (1984). Nuclear immunofluorescence by a monoclonal antibody against microtubule-associated protein 1 as it is associated with cell proliferation and transformation. Exp. Cell Res. 155, 33-42.

Singer, S.J., Ball, E.H., Geiger, B. and Chen, W-T. (1982). Immunolabeling studies of cytoskeletal associations in cultured cells. Cold Spring Harbor Symp. Quant. Biol. 46, 303-316.

Wiche, G. (1985). High-molecular-weight microtubule associated proteins (MAPs): a ubiquitous family of cytoskeletal connecting links. Trends Biochem. Sci. 10, 67-70.

Wiche, G. and Baker, M.A. (1982). Cytoplasmic network arrays demonstrated by immunolocalization using antibodies to high molecular weight proteins present in cytoskeletal preparations from cultured cells. Exp. Cell Res. 138, 15-29.

Wiche, G., Herrmann, H., Leichtfried, F.E. and Pytela R. (1982). Plectin: a high molecular weight polypeptide component that copurifies with intermediate filaments of the vimentin type. Cold Spring Harbor Symp. Quant. Biol. 46, 475-482.

Wiche, G., Krepler, R., Artlieb, U., Pytela, R. and Denk, H. (1983). Occurrence and immunolocalization of plectin in tissues. J. Cell Biol. 97, 887-901.

Wiche, G., Briones, E., Koszka, C., Artlieb, U. and Krepler, R. (1984). Widespread occurrence of polypeptides related to microtubule-associated proteins (MAP-1 and MAP-2) in non-neuronal cells and tissues. EMBO. J. 3, 991-998.

Wiche, G., Herrmann, H., Dalton, J.M., Foisner, R., Leichtfried, F.E., Lassmann, H., Koszka, C. and Briones, E. (1986). Molecular aspects of MAP-1 and MAP-2: Microheterogeneity, in vitro localization and distribution in neuronal and non-neuronal cells. Ann. N.Y. Acad. Sci. 466, 180-198.

NEW METHODS TOWARD PURIFICATION AND STRUCTURAL STUDIES OF MICROTUBULE ASSOCIATED PROTEINS

J.C.Vera, C.I.Rivas and R.B.Maccioni

Department of Biochemistry, Biophysics and Genetics, University of Colorado Health Sciences Center, Denver, Colorado 80262, USA

ABSTRACT

Because of limitations inherents in the methodology utilized to purify MAPs, we do not know practically anything about the biochemical and functional properties of MAPs present in most non-neuronal cells and tissues. In a attempt to overcome these limitations, we have searched for new experimental approaches toward the purification and characterization of MAPs. As a result of this, we have developed an affinity purification method that take advantage of two well known properties of MAPs: heat stability and binding to calmodulin. This method has been succesfully applied to obtain MAPs from tissue of neuronal and non-neuronal origins. The results obtained confirm the wide distribution of MAPs in different systems and provide reliable information on the *in vivo* presence of different isoforms of MAPs.

INTRODUCTION

Microtubules isolated by temperature-dependent cycles of polymerization-depolymerization contain, besides tubulin, several other proteins called the microtubule-associated proteins (MAPs). Some of these proteins are well characterized, specially those isolated by the *in vitro* reversible polymerization of tubulin from brain extracts. In this method, microtubules are induced to polymerize in warm buffers and after collecting them by centrifugation they are subjected to cold induced depolymerization. Proteins that copurify in constant stiochiometry with tubulin and whose association with microtubules depend upon the formation of the polymer have been identified (Borisy et al,1975; Weingarten et al,1975). Hence, in the reversible assembly procedure, a MAP is a microtubule protein that can be sedimented by centrifugation under polymerization conditions in warm, but not under depolymerization conditions in the cold.

The MAPs isolated from mammalian brains consist of multiple components that can be classified in three groups: MAP-1, MAP-2 and tau (Olmsted,1986). MAP-1 (Mr: 330,000-350,000) contains at least three prominent electrophoretic species (MAP-1A,B and C) when analyzed by high resolution polyacrylamide gel electrophoresis (Bloom et al,1984; Herrmann et al,1985). Using the same methodology, the presence of at least two MAP-2 bands (MAP-2A and B, Mr: 270,000-300,000) was established (Kim et al,1979; Herrmann et al,1985). Tau protein isolated from bovine brains consists of four polypeptides with molecular weights of 63,000, 59,000, 56,000 and 53,000 (Cleveland et al,1977a,b). Within each family of proteins, the individual subcomponents are highly homologous as indicated by one and two-dimensional peptide mapping, aminoacid composition determinations and immunological analyses (Cleveland et al,1977a,b; Herrmann et al,1984,1985). MAP-1, MAP-2 and tau promote the assembly of microtubules *in vitro* and decrease the critical concentration of tubulin needed to polymerize into microtubules (Cleveland et al,1977a,b; Herzog and Weber,1978; Kuznetsov et al,1981a,b). A new MAP have been recently described, MAP-3,that includes two components (Huber et al,1985).

For most tissue and cellular sistems from non-neuronal origin, attempts to purify MAPs were largerly unsuccessful because the self-assembly of cytoplasmic microtubules does not proceed to a significant extent presumably due to the low concentration of tubulin in non-neuronal cell types. A few exceptions were human HeLa and differentiated mouse neuroblastoma cells (Bulinski and Borisy,1979,1980; Olmsted and Lyon,1981). Finally, a method was devised (Vallee,1982) in wich microtubules are induced to assemble with the aid of taxol, a drug that promotes the assembly of microtubules *in vivo* and *in vitro*. In the taxol procedure, a protein is considered a MAP if it can be sedimented by centrifugation under assembly conditions in a taxol dependent fashion, but not in the absence of taxol.

When this method was applied to brain tissue, the same microtubular components (tubulin and MAPs) were obtained as compared with the temperature-driven reversible assembly (Vallee,1982). Using taxol-induced polymerization of microtubules, MAPs have been identified in a wide variety of sistems, including different brain regions and cells and tissues of non-neuronal origins (Vallee,1982; Vallee and Bloom,1983; Wiche et al,1984; Bloom et al,1985a,b). These studies have confirmed previous immunological analyses that indicated a widespread ocurrence of MAP-like proteins in neuronal as well as in non-neuronal cells and tissues, and their possible tissue especificity (Wiche,1985; Olmsted,1986).

NEW EXPERIMENTAL APPROACHES ARE NEEDED

The application of the temperature-dependent and taxol-induced polymerization of tubulin to the identification and purification of MAPs has increased our understanding of their structural diversity. However, both methods present some disadvantages in that they are relatively slow and depend on the co-polymerization of MAPs with microtubules. It has been reported that under these purification conditions MAPs are exposed to unwanted enzimatic modifications, including proteolysis and phosphorylation (Sloboda et al, 1975; Sandoval and Weber,1978; Lindwall and Cole,1984). Hence, differents forms of tau polypeptides have been observed (Lindwall and Cole,1984). According to this, an important consideration in the development of new methods for the purification of MAPs is the need to avoid the unnecessary exposure of MAPs to the action of the different enzymes present in the initial homogenate. This is especially important if the aim is to obtain MAPs to be used in comparative studies. Furthermore, depending on the cell lysis method used prior to the assembly-disassembly of microtubules the relative proportions of MAP-1 and MAP-2 vary (Murphy and Borisy,1975; Karr et al,1979). On the other hand, the co-polymerization procedure, even in the presence of taxol, limits many studies of MAPs to sources with a sufficient high concentration of tubulin to allow the polymerization procedure to be performed. As a result of this, we do not know practically anything about the biochemical characteristics of MAPs from non-neuronal tissue, or if they share the capacity to promote tubulin polimerization with their brain counterparts. Clearly, the unambiguous identification and biochemical characterization of the MAP complement of non-neuronal tissue is an essential step toward understanding the possible modulatory role of MAPs within the dynamics and structural aspects of the cytoskeleton.

AN AFFINITY PURIFICATION PROCEDURE BASED ON TWO WELL KNOWN PROPERTIES OF MICROTUBULE ASSOCIATED PROTEINS; HEAT STABILITY AND BINDING TO CALMODULIN

A useful criterion widely utilized for the classification of MAPs has been the resistence of MAP-2 and tau to denaturation at elevated temperatures (Herzog and Weber,1978). Preparations freed of MAP-1 can be easily obtained by using a thermoprecipitation step in which three-cycled microtubule protein is heated at 85-100 °C. Both MAP-2 and tau remain solubles and MAP-1 is denatured and precipitate (Herzog and Weber,1978). However, under certain conditions MAP-1 is also heat stable (Vallee,1985). In regard with the possible detrimental effect of the boiling step on the structure-activity of MAP-2 and tau, these proteins retain their capacity to induce tubulin polymerization (Herzog and Weber,1978). Furthermore, physicochemical studies indicate that boiled MAP-2 is indistinguishable from MAP-2 prepared without the boiling step (Hernandez et al,1986).

The assembly of tubulin in crude brain extracts is inhibited by micromolar Ca^{2+} concentrations, whereas the assembly of purified preparations of tubulin is sensitive to millimolar Ca^{2+} concentrations (Weisenberg,1972; Berkowitz and Wolf,1981). It has been shown that Ca^{2+} exerts its inhibitory effect in two different ways: an intrinsec calcium ion effect on the tubulin dimer and a calmodulin mediated effect (Marcum et al,1978; Berkowitz and Wolff,1981). The calmodulin-dependent inhibition require the presence of MAPs. Both

MAP-2 and tau appear to bind calmodulin specifically with similar affinity (Lee and Wolf,1984; Kumagai et al,1986). The possibility that MAP-1 could be able to bind calmodulin is suggested by its functional similarity to both MAP-2 and tau, but this is a matter that has not been examined at all.

We have relied mainly on these two properties of MAPs, heat stability and binding to calmodulin, to develop novel methods of purification without the requirement of a certain treshold concentration of microtubule protein (and hence, MAPs) during the actual purification. We decided to take advantage of the heat stability of MAPs using the boiling method as an early purification step, to inactivate unwanted enzimatic activities but without affecting the _in vitro_ activity of MAPs. This step is followed by affinity chromatography onto a column of calmodulin-Sepharose to discard possible heat-stable contaminating proteins. This purification procedure was succesfully applied to obtain functional MAPs from brain and also from tissue of non-neuronal origin (Vera et al,1987).

PURIFICATION OF FUNCTIONAL MICROTUBULE ASSOCIATED PROTEINS FROM BRAIN TISSUE

When boiled brain proteins were fractionated in a column of calmodulin-Sepharose and the especifically retained material was eluted and analyzed by polyacrylamide gel electrophoresis, a restricted number of polypeptides was observed (Fig 1A). These include protein bands that migrate at positions in the gel corresponding to those expected for MAP-1, MAP-2 and tau (see Fig 1A). A close examination of the electrophoretic pattern revealed the presence of multiple subcomponents according to the expected distribution for each family of proteins: 3 isoforms of MAP-1, 2 forms of MAP-2 and at least 4 molecular species of tau. The different subcomponents of MAPs were efficiently resolved using a linear polyacrylamide gradient.

Unexpected at this point was the presence of polypeptides with characteristics of MAP-1 because it has been described that these proteins are denatured during the boiling procedure (Herzog and Weber,1978). However, the heat denaturation step is normally carried out after at least two or three cycles of temperature-induced polymerization-depolymerization, with protein preparations that consist fundamentally of tubulin and MAPs. We have to point out that our heat denaturation step was carried out directly in the initial homogenate without previous tubulin polymerization. The different results obtained suggest that MAP-1 polypeptides could be modified during the previous assembly cycles altering some of its physicochemical properties. It is also possible that intrinsec factors present in the initial homogenate can modulate the heat stability of MAP-1, factors which are likely to be lost during the purification procedure. Evidence that the heat stability of MAP-1 is a function of the composition of proteins present during exposure to elevated temperature has been presented (Vallee,1985), but this depend on the previous treatment of the protein preparation (Vallee,1985; Kuznetsov et al,1981a,b). Hence, caution must be excercised on the use of heat stability as a criterion for the classification of MAPs.

A pair of closely spaced protein bands with aproximate molecular weight of 180,000 were also observed after gel electrophoresis (Fig 1A). In principle, these could be the MAP-3 polypeptides (Huber et al,1985). However, it has not been possible to detect the presence of MAP-3 by Coomassie ble staining after polyacrylamide gel electrophoresis (Huber et al,1985,1986). Furthermore, the identity of MAP-3 as defined using a monoclonal antibody raise several interrogants. Specially intriguing is their immunological relationship to MAP-1 (Huber et al,1986).

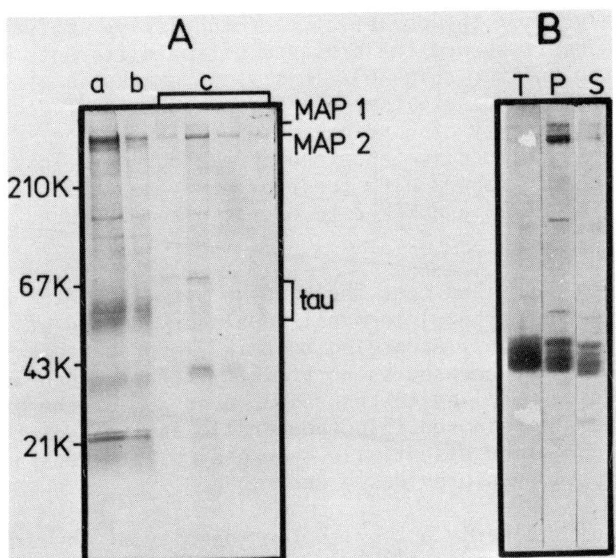

FIG 1A. Affinity chromatography of boiled brain proteins on calmodulin-Sepharose. Gel electrophoresis of the (a) initial homogenate, (b) non-retained material and (c) different fractions of the material eluted with EGTA. The position of MAP1, MAP2 and tau was determined using the corresponding brain proteins.

FIG 1B. Incorporation of MAPs in the microtubule polymer. Gel electrophoresis of phosphocellulose purified tubulin (T), pellet (P) and supernatant (S) obtained after MAPs-induced polymerization of tubulin.

The proteins isolated using the metodology described in the previous section share four properties with the classically purified MAPs: heat stability, binding to calmodulin, molecular weights and presence of multiple subcomponents. The most remarkable property of brain MAPs is their capacity to induce the polymerization of phosphocellulose purified tubulin in a temperatura dependent manner, under polymerization conditions (Cleveland et al,1977a,b;Herzog and Weber,1978; Kuznetsov et al,1981a,b). Under standard polymerization conditions, phosphocellulose purified tubulin do not show any noticeable assembly. On the other hand, when a constant amount of tubulin was assambled in the presence of increasing amounts of the affinity-purified proteins, a clear increase in the levels of polymerized protein was observed. As expected, the final level of polymerized protein was proportional to the amount of affinity proteins added, as assessed by using turbidimetry at 350 nm and the centrifugation method (Maccioni et al,1985). The polymerization was cold reversible. Analyses of the pellet of microtubules by polyacrylamide gel electrophoresis showed the presence of the differents MAPs associated with the polymer (Fig 1B). This is a remarkable observation because it clearly indicate that all the components of each MAP family share the property of binding to tubulin after the boiling step followed by calmodulin-Sepharose chromatography. This is especially important in regard with the previously reported differential response of MAP-1 and MAP-2 to elevated temperature (Herzog and Weber,1978).

It has been described that the binding site for the MAPs is located in the 4-kDa carboxyl terminal regulatory domain of tubulin (Serrano et al,1984a,b). According to this, we prepared a 4 kDa -Sepharose 4B column to be used as an affinity matrix. When our preparation of MAPs was applied to the resin, over 80% of the protein sample applied was retained. Electrophoretic analysis of the retained fraction confirmed essentially the results obtained in the polymerization experiments previously described.

These results clearly establish the identity of the proteins purified using this new experimental approach as tipical MAPs, with the additional advantage of extending our knowledge on the properties of the less studied MAP-1. In addition, these data show the great functional similarity shared by the brain MAPs, with at least five characteristics, physicochemical and functional, that can be used as a powerful criterium to identify them (Table 1).

MICROTUBULE ASSOCIATED PROTEINS FROM TISSUE OF NON NEURONAL ORIGIN

The presence of MAPs in tissue of non-neuronal origin has been documented fundamentally using immunological methods (Wiche, 1985; Olmsted,1986). However, our knowledge about the biochemical properties of non-neuronal MAPs is restricted to a very limited number of examples (Wiche,1985; Olmsted,1986). In addition, contradictory results in regard to the presence of different isoforms of MAPs have been reported by using several experimental approaches (Wiche, 1985 Olmsted,1986). Thus, another important consideration implicit in our interest in to develop new methods for the purification of MAPs was the possibility of their application to tissue of non-neuronal origin.

TABLE 1: PROPERTIES OF BRAIN MICROTUBULE ASSOCIATED PROTEINS[a]

1: Heat stability (100 °C for 10 minutes)
2: Specific binding to calmodulin (requires Ca^{2+})
3: Presence of multiple subcomponents (3 for MAP-1, 2 for MAP-2 and at least 4 for tau)
4: Promotion of tubulin polymerization in a temperature dependent fashion
5: Binding to the 4 kDa carboxyl terminal regulatory domain of tubulin

(a) MAPs purified according to our method using the boiling step followed by affinity chromatography on calmodulin-Sepharose

We applied our procedure to the purification of MAPs from mouse brain, liver and lung. Under the standard chromatographic conditions, the boiled homogenates from brain, liver and lung showed the presence of proteins with the capacity to bind specifically to calmodulin. As expected, the electrophoretic pattern of the brain proteins closely resemble to those described as belonging to MAPs. The similarities are reflected in equivalent relative molecular weigth and presence of multiple subcomponents. Most interestingly,both liver and lung protein preparations showed a protein distribution, in polyacrylamide gels, similar but not identical to the protein pattern of the brain proteins. When tested in an *in vitro* assay, the three protein preparations showed the capacity to induce the polymerization of tubulin in a temperature dependent fashion. Electro phoretic analyses indicated that these proteins were incorporated in the microtubule polymer. The co-polymerized proteins included a high molecular weight fraction from the liver and lung preparations This is especially important because this is, to our knowledge, the first report of active MAP-like proteins in this kind of tissues, and corroborate previous immunological evidences of their presence. This method was also succesfully used to obtain MAPs from teratocarcinoma cells.

ADVANTAGES OF A NEW PURIFICATION PROCEDURE

According to the results described in the previous sections,our purification method presents several advantages as compared with the traditional ones. The early boiling step appear to be efficient in inactivating endogenous proteases as indicated by the practically complete absence of degradation products from MAP-1 or MAP-2 (Fig 1). Under the same considerations, the presence of multiple subcomponents of MAP-1, MAP-2 and tau appears to be a reliable representation of the *in vivo* situation, which is an important point. Furthermore, the metod has also permited to obtain novel information about the less known MAP-1 and confirmed those described for MAP-2 and tau, enlightening the functional relationships between the different MAPs. The most important advantage of this method is that it can be applied to obtain functional MAPs from non-neuronal tissue in sufficient amounts to carry out biochemical and structural-functional analyses. Actually, on the basis on the use of affinity chromatography on calmodulin-Sepharose we are developing new experimental procedures to separate the individual subcomponents of MAPs, with promising preliminary results.

CONCLUDING REMARKS

During the last few years, the study of MAPs has been largerly directed toward their biochemical characterization and documenting where these proteins occur in cells. Recent evidences indicate that they exhibit a rather wide cellular distribution and are tissue especific. Furthermore, data obtained from studies on the distribution and activity of MAPs at various stages of brain development suggest that the expression of MAPs is under regulation at both transcriptional and translational levels. MAPs are also subject to post translational modifications (phosphorylation by endogenous kinases) with functional consequences in their capacity to induce tubulin polymerization(Wiche,1985; Olmsted,1986). Relative to neuronal MAPs, very little is known about MAPs from non-neuronal tissues. Our knowledge is based fundamentally on immunological studies and taxol-driven assembly of microtubules (Olmsted,1986).On the other hand, the presence of individual tubulin isotypes, until now functionally indistinguishables, has led to the suggestion that tubulin assembly characteristics could be modulated *in vivo* by other especific microtubule components (Cleveland,1987). Hence, the structural diversity of MAPs could be functionally important, with different isoforms bearing the capacity of differential binding to particular tubulin isotypes. According to this, a detailed knowledge of the distribution, biochemical characteristics and evolution of MAPs is a basic requisite toward our understanding of the possible multifacetic aspects of MAPs functions. These facts point to the importance of developing new experimental approaches aimed to identify and purify MAPs from different types of cells and tissues, especially those from non-neuronal origins.

ACKNOWLEDGEMENTS

We gratefully acknowledge the support of the Council for Tobacco Research, USA (Grant 1913), to this investigation.

REFERENCES

Berkowitz SA and Wolf J (1981) Intrinsec calcium sensitivity of tubulin polymerization.J.Biol.Chem.257,11216-11223

Bloom GS, Luca FC and Valle RB (1985a) Microtubule associated-protein 1B: identification of a major component of the neuronal cytoskeleton Proc.Natl.Acad.Sci.USA 82,5404-5408

Bloom GS, Luca FC and Vallee RB (1985b) Identification of high molecular weight microtubule-associated proteins in anterior pituitary tissue and cells using taxol-dependent purification combined with microtubule-associated protein especific antibodies Biochemistry 24,4185-4191

Bloom GS, Schenfeld TA and Vallee RB (1984) Widespread distribution of the major polypeptide component of MAP 1 (microtubule associated protein 1) in the nervous sistem J.Biol.Chem.98, 320-330

Borisy GG, Marcum JM, Olmsted JB, Murphy DB and Johnson KA (1975) Purification of tubulin and of associated high molecular weight proteins from porcine brain and characterization of microtubule assembly <u>in vitro</u> Ann.NY Acad.Sci.253,107-132

Bulinski J and Borisy GG (1979) Self-assembly of microtubules in extracts of cultured HeLa cells and the identification of HeLa microtubule-associated proteins Proc.Natl.Acad.Sci.USA 76, 293-297

Bulinski J and Borisy GG (1980) Microtubule associated proteins from cultured HeLa cells. Analysis of molecular properties and effects on microtubule polymerization J.Biol.Chem. 255, 11570-11576

Cleveland DW, Hwo SY and Kirschner MW (1977a) Purification of tau a microtubule associated protein that induces assembly of microtubules from purified tubulin J.Mol.Biol.116,207-226

Cleveland DW, Hwo SY and Kirschner MW (1977b) Physical and chemical properties of purified tau factor and the role of tau in microtubule assembly J.Mol.Biol.116,227-248

Hernandez MA, Avila J and Andreu JM (1986) Physicochemical characterization of the heat-stable microtubule associated protein

MAP-2 Eur.J.Biochem.154,41-48

Herrmann H, Dalton JM and Wiche G (1985) Microheterogeneity of microtubule associated proteins, MAP 1 and MAP 2, and differential phosphorylation of individual subcomponents J.Biol.Chem. 260,5797-5803

Herrmann H, Pytella R, Dalton J and Wiche G (1984) Structural homology of microtubule associated proteins 1 and 2 demostrated by peptide mapping and immunoreactivity J.Biol.Chem.259, 612-617

Herzog W and Weber K (1978) Fractionation of brain microtubule associated proteins. Isolation of two different proteins which stimulate tubulin polymerization _in vitro_ Eur.J.Biochem 92,1-8

Huber G, Alaimo-Beuret D and Matus A (1985) MAP 3: Characterization of a novel microtubule-associated protein J.Cell Biol. 100,496-507

Huber G, Pehling G and Matus A (1986) The novel microtubule associated protein MAP 3 contributes to the _in vitro_ assembly of brain microtubules J.Biol.Chem.15,2270-2273

Karr TL, White HD and Purich DL (1979) Characterization of brain microtubule proteins prepared by selective removal of mitocondrial and synaptonemal components J.Biol.Chem.254,6107-6111

Kim H, Binder LI and Rosembaun JL (1979) The periodic association of microtubule-associated proteins with brain microtubules _in vitro_ J.Cell Biol.80:266-276

Kumagai H, Nishida E, Kotani S and Sakai H (1986) On the mechanism of calmodulin-induced inhibition of microtubule assembly _in vitro_ J.Biochem.99,521-525

Kuznetsov SA, Rodionov VI, Gelfand VI and Rosenblat VA (1981a) Purification of high-Mr microtubule proteins MAP1 and MAP2 FEBS lett.135,237-240

Kuznetsov SA, Rodionov VI, Gelfand VI and Rosenblat VA (1981b) Microtubule associated protein MAP1 promotes microtubule assembly _in vitro_ FEBS lett.135,241-244

Lee Y and Wolf J (1984) Calmodulin binds to both microtubule-associated protein 2 and tau proteins J.Biol.Chem.259,1226-1230

Lindwall G and Cole RD (1984) The purification of tau protein and the occurrence of two phosphorilation states of tau in brain J.Biol.Chem.259,12241-12245

Maccioni RB, Serrano L, Avila J and Cann J (1985) Characterization and structural aspects of the enhanced assembling capacity of tubulin after removal of its carboxyl-terminal domain Eur.J. Biochem.156,375-381

Marcum JM, Dedman JR, Brinkley BR and Means AR (1978) Proc.Natl. Acad.Sci.USA 75,3771-3775

Murphy DB and Borisy GG (1975) Association of high-molecular weight proteins with microtubules and their role in microtubule assembly _in vitro_ Proc.Natl.Acad.Sci.USA 72,2696-2700

Olmsted JB (1986) Microtubule-associated proteins Annu.Rev.Cell Biol.2,421-457

Olmsted JB and Lyon HD (1981) A microtubule associated protein specific to differentiated neuroblastoma cells J.Biol.Chem. 256, 3507-3511

Sandoval IV and Weber K (1978) Calcium-induced inactivation of microtubule formation in brain extracts. Presence of a calcium dependent protease acting on polymerization-stimulating micro tubule-associated protein Eur.J.Biochem.92,463-470

Serrano L, Avila J and Maccioni RB (1984a) Controlled proteolysis of tubulin by subtilisin. Localization of the site for MAP2 interaction Biochemistry 23,4675-4681

Serrano,L, DeLaTorre J, Maccioni RB and Avila J (1984b) Involvement of the carboxy-terminal domain of tubulin on microtubule assembly and regulation Proc.Natl.Acad.Sci.USA 81,5989-5993

Sloboda RD, Rudolph SA, Rosembaun JL and Greengard P (1975) Cyclic AMP-dependent endogenous phosphorylation of a microtubule-associated protein Proc.Natl.Acad.Sci USA 72,177-181

Vallee RB (1982) A taxol-dependent procedure for the isolation of microtubules and microtubule-associated proteins (MAPs) J.Cell Biol.92,435-442

Vallee RB (1985) On the use of heat stability as a criterion for the identification of microtubule-associated proteins (MAPs) Biochem.Biophys.Res.Commun.133,128-133

Vallee RB and Bloom GS (1983) Isolation of sea urchin egg microtubules with taxol and identification of mitotic spindle microtubule-associated proteins with monoclonal antibodies Proc.Natl Acad.Sci.USA 80,6259-6263

Vera JC, Rivas CI and Maccioni RB (1987) Affinity purification of functional microtubule-associated proteins from neuronal and non-neuronal tissue (Submitted)

Weingarten MD, Lockwood AH, Hwo SY and Kirschner MW (1975) A protein factor essential for microtubule assembly Proc.Natl.Acad. Sci.USA 72,1858-1862

Weisenberg R (1972) Microtubule formation *in vitro* in solutions containing low calcium concentrations Science 177,1104-1105

Wiche G (1985) High molecular weight microtubule-associated proteins (MAPs). A ubiquitous family of cytoskeletal connecting links Trends Biochem.Sci.10,67-70

Wiche G, Briones E, Koszka C, Artlieb U and Krepler R (1984) Widespread ocurrence of polypeptides related to neurotubule-associated proteins (MAP1 and MAP2) in nonneuronal cells and tissues EMBO J.3,991-998

REGULATION BY CALCIUM OF THE cAMP-DEPENDENT PROTEIN KINASE-MAP2 INTERACTION

M.A. Hernández, L. Serrano and J. Avila. Centro de Biología Molecular (CSIC-UAM). Canto Blanco. 28049 Madrid. Spain.

Microtubule associated protein, MAP_2, serves as an anchorage point for the cytoplasmic typeII cAMP-dependent kinase (PKA) present in brain (1,2). MAP_2 associates this kinase through its projection domain (1). This interaction could be regulated by calcium since previously we have found that the projection region of MAP_2 binds calcium (1-2 moles) (Kd ≃ 10 μM) (submitted for publication). Here, we present evidences for calcium modulation of the MAP_2-PKA interaction.

(I) Fractionation of microtubular proteins and protein kinase activities in the presence or absence of calcium. In order to analyze the effect of calcium on the MAP_2-PKA interaction, microtubule protein was fractionated by gel filtration in the presence or absence of this cation (10 μM or 1 mM Ca^{2+}). As shown in Fig. 1, when the cAMP-stimulated protein kinase activity was assayed, different activity peaks were obtained in the presence or absence of calcium. These differences were not due to changes in the elution of microtubular protein (Fig. 1B). In the absence of calcium, the maxima of kinase activity peaks were found in the same position

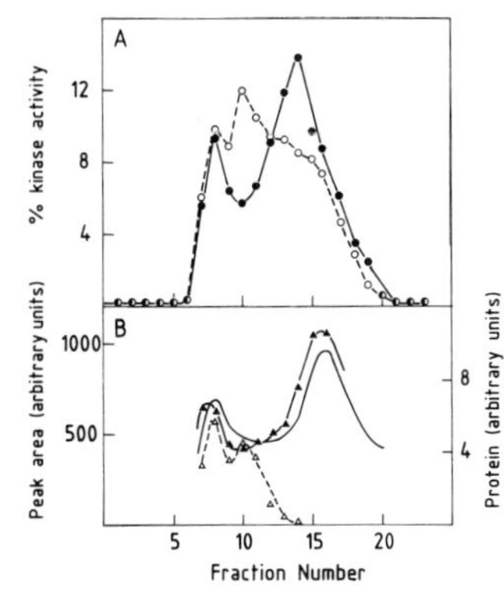

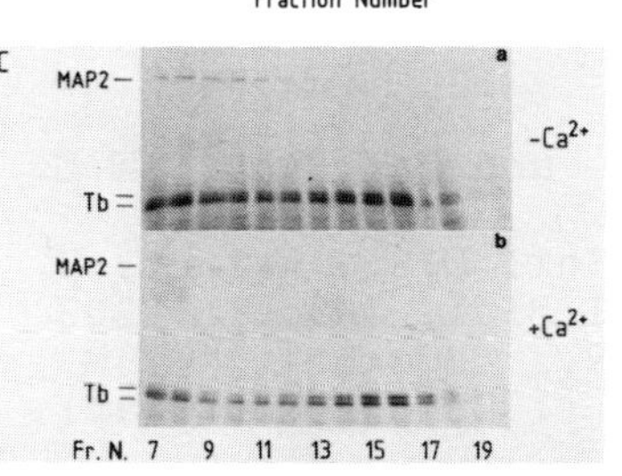

Fig. 1. Sepharose 4B-CL gel filtration of microtubule protein. Microtubule protein (0.25 mg) applied to a 0.5x20 cm Sepharose 4B-CL column preequilibrated with 0.1 M Mes pH 6.4, 0.1 mM $MgCl_2$ plus 10 μM $CaCl_2$ (●—●) or 1 mM EGTA (o---o). Panel (A) shows the cAMP-stimulated protein kinase activity of the fractions and (B) the total amount of protein in each fraction (—) together with the amount of MAP_2 (Δ--Δ) and tubulin (▲—▲) determined by densitometry of the gels shown in panel C. (C) Gel electrophoresis showing the proteins present in each fraction from the column in the presence (b) or absence of calcium (a).

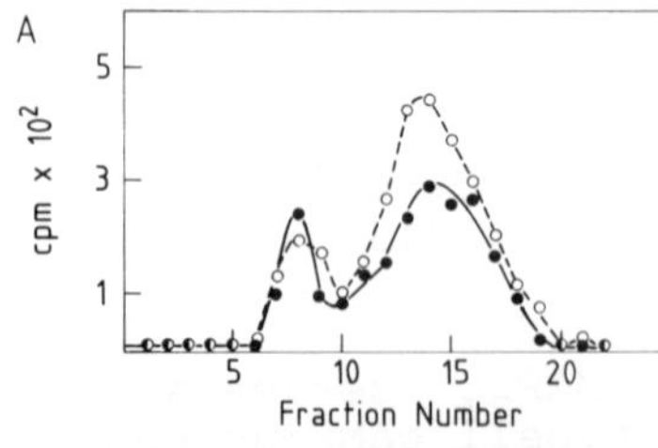

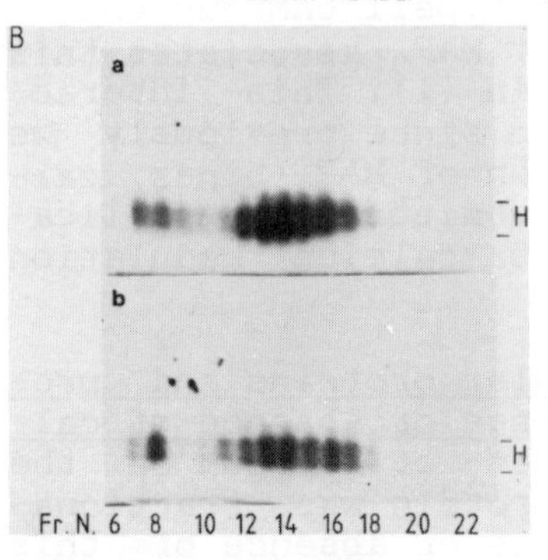

Fig. 2. Protein kinase assays. The protein kinase activity for each fraction from the column shown in Fig. 1 was assayed on histones (H) in the presence (o--o) or absence (●—●) of 10 μM cAMP. After autoradiography, the radiolabeled bands were excised and counted (A). Panel (B) shows the autoradiographs of the phosphorylated histones in the presence (a) or absence (b) of cAMP.

as the MAP_2 peaks (Fig. 1A and 1B) as previously described (1). When calcium was present, one of the maxima of kinase activity changed to a position where MAP_2 was almost absent. However, the other maximum of kinase activity remained in the void volume.

(II) Analysis of the protein kinase activities. Since it has been suggested that there may be two kinases associated to microtubules, one dependent and the other independent of cAMP (3), an experiment was performed to determine which one corresponded to that shifted by calcium. As shown in Fig. 2, when the protein kinase activity from the calcium column was assayed, the void volume activity was not activated by cAMP (cAMP-independent). On the other hand, the peak of kinase activity shifted by calcium was cAMP-dependent. The above results indicate that calcium dissociates the MAP_2-PKA interaction either by a direct or indirect effect of this cation on MAP_2 or the regulatory RII subunit of PKA. To analyze this point the above experiments were repeated but this time purified MAP_2 and PKA were used. In this case the interaction of PKA with MAP_2 was not affected by the presence of calcium up to 0.1 mM (data not shown). This result argues for a mediated role of calcium in the MAP_2-PKA interaction.

REFERENCES

(1) Valle, R.B., Di Bartolomeis, M.J. and Theurkauf, W.E. (1981) J. Cell Biol. 90, 568-576.
(2) De Camilli, P., Moretti, M., Donini, D.S., Walter, U. and Lohmann, S.M. (1986). J. Cell Biol. 103, 189-203.
(3) Vallano, M.L., Goldenring, J.R., Buckholz, T.M., Larson, R.E., De Lorenzo, R.J. (1985) Proc. Natl. Acad. Sci. USA 82, 3302-3206.

POLYMERIZATION OF MICROTUBULE-ASSOCIATED PROTEIN TAU FACTOR

E. Montejo, A. Nieto and J. Avila. Centro de Biología Molecular (CSIC-UAM) Cantoblanco. 28049-Madrid. Spain

Alzheimer's disease, one of the most frequent forms of senile dementia, is histologically characterized by the appearance of neurofibrillary tangles (NFT) (1), which are composed by aggregates of paired helical filaments (PHF). At present it is not known which is the protein component of those PHF but such a component should have the ability to assemble. Recently it has been shown that tau factor, one of the microtubule associated proteins, is involved in the formation of PHFs (2-4). In this context we have carried out _in vitro_ experiments with purified tau factor to test its ability to polymerize into filaments.

(I) _Polymerization studies_. Highly purified tau factor was extensively dialyzed against 6M urea in a buffer containing 0.1M MES pH 6.4; 2 mM EGTA and 2 mM $MgCl_2$ and after the removal of urea, it was checked its ability to polymerize. In these conditions, a high proportion of filaments with a diameter of 10-20 nm, ressembling those found in PHF were observed. These filaments are not seen with untreated tau protein. The polymers were further characterized by immunoelectronmicroscopy using a monospecific tau antibody and protein A conjugated with colloidal gold. The specific staining with gold particles of the filaments confirms that they are composed of tau protein (Fig. 1).

(II) _Possible modification of tau protein_. Besides to the denaturing effect of urea it can also produce deamination of glutamine or

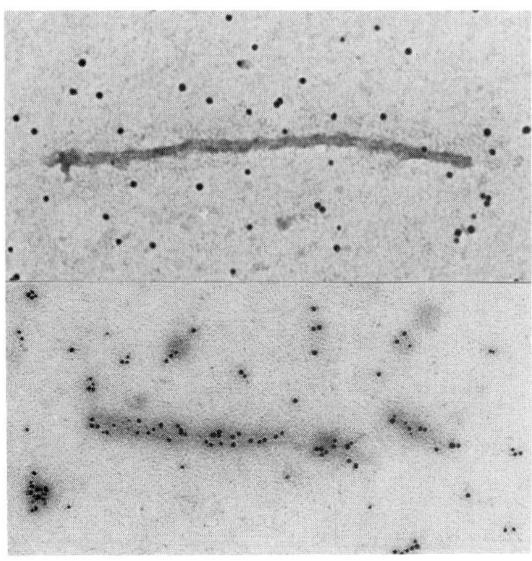

Fig. 1. _Immunoelectronmicroscopy of tau polymers_. Urea-treated tau preparations were incubated with protein A conjugated with colloidal gold and in the absence (upper) or the presence (lower) of tau antibody.

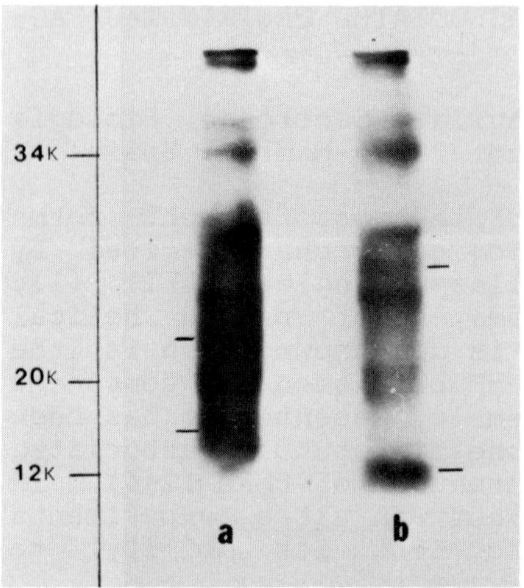

Fig. 2. S.aureus V8 protease digestion of tau factor. Lane a corresponds to untreated tau and lane b to urea treated tau. Molecular weight markers are indicated.

asparagine residues. To test this possibility we carried out digestions with S. aureus V8 protease (an enzyme which cleaves preferentially at glutamic or aspartic groups) (5) of treated and untreated tau protein. The results are shown in Fig. 2 where a larger proportion of low molecular weight fragments for the treated tau compared with the untreated form, can be observed. This suggests the presence of a higher number of glutamic or aspartic residues in the treated sample. In agreement with that, we have also found that urea treated tau is a better substrate for transglutaminase, an enzyme which requires accesible glutamic residues.

Although the presented data suggest that a deamination takes place during the urea treatment which could be involved in the formation of tau polymers; other factors such as conformational changes could also be responsible for the formation of these filaments.

REFERENCES

(1) Kidd, M. (1963) Nature 197, 192-193.

(2) Grundke-Iqbal, F., Iqbal, K., Quinlan, M. Tung, Y.C., Zaidi, M. and Wisniewski, H.M. (1986) J. Biol. Chem. 261, 6084-6089.

(3) Wood, J.G., Mirra, S.S., Pollock, N.J. and Binder, L.I. (1986) Proc. Natl. Acad. Sci. USA 83, 404-4043.

(4) Kosik, K., Joachim, C.L. and Selkoe, D. (1986) Proc. Natl. Acad. Sci. 83, 4044-4048.

(5) Houmard, J. and Drapeau, G.R. (1986) Proc. Natl. Acad. 69, 3506-3509.

AXONAL LOCALIZATION OF MAP2 IN SITU IN EMBRYONIC TECTAL PLATE AT THE EARLIEST STAGE OF NEURONAL DIFFERENTIATION.

S.Binet, A.Fellous*, R.Ohayon*, E.Cohen and V.Meininger.

Laboratoire d'Anatomie, Faculté de Médecine,45 rue des St-Péres, 75270 PARIS CEDEX 06,France and * INSERM, Hôpital Bicêtre,94275, Le Kremlin-Bicêtre, France.

Neuronal polarity is clearly related with the development of two distinct classes of processes, axon and dendrites. This anisotropy seems to arise from molecular asymmetries of the microtubules (1), particularly of the microtubule-associated-proteins (MAPs). Among these proteins, MAP2 is neuron specific and seems predominantly located in dendrites, both in adult and differentiating nervous system (2). However, all studies have been performed at late stages of differentiation, i.e. after the dendritic growth. Here we analyzed in situ the apparition and localization of this protein in the embryonic tectal plate, before the appearance of the dendritic processes, using a polyclonal anti-MAP2 antibody and 0.5 microns semi-thin sections obtained after PEG embedding (3).

(I) ANTIBODY CHARACTERIZATION.

MAPs were obtained according to (4) and MAP2 separated by SDS-PAGE. MAP2 band was injected to a rabbit. Specificity of the antibody was checked by immunoblotting against whole rat brain extracts. As seen in Fig.1, the anti-MAP2 antibody has a major activity against two polypeptide bands (►) which have the same electrophoretic mobilities as the two components of the standard MAP2 (5). Several other bands (►) were also faintly stained. As demonstrated (6), these bands are degradation products of the parent molecule, cross-reacting both with monoclonal and polyclonal anti-MAP2 antibodies.

A specific anti-tubulin polyclonal antibody (7) (kindly provided by J.de Mey),was also used to localize the microtubules in the different types of cells present in the tectal plate at stage E10, at ten days post-mating.

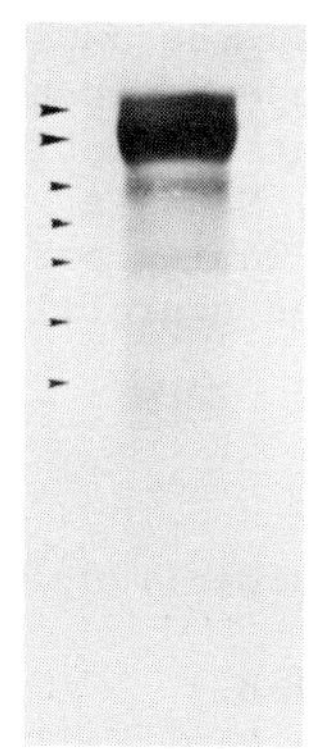

Figure 1:immunoblot of the anti-MAP2 antibody.

(II) ANTIBODIES LOCALIZATION-TUBULIN, MAP2.

As previously demonstrated (8), at this stage, microtubules display two major localizations: the ventricular processes of the bipolar neuro-epithelial cells, near the ventricular surface, and the axonal processes of the young neurons, tangentially orientated and located near the pial, or apical, surface.

Cold-treatment (0°C,90min) depolymerizes all the microtubules of the bipolar cells and in the axonal profiles, it leaves intact short fragments corresponding to the cold-stable pool of microtubules (8).

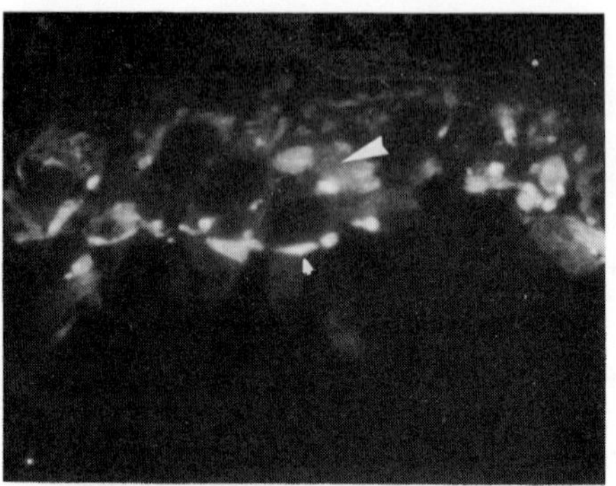

Figure2: immunofluorescence with the anti-MAP2 antibody.Apical zone of the tectal plate.E10.

Fig.2 shows the immunofluorescence staining in the intermediate zone obtained with the anti-MAP2 antibody. No staining was observed in the bipolar neuro-epithelial cells. In the young neurons, a diffuse and faint fluorescence was present in the cell bodies (➤), whereas axonal profiles (▲) were densely stained. These aspects are nearly identical to those obtained in the axonal profiles with the anti-tubulin antibody.

After cold-treatment, the axonal staining disappeared, but a faint fluorescence remained in the cell bodies.

The localization of MAP2 in the cell bodies of the differentiating young neurons confirm that this protein is neuron specific and is a good marker of the neuronal differentiation (9).

The localization of the anti-MAP2 antibody in the axonal profile raises many questions. First, is this staining specific? Specificity of the antibody (Fig.1), identical results obtained with a monoclonal anti-MAP2 antibody (results non shown) favor this hypothesis. Second, where is localized the protein? Comparison with the aspects obtained with the anti-tubulin antibody (8), the disappearance of the axonal staining after cold-treatment when neurofilaments are still observed (8) suggest that before the dendritic growth, MAP2 is transiently present in the growing axons, in the cold-labile part of the microtubules, without the functional links with the neurofilament described in adult.

REFERENCES.

(1) Caceres,A., Banker,G.A. and Binder,L. (1986) J.Neurosci. 6, 714-722.

(2) Burgoyne,R.D. (1986) Comp.Biochem.Physiol. 83 B, 1-8.

(3) Wolosewick,J.J., de Mey,J. and Meininger,V. (1983) Biol.Cell 49, 219-226.

(4) Fellous,A., Francon,A., Lennon,M. and Nunez,J. (1977) Eur.J. Biochem. 78, 167-174.

(5) Kim,H., Binder,L.I. and Rosenbaum,J.L. (1979) J.Cell Biol. 80, 266-276.

(6) Huber,G. and Matus,A. (1984) J.Neurosci. 4, 151-160.

(7) de Brabander,M., Geuens,G., de Mey,J. and Joniau,M. (1979) Biol. Cell 34, 83-96.

(8) Cohen,E., Binet,S. and Meininger,V. (1987) Dev.Brain Res. (in press)

(9) Izant,J.and Mc.Intosh,J. (1980) P.N.A.S. 77,4741-4745

SYNTHESIS AND PHOSPHORYLATION OF MAP'S DURING IN VITRO DIFFERENTIATION OF EMBRYONAL CARCINOMA CELLS, A MODEL FOR NEURONAL DIFFERENTIATION.

A. Fellous°, R. Ohayon° and H. Jakob*
°Unité INSERM, 78 avenue du Gl Leclerc, 94275 Kremlin-Bicêtre ; *Unité de Génétique cellulaire du Collège de France et de l'Institut Pasteur, 25 rue du Dr. Roux, 75724 Paris Cedex 15.

INTRODUCTION. The differentiation of neuronal cells can be studied in primary neurons, but more easily in model systems of cell lines in vitro. When differentiation of certain embryonal carcinoma cells (EC) is triggered either spontaneously or by drugs, they can give rise to differentiated derivatives resembling closely to neurons and glial cells (1). Morphology of these neurites can be diverse, presenting axon and dendritic like processes. We studied the localization, synthesis and phosphorylation of two brain specific MAP's during this process of differentiation. The results were compared with those found for brain tissues (2).

RESULTS. I. Differential subcellular localization of MAP_2, MAP_1 and GFA as revealed by indirect immunofluorescence studies. To promote differentiation of PCC7-S-1009 embryonal carcinoma cells, retinoic acid (RA) 10^{-7}M and 10^{-3}M dibutyryl adenosine-3'-5' cyclic monophosphate (dbcAMP) were added for various length of time to the cultures. Immunofluorescence studies performed according to (3) show that with anti-MAP_2 antibody, only a few processes were stained on day 2, despite the fact that numerous neurite outgrowths can be seen. On day 6, treated cells extend an elaborate network of neurites : numerous processes showed a bright staining. In contrast, all the processes were stained by a monoclonal antibody against MAP_1. This result is consistent with the conclusion of several authors (2) studying brain differentiation, who show that MAP_2 is a dendritic marker and MAP_1 is present in axons as well as in dendrites.

On day 6, some cells showed a significant staining with anti-GFA antibody, suggesting that some cells do differentiate to glial derivatives.

II. Biosynthesis and phosphorylation of MAP_2 and MAP_1. To determine whether the presence of MAP_2 and MAP_1 reflects new biosynthesis or a subcellular redistribution of these two MAP's during 1009 cell differentiation, the cells were pulse-labelled with ^{35}S-methionine. Immunoprecipitation of the cell extracts was performed with anti-MAP_2 and anti-MAP_1 antibodies. Fig. A shows an increase in synthesis of MAP_2 and MAP_1 with time. In contrast to results obtained with other transformed cells, their apparent molecular weights are identical to those found in brain tissues. Fig. B demonstrates that MAP's are phosphorylated in treated 1009 cells, with an increase in the degree of phosphorylation.

CONCLUSIONS. Differentiation of 1009-EC cells obtained by treatment with RA + dbcAMP stimulates the synthesis of MAP_2 and MAP_1 similarly to brain MAP's. The following similarities : 1) a rate of synthesis which follows the process of differentiation, 2) a restricted localization of MAP_2 and a widespread distribution of MAP_1, 3) a same apparent molecular weight, 4) the ability to be phosphorylated, provide further support to the idea that some EC cells are a valid tool for the study of neuronal differentiation.

REFERENCES

(1) Paulin, D., Jakob, H., Jacob, F., Weber, K. and Osborn, M. (1982) Differentiation 22, 90-99.
(2) Vallee, R.B., Bloom, G.S. and Luca, F.C. (1984) in The Molecular Biology of the Cytoskeleton. Borisy, G.G., Cleveland, D.W. and Murphy, D.B. eds, 111-130.
(3) Denis-Donini, S., Glowinski, J. and Prochiantz, A. (1984) Nature 307, 641-643.
(4) Greene, L.A., Liem, R.K.H. and Shelanski, M.L. (1983) J.Cell. Biol. 96, 76-83.

<u>Immunoprecipitation of ^{35}S methionine and ^{32}P 1009 labelled cells</u>

Untreated and treated cells were pulse-labelled with ^{35}S methionine for (12 hours) or with ^{32}P orthophosphate (for 3 hours). Extractions and immunoprecipitation were performed as described (4).

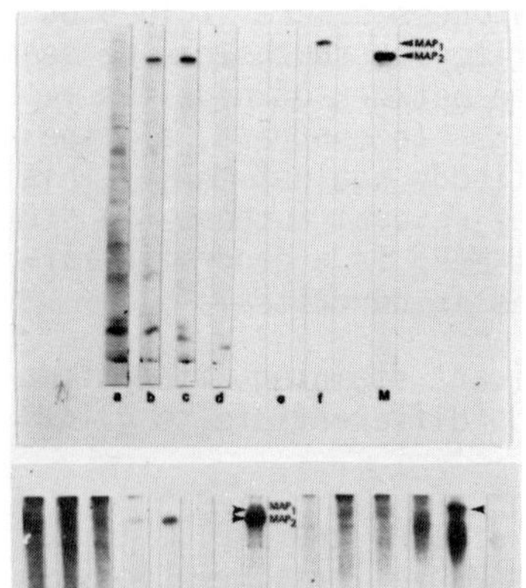

A. Immunoprecipitation of ^{35}S-labelled extracts with anti-MAP_2 antibody (lanes a-d) and anti-MAP_1 (lanes e-f). Lanes a and e : untreated cells ; b : 4 days treated cells ; c and f : 6 days ; d : 8 days ; lane M : in vitro phosphorylated MAP_2 from brain.

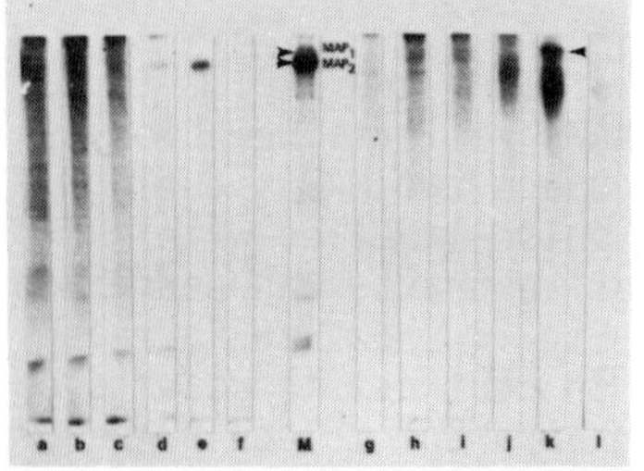

B. Immunoprecipitation of ^{32}P labelled cell extracts with anti-MAP_2 (lanes a-f) and anti-MAP_1 (lanes g-l). Lanes a and g : untreated cells; b and h : 1 day treated cells ; c and i : 2 days ; d and j : 4 days ; e and k : 6 days ; f and l : 8 days; lane M : brain ^{32}P labelled MAP_2.

Trypontin, a novel microtubule-membrane linker in *Trypanosoma brucei*

T. Seebeck[1]), V. Küng[1]), J. Stieger[1]), M. Müller[2]) and T. Wyler[3])

[1]) Institute for General Microbiology, [2]) Institute for Biochemistry and Molecular Biology, [3]) Department for Developmental Biology, University of Bern, Baltzerstrase 4, CH-3012 BERN, Switzerland

Many eukaryotic cells contain membrane-associated cytoskeletal structures (membrane skeletons), either in addition to, or in place of, a transcellular cytoskeleton. The membrane skeleton has been studied in most detail in the human erythrocyte (1), but analogous structures have been observed in many other cell types. The membrane skeleton of the parasitic hemoflagellate *Trypanosoma brucei* consists of a dense array of singlet microtubules, which are tightly connected to the overlaying cell membrane (2). Present knowledge suggests that this membrane skeleton is the sole cytoskeletal component of the cell body. Here we describe the identification of a 60 kDa protein which may be involved in this microtubule-membrane contact. Due to this presumed bridgeing function, the name **Trypontin** is proposed for this novel type of microtubule-associated protein.

Microtubule-membrane links can be disrupted by phenothiazines. Early observations have shown that micromolar concentrations of many compounds of the phenothiazine family, a group of major tranquilizers extensively used in clinical medicine, rapidly disrupt the layer of membrane-bound microtubules in cultured trypanosomes (3).

Hemoflagellates may contain a specific protein which is involved in the phenothiazine-sensitive link between microtubules and membranes. Affinity chromatography of proteins from microtubule-membrane complexes on phenothiazine affinity columns led to the identification of a 60 kDa protein which is strongly retained by this matrix. Monoclonal and polyclonal antibodies were used to demonstrate that this protein is specific for kinetoplastid flagellates and is absent in a variety of other protozoa, as well as in higher organisms (4). Furthermore, the protein is not strongly conserved between the different hemoflagellates. This is corroborated by crosshybridization analyses of DNAs from various organisms using a DNA sequence for the 60 kDa protein which has been isolated from a lambda gt11 library of *Trypanosoma brucei*.

The 60 kDa protein (Trypontin) is a microtubule-associated protein. Copolymerization experiments, as well as binding experiments with preformed microtubules have established that trypontin is a *bona fide* microtubule-associated protein. Cell fractionation experiments further confirm its colocalization with the membrane-bound microtubules.

The 60 kDa protein (Trypontin) is a membrane-binding protein. Retention on hydrophobic matrices as well as charge-shift electrophoresis have indicated that trypontin contains strongly hydrophobic domains. Direct binding experiments of purified trypontin to artificial liposomes have confirmed this and have demonstrated that binding is strongly dependent on the composition of the liposomes.

Trypontin can crosslink microtubules and liposomes. Light and electron microscopy as well as sedimentation studies have demonstrated that trypontin can crosslink artificial liposomes, as well as trypanosomal membrane vesicles, to microtubules and to cytoskeletons. These experiments strongly suggest that trypontin is cogently involved in the microtubule-membrane contact *in vivo* and hence may represent a highly suitable model protein for studying such interactions in more detail.

REFERENCES

(1) Elgsaeter,A., Stokke,B.T., Mikkelsen,A. and Branton,D. 1986. The molecular basis of erythrocyte shape. Science **234** , 1217-1223.

(2) Angelopoulos,E. (1970) Pellicular microtubules in the family trypanosomatidae. J. Protozool. **17** , 39-51.

(3) Seebeck,T. and Gehr,P. (1983) Trypanocidal action of neuroleptic phenothiazines in *Trypanosoma brucei*. Mol. Biochem. Parasitol. **9** , 197-208.

(4) Stieger,J. and Seebeck,T. (1986) Monoclonal antibodies against a 60 Kd phenothiazine-binding protein from *Trypanosoma brucei* can discriminate between different trypanosome species. Mol. Biochem. Parasitol. **21** , 37-45.

CELLULAR AND MOLECULAR BIOLOGY OF CYTOSKELETON DURING DEVELOPMENT

REORGANIZATION OF TUBULIN POOLS DURING EARLY MOUSE EMBRYOGENESIS

Juan Aréchaga, Coralia Rivas, Juan C. Vera and Ricardo B. Maccioni

Department of Biochemistry, Biophysics & Genetics, University of Colorado Health Sciences Center, Denver, CO 80262

ABSTRACT

Microtubules appear to be involved in the morphological changes of blastomers during preimplantation embryogenesis, namely in cellular polarization. We have examined the intracellular free and assembled tubulin pools as an attempt to understand the molecular aspects of microtubule involvement in early mouse embryogenesis. After embryo's homogeneization at 4°C tubulin levels in the soluble fraction did not change significantly throughout the precompacted stages but decreased in compacted morulae and blastocyst to 36-40% of that of preliminary stages. On the other hand, tubulin in NP-40 solubilized cellular fractions was significantly higher in compacted morulae. Sucrose gradients centrifugation at 32°C of supernatants and solubilized pellets obtained after homogeneization of embryos in warm buffers also indicated a preferential distribution of assembled tubulin in the membrane domain of blastomers of compacted embryos. Indirect immunofluorescence studies using antitubulin antibodies showed topographic differences in the distribution of tubulin at blastocyst stage with a preferential organization of assembled tubulin in the inner cell mass.

INTRODUCTION

Compaction (Pratt *et al.*, 1982) and cavitation (Wiley and Eglitis, 1981) are the two major morphological changes in preimplanted mammalian embryos. During these important processes a homogeneous population of blastomeres at 8-cell stage lose normally their totipotency (Tarkowski and Wroblewska, 1967) and segregate themselves after the next two cleavages resulting in specification for inner cell mass (from which the embryo arises) and trophectoderm (which forms only extra-embryonic structures) (Gardner, 1981). In a short period of time (24 hours approximately between 8-cell non-compacted morula and cavitated blastocyst in the mouse) a complex mixture of biological events occurs simultaneously, making it extraordinarily difficult to obtain actually a complete understanding of the above mentioned processes. Therefore, general hypotheses or partial mechanisms have been invoked to explain blastomere determination in relation to the possible role of positional information (Tarkowski and Wroblewska, 1967; Hillman *et al.*, 1972; Kelly, 1977), blastocoelic fluid environment (Pedersen and Spindle, 1980), temporal and regional X-chromosome differentiation (Monk, 1981), polarity and phenotypic differences after cell divisions (Johnson and Ziomek, 1981; Ziomek and Johnson, 1982) and changes in cell surface molecules (Richa *et al.*, 1981; Johnson, 1986). However, only careful analysis and correlations of the many cytological and biochemical modifications during compaction and cavitation can lead to a complete interpretation of this crucial period of development.

Through studies with different kinds of drugs which selectively affect the cytoskeleton (colchicine, colcemid, taxol, nocodazol) indirect evidence implicating the microtubule system in compaction and cavitation has been obtained (Ducibella and Anderson, 1975; Azim et al., 1980; Ducibella, 1982; Maro and Pickering, 1984; Sutherland and Calarco-Guillam, 1983). On the other hand, it has also been shown that the rate of tubulin synthesis increases significantly between 8-cell and blastocyst stages (Abreu and Brinster 1978), and studies with monospecific antibodies using indirect immunofluorescence have indicated the presence of tubulin in all cells in the blastocyst (Paulin et al., 1980). Moreover, it is a current observation that calcium-free media reverses the compaction process (Ducibella and Anderson, 1975), which could be related to the very well known requirement for calcium ions in the regulation of microtubule assembly in the cells.

As an attempt to further understand the molecular aspects of the role of microtubules in mammalian preimplantation embryos, we have examined the intracellular tubulin pools and the intracellular distribution of their assembled forms. Our studies suggest that an intracellular redistribution of microtubules from cytoplasm to the membrane domain occurs with the onset of the compaction process. Moreover, some changes have been found by indirect immunofluorescence in the distribution of the amount of tubulin during cavitation of mouse blastocyst. These changes may respond to demands for cytoskeleton stabilization during compaction and suggest a commitment of the inner cell mass during blastulation..

RESULTS

Tubulin pools have been analyzed during early mouse embryogenesis by microdeterminations of colchicine binding of soluble and NP-40 solubilized cellular fractions from preimplanted mouse embryos at different developmental stages. The strong interaction nearly irreversible of colchicine and tubulin with a stoichiometry of one mole of drug per mole of tubulin allows to use the binding assay for quantitative determination of tubulin in different animal tissues (Wilson, 1975). Initial values for colchicine binding were obtained by extrapolation to zero of the time dependent decay curves of colchicine binding of embryonic extracts. Colchicine binding assay in mouse embryos homogeneized in cold shows that there is a slight increase from 1.7 to 2.4 pmoles/μg of protein in the colchicine binding activity as 2-cell embryos develop to the 8-cells stage, followed by a drastic decrease occurring during the process of compaction. After that, the colchicine binding activity remains stable until late blastocyst stage. Quite the reverse happens with the NP-40 solubilized fraction; the colchicine binding increases from 4.2 to 5.6 pmol/μg of protein during compaction (see Fig. 1). The specific colchicine binding activities were higher in the particulate fraction as compared with high speed supernatants throughout the entire preimplantation embryogenesis. Interestingly, the total colchicine binding activity (considering both soluble and particulate fractions) was essentially constant through the different embryonic stages analyzed.

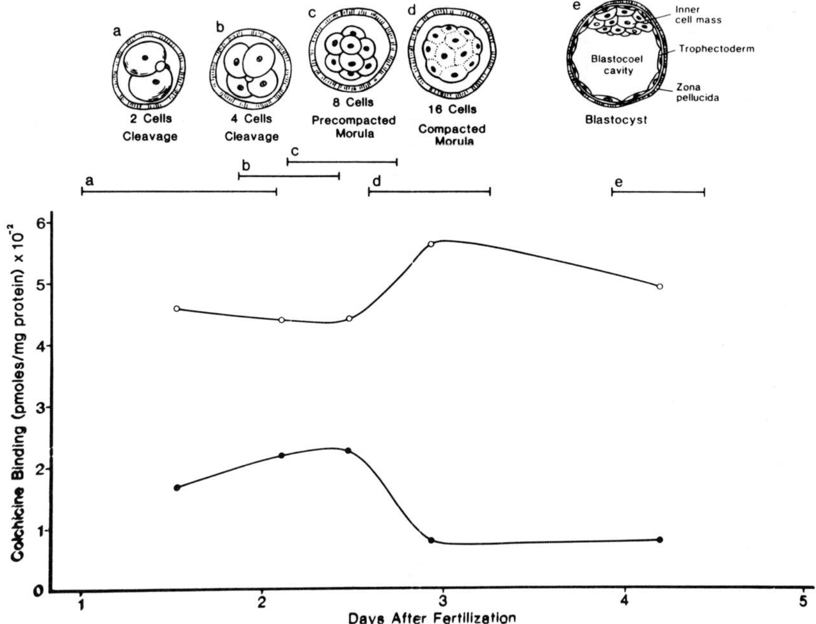

Figure 1: Colchicine binding activity of the supernatants (●) and the solubilized pellets (O) of embryos from different developmental stages homogeneized at 4^{0}. Each point represent the actual colchicine binding activity, corrected for time dependent inactivation by extrapolation to zero time the plots of time course binding decay (Maccioni and Mellado, 1981). The upper part shows a representation of the embryonic stages analyzed. Colchicine binding assays were performed according to the procedure of Sherline et al., J. Biol. Chem. 250:5481-5486 (1975). Protein concentrations were determined by the Lowry method (Lowry et al. ,1951).

In order to define the internal distribution of assembled tubulin in embryonic cells we have performed the homogenezation at 37°C and under microtubule stabilizing conditions (see Fig. 2). The embryonic extracts were subjected to warm low-speed centrifugation and the resuspended pellet and supernatants centrifuged at high speed on sucrose cushions. The colchicine binding was determined after NP-40 solubilization and incubation at 4°C of cellular fractions from the resuspended pellet fractionated by ultracentrifugation. The determinations of colchicine binding of the different fractions from the sucrose gradient of the embryonic supernatants were also performed after incubation at 4°C to depolymerize assembled tubulin.
Colchicine binding activity per microgram of total protein in the soluble fraction appears to be essentially the same in precompacted and compacted embryos while a slight increase was observed in blastocysts. However, in

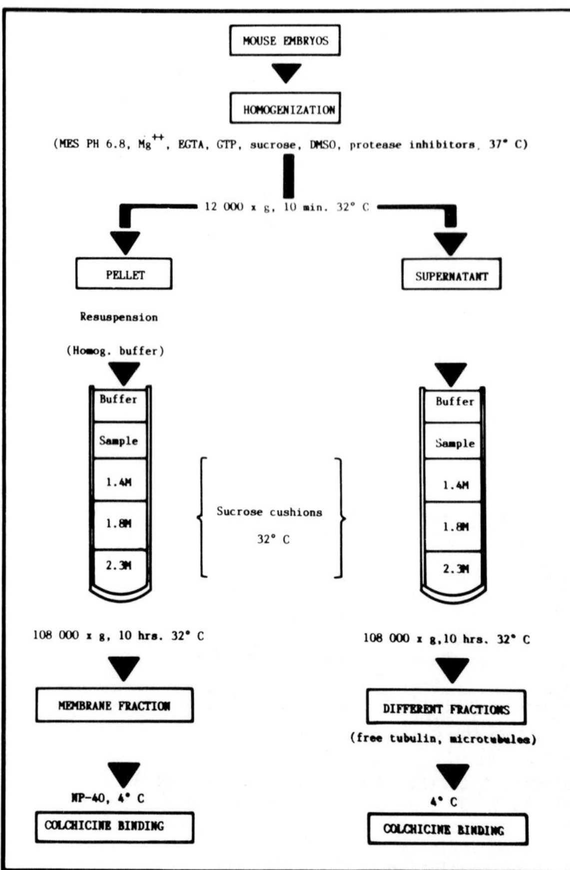

Figure 2: Protocol used for fractionation of embryos under microtubule stabilizing conditions.

Cell extracts from mouse embryos cultured in Whitten's medium (Whitten, 1971) were washed in Hank's balanced solution and homogeneized at 37° in a microtubule stabilization buffer (0.05 M Mes,pH 6.8, 1.5mM $MgCl_2$, 0.5 mM EGTA, 1mM GTP, 0.3M sucrose and 5% DMSO) containing the protease inhibitors aprotinin, soybean trypsin inhibitor and PMSF. Microtubules and free tubulin pools from both the cytoplasmic and membrane domains were separated by differential centrifugation followed by sucrose gradient centrifugation.
Colchicine binding was carried out according to Sherline et al. (1975)

TABLE 1: Subcellular distribution of colchicine binding activity during preimplantation mouse embryogenesis(1)

Embrionic stage	Membrane fraction (pmol colchicine/ ug protein)	Cytoplasmic fraction (pmol colchicine/ ug protein)
Precompacted	0.28 (0.43)	0.44
Compacted	0.74 (0.76)	0.42
Blastocyst	0.56 (0.72)	0.52

(1) Homogeneizationof embryos at 37° in microtubule stabilization buffer (0.05 M Mes pH 6.8, 1.5mM $MgCl_2$, 0.5mM EGTA, 1 mM GTP, 0.3 M sucrose and 5% DMSO). Other experimental conditions as indicated in protocol of Fig. 2.

the membrane fraction there was a significant increase of colchicine binding activity and therefore on the amount of tubulin (under microtubule stabilizing conditions) during the transition from precompacted embryos to the compacted stage. (Table 1). A small decrease in the binding activity occured when passing from the compacted embryos to blastocysts.

Indirect immunofluorescence studies with tubulin antibodies showed a diffuse pattern in all embryonic stages until blastulation. At the latter stage the inner cell mass became more stained as compared with trophectoderm cells (see Fig. 3).

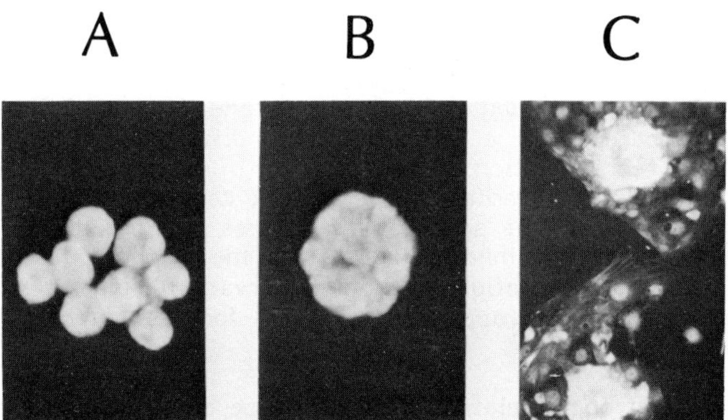

Figure 3: Indirect immunofluorescence staining of microtubules in mouse embryos at different stages of development. Embryos were incubated with anti-tubulin polyclonal antibody followed by treatment with fluorescein-labeled goat anti-rabbit immunoglobulin. (A) Eight-cell pre-compacted morula. (B) Compacted morula and (C) Hatched blastocysts.

DISCUSSION

The present study shows that the process of compaction in mouse embryos is associated with an increase of assembled tubulin pools in the membrane fraction of embryonic cells. In addition, these studies indicate a preferential distribution of assembled tubulin in the inner cell mass of blastocysts as compared with the trophectoderm.

A previous report (Abreu and Brinster, 1978) has shown that tubulin synthesis in mouse embryos increase from 1% to 19% (19-folds) of total protein synthesis between 1-cell embryo and blastocyst stages, with very low rates during the three initial cleavages and peaking at 8-cell stage (where the process of compaction begins). The differential distribution of assembled tubulin between the cytoplasmic and the membrane domains occuring during compaction suggests an internal reorganization of tubulin pools associated to the significant biological changes of this period of development. Earlier electron microscopy studies (Ducibella and Anderson, 1975; Wiley and Eglitis, 1981) support this conclusion because the observation of a higher density of microtubules in the membrane domain of flattened blastomeres as compared with a more scattered distribution in previous stages. Therefore, reorganization of microtubules in mouse embryos could be related to the needs of cytoskeleton in distinct cellular domains during compaction and cavitation. The apparent differences in the distribution of polymeric tubulin between the inner cell mass and the trophectoderm (as it seems from indirect immunofluorescence observation) indicates that assembled tubulin could be an appropriate marker for cell differentiation at this stage of development.

Cytoskeleton is strongly related to maintenance of cellular shape and therefore morphological changes during embryonic development are likely to be correlated to changes in the organization or dynamics of their structural components. In previous studies the intercellular flattening mechanism during compaction was associated mainly with the actin microfilament system (Sutherland and Calarco - Guillam, 1983; Johnson and Maro, 1984) and with spectrin (Reima and Lethonen, 1985). The cell surface polarization process seems to be related essentially with the microtubular network (Maro and Pickering, 1984, Houliston et al, 1987), but both systems are intimately related by multiple interactions as can be gleaned from in vitro experiments with different drugs which affect these components selectively (Pratt et al, 1982). Cavitation involves qualitative or quantitative differences in the distribution of cytoskeletal components including cortical localization of microtubules (Wiley and Eglitis, 1981). Thus in this report we show a prefential distribution of microtubules in the inner cell mass and other authors have shown an specificity for trophectoderm of keratin-type filaments. (Lehtonen, 1985). Relatedly, changes in the enzymatic systems like transglutaminase (Maccioni and Arechaga, 1986), which favor cytoskeleton stabilization are important in these stages of development. Therefore, structural and functional studies on cytoskeleton components during embryogenesis appears to be a relevant area of research aimed to elucidate the regulatory aspects of the cytoskeleton involvement in early development.

ACKNOWLEDGEMENTS

This research has been supported by grants from The Council for Tobacco Research, U.S.A., Milheim Foundation and CAICYT.

REFERENCES

1. Abreu, S. and Brinster, R. L. (1978). Synthesis of tubulin and actin during the preimplantation development of the mouse. Exp. Cell Res. 114, 135-141

2. Azim, M., Surani, H., Barton, S.C. and Burling, A. (1980). Differentiation of 2-cell and 8-cell mouse embryos arrested by cytoskeletal inhibitors. Exp. Cell Res. 125, 275-286

3. Ducibella, T. (1982). Depolymerization of microtubules prior to compaction: development of cell spreading is not inhibited. Exp. Cell Res. 138, 31-38

4. Ducibella, T. and Anderson, E. (1975). Cell shape and membrane changes in the eight-cell mouse embryo: prerequisites for morphogenesis of the blastocyst. Dev. Biol. 47, 45-58

5. Gardner, R.L. (1981). In vivo and in vitro studies on cell lineage and determination in early mouse embryo. IN "Cellular Controls in Differentiation" (C.W. Lloyd, and D.A. Reed, eds.), pp. 257-278. Academic Press, London

6. Hillman, N., Sherman, M.I., Graham, C. (1972). The effect of spatial arrangement on cell determination during mouse development. J. Embryol. exp. Morph. 28, 263-278

7. Houliston E., Pickering, S.J. and Maro B. (1987) Redistribution of microtubules and pericentriolar material during the development of polarity in mouse blastomeres. J. Cell Biol. 104:1299-1308.

8. Johnson, L.V. (1986). Wheat germ agglutinin induces compaction and cavitation-like events in two-cell mouse embryos. Dev. Biol. 113, 1-9

9. Johnson, M. H. and Maro, B. (1984). The distribution of cytoplasmic actin in mouse 8-cell blastomeres. J. Embryol. exp. Morph. 82, 431-439

10. Johnson, M.H., and Ziomek, C.A. (1981). The foundation of two distinct cell lineage within the mouse morula. Cell 24, 71-80

11. Kelly, S.J. (1977). Studies of the developmental potential of 4- and 8-cell stage mouse blastomeres. J. Exp. Zool. 200, 365-376

12. Lehtonen, E. (1985). A monoclonal antibody against mouse oocyte cytoskeleton recognize cytokeratin-type filaments. J. Embryol. exp. Morph. 90, 197-209

13. Lowry, O.H., Rosenbrough, N.J., Farr, A.L., Randal, R.J., (19 1). Protein measurements with the Folin phenol reagent. J. Biol. Chem. 193, 265-275

14. Maccioni, R. B. and Arechaga, J. (1986). Transglutaminase involvement in early embryogenesis. Exp. Cell Res. 167:266-270

15. Maccioni, R. B. and Mellado, W. (1981) "Characteristics of the *in vitro* assembly of tubulin from brain of *Cyprinus carpio*". Comparative Biochemistry and Physiology 70B, 375-380

16. Maro, B. and Pickering, S.J. (1984). Microtubules influence compaction in preimplanted mouse embryos. J. Embryol. exp. Morph. 84, 217-232

17. Monk, M. (1981). A stem-line model for cellular and chromosomal differentiation in early mouse development. Differentiation. 19, 71-76

18. Paulin, D., Babinet, C., Weber, K. and Osborn, M. (1980). Antibodies as probes of cellular differentiation and cytoskeletal organization in the mouse blastocyst. Exp. Cell Res. 130, 297-304

19. Pedersen, R., and Spindle, A.J. (1980). Role of the blastocoele microenvironment in early mouse embryo differentiation. Nature 284, 550-552

20. Pratt, H.P.M., Ziomek, C.A., Reeve, W.J.D. and Johnson, M.J. (1982). Compaction of the mouse embryo: an analysis of its components. J. Embryol. exp. Morph. 70, 113-132

21. Reima, I. and Lethonen, E. (1985). Localization of nonerythroid spectrin and actin in mouse oocytes and preimplantation embryos. Differentiation 30, 68-75

22. Richa, J., Damsky, C.H., Buck, C.A., Knowles, B.B. and Solter, D. (1981). Cell surface glycoproteins mediate compaction, trophoblast attachment, and endoderm formation during early mouse development. Dev. Biol. 108, 513-521

23. Sherline, P., Bowdin, C.K. and Kipnis, D. M. (1974). A new cholchicine binding assay for tubulin. Anal. Biochem 62, 400-407

24. Sutherland, A. and Calarco-Gullam, P.G. (1983). Analysis of compaction in the preimplantation mouse embryo. Dev. Biol. 100, 328-338

25. Tarkowski, A.K. and Wroblewska, J. (1967). Development of blastomeres of mouse eggs isolated at 4 and 8-cell stage. J. Embryol. exp. Morph. 18, 155-180

26. Whitten, W.K. (1971). Nutrient requirements for the culture of preimplantation mouse embryos *in vitro.* Adv. Bioscien. 6, 129-140

27. Wiley, L.M., Eglitis, M.A. (1981). Cell surface annd cytoskeletal

elements: cavitation in the mouse preimplantation embryo. Dev. Biol. 86, 493-501

28. Wilson, L. (1975). Microtubules as drug receptors: pharmacological properties of microtubule proteins. Ann. N.Y. Acad. Sci. 253, 213-231

29. Ziomek C.A. and Johnson, M.H. (1982). The role of phenotype and position in guiding the fate of 16-cell mouse blastomeres. Dev. Biol. 91, 440-447

CALCIUM-DEPENDENT DEGRADATION OF GLIAL FIBRILLARY ACIDIC PROTEIN AND VIMENTIN IN RAT CORTICAL ASTROCYTES IN CULTURE

J. Ciesielski-Treska, J.F. Goetschy, G. Ulrich and D. Aunis
Groupe de Neurobiologie Structurale et Fonctionnelle, INSERM U-44, Centre de Neurochimie du CNRS, 5, rue Blaise Pascal, 67084 Strasbourg Cedex, France

Glial fibrillary acidic protein (GFAP) and vimentin are the principal constituents of intermediate filaments (IF) found in astrocytes in culture (1). We have shown previously that during the extraction of astrocyte cytoskeletal proteins in the presence of calcium, there was a marked increase of the degradation of GFAP and vimentin (2). In this report, we present evidence that the proteolysis of astrocyte IF proteins is also stimulated in intact cells by ionomycin, a calcium ionophore. One of the problems in studying the molecular mechanisms underlying the proteolysis of IF proteins is the difficulty of manipulating the concentration of calcium in the cell cytoplasm. To overcome this, monolayer astrocyte cultures were exposed to digitonin which renders the plasma membrane permeable to ions and small molecules (3). Using digitonin-permeabilized cells we demonstrate that the proteolysis of GFAP and vimentin is stimulated by calcium in the micromolar and millimolar concentration range.

INTACT CELLS. The extent of degradation of GFAP and vimentin has been determined by quantitative electroimmunoblotting (see [2]). Figure 1A shows that in control culltures, about 90 % of anti-GFAP immunoreactivity was found in the 50 KD protein band which corresponds to intact non-degraded polypeptide.

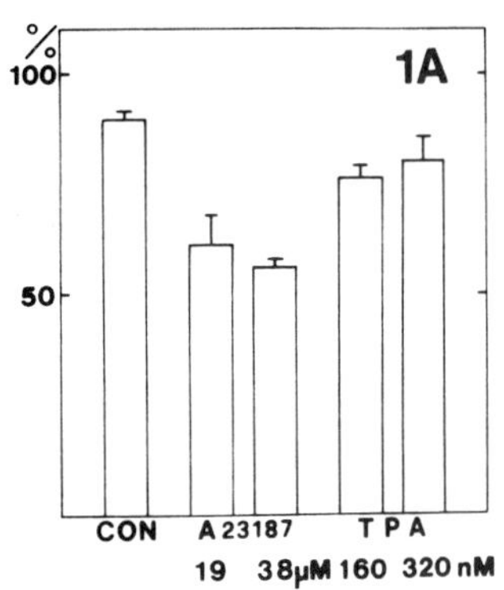

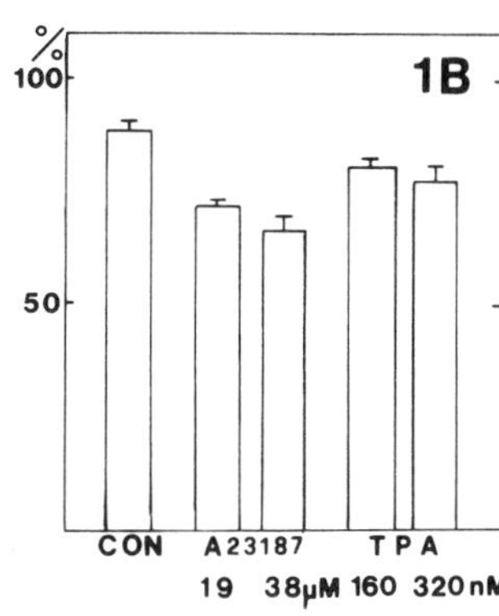

Fig. 1. Effects of calcium ionophore and phorbol ester analogue on degradation of GFAP (A) and vimentin (B). Monolayer cultures (10 cm Petri dish) were incubated for 5 min in medium containing 2 mM calcium with A23187 and TPA at indicated concentration, with vehicle (0.01 % dimethyl sulfoxide). Controls (con) were incubated with vehicle alone. The incubation medium was removed, the cells were

washed with PBS/EGTA and scraped in extraction buffer containing leupeptin (100 µM), PMSF (2 mM), EGTA (1 mM) as described (2). Electroimmunoblots of nonionic detergent-insoluble fractions were scanned using a reflectance densitometer (Camag, Switzerland) and intact GFAP and vimentin were determined as relative peak areas corresponding to 50 KD and 57 KD bands respectively. Data are expressed as the mean ± SEM from six separate experiments.

In cultures incubated with calcium ionophore, the 50 KD GFAP-positive band decreased by 20 to 30 %. Incubation of the cells with 12-O-tetradecanoylphorbol-13-acetate (TPA), which stimulates protein kinase C in astrocytes (4), also reduced the 50 KD GFAP-positive band. As can be seen in Fig. 1B calcium ionophore and TPA also decreased the proportion of intact vimentin, although under the same experimental conditions, the extent of proteolysis of vimentin was less than that of GFAP.

DIGITONIN-PERMEABILIZED CELLS. We have adapted the digitonin permeabilization technique (3) for application to monolayer cultures of astroglial cells. Incubation for 15 min with 15 µM digitonin at 37°C was sufficient to permeabilize the plasma membrane but caused little leakage of protease activities from the cells. Fig. 2A shows that the degradation of GFAP becomes apparent at concentrations of free calcium higher than 10 µM. At 1 mM concentration, calcium induces a 45 to 50 % reduction in the 50 KD GFAP-positive band.

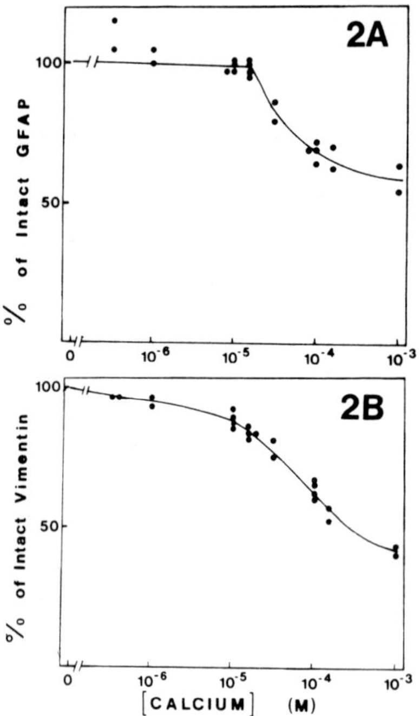

Fig. 2. Calcium-dependent proteolysis of GFAP and vimentin. Monolayer cultures were incubated with digitonin in potassium glutamate medium (3) containing indicated free calcium concentrations and then treated as for Fig. 1. Data are expressed as the percent of intact GFAP and vimentin determined in calcium-containing media relative to the percent of intact GFAP and vimentin determined in calcium-free medium. Each point represents the average of three separate experiments.

Calcium-dependent proteolysis of vimentin is demonstrated in Fig. 2B. Low levels of free calcium (10 to 20 µM) were sufficient to induce a 20 % reduction in the 57 KD vimentin band. The extent of degradation of vimentin increased in parallel with increasing free calcium concentrations and reached 60 % in cells exposed to 1 mM calcium.

Table 1. Effect of leupeptin on proteolytic activity in digitonin-permeabilized cells

AGENT (15 min, 37°C)	FINAL CONCENTRATION (μM)	% OF INTACT GFAP	% OF INTACT VIMENTIN
NO DRUG		96 ± 9.6	82 ± 11.2
DIGITONIN (D)	15	72 ± 5.2	67 ± 9.4
D (15 μM) + LEUPEPTIN	10	94 ± 2.2	80 ± 3.1
	100	95 ± 2.0	80 ± 5.5
D (15 μM) + Ca^{2+}	50	65 ± 9.4	51 ± 10.8
D(15 μM) μ Ca^{2+} (50 mM) + LEUPEPTIN	10	90 ± 7.1	82 ± 3.4

Table 1 shows that μM concentrations of leupeptin were effective in blocking the calcium-dependent proteolysis of GFAP and vimentin. The proteolysis was also inhibited by the calcium-chelating agents EGTA and EDTA but pepstatin or PMSF had no inhibitory effects. The proteolytic activity detected in cells incubated with digitonin in calcium-free medium is also sentitive to leupeptin.

For reaction conditions, see Fig. 2.

It will be necessary to ascertain to what extent this is related to activation of proteases by endogeneous calcium or to the activation of lysosomal thiol proteases.

In conclusion the results show that astrocyte IF proteins are sensitive to proteolytic degradation stimulated by low and high calcium concentrations. Proteolysis of GFAP and vimentin may reflect the activity of low- and high-calcium requiring forms of neutral proteases identified in the brain (5). The differing sensitivities of GFAP and vimentin to calcium may imply that IF proteins mediate certain effects of calcium on intracellular function in astrocytes.

REFERENCES

(1) Goetschy, J.F., Ulrich, G., Aunis, D. and Ciesielski-Treska, J. (1986) J. Neurocytol. 15, 375-385.
(2) Ciesielski-Treska, J., Goetschy, J.F. and Aunis, D. (1984) Eur. J. Biochem. 138, 465-471.
(3) Wilson, S.P. and Kirshner, N. (1983) J. Biol. Chem. 258, 4994-5000.
(4) Mobley, P.L., Scott, S.L. and Cruz, E.G. (1986) Brain Res. 398, 366-369.
(5) Hamakubo, T., Kannagi, R., Murachi, T. and Matus, A. (1986) J. Neurosci. 6, 3103-3111.

FUNCTIONAL VERSATILITY OF MAMMALIAN ß-TUBULIN ISOTYPES

N.J. Cowan, S.A. Lewis, S. Sarkar and W. Gu

New York University School of Medicine, New York, NY 10016, USA

ABSTRACT

The structure and patterns of expression of five mammalian α-tubulin isotypes and six mammalian ß-tubulin isotypes has been determined. Comparison of ß-tubulin isotype amino acid sequences between mouse and human shows that in four out of six cases, conservation is absolute. Several of these isotypes are expressed ubiquitously; others are restricted in their expression to specific tissues (brain, testis and bone marrow) that contain cell types with specialized kinds of microtubule structure. To investigate the contribution (if any) of different ß-tubulin isotypes to microtubule function, we raised antisera that are capable of distinguishing among these isotypes. Animals were first rendered immunologically tolerant to common (i.e., shared) epitopes; a specific response was then elicited using the immunogen of choice. Both tolerogens and immunogens were presented as cloned fusion proteins expressed in a bacterial host. Immunofluorescence experiments with these sera show no evidence of segregation of ß-tubulin isotypes among functionally distinct microtubule structures. Therefore, differences between ß-tubulin isotypes that are maintained by selective pressure influence the function of these isotypes not through their segregation, but rather in isotypically heterogeneous microtubules, possibly via interactions with isotype-specific microtubule associated proteins (MAPs).

INTRODUCTION

Microtubules are long, dynamic polymers that are involved in a number of essential functions in both immature and differentiated cells. A basic problem in the study of microtubules is the extent (if any) to which multiple tubulin polypeptides might participate in the establishment of diverse microtubule functions. One possibility is that different forms of α- and ß-tubulin could define the basis of functionally distinct populations of microtubules (Fulton and Simpson, 1976). In addition, post-translational modifications or modulation of different MAPs could also contribute to the remarkable diversity of microtubule structures in higher eukaryotes.

In mammals, the α- and ß-tubulins are encoded by multigene families (Cleveland _et al_., 1980). Thus, the genetic potential exists whereby distinct polypeptides could in principle contribute to the diversity of microtubule function. Alternatively, the dominant or restricted expression of some α- and ß-tubulin isotypes in specific tissues could reflect the need for quantitative regulation of tubulin synthesis in a situation where one tubulin gene product is capable of assembling into several functionally distinct kinds of microtubule structure. In lower eucaryotes and in *Drosophila melanogaster*, genetic experiments have shown that the function of one tubulin subunit can be substituted by another tubulin gene product (Shatz *et al*., 1986; Adachi *et al*., 1986; May *et al*., 1985; Kemphues *et al*., 1982). However, the greater variety of α- and ß-tubulin isotypes in mammals, their conserved patterns of expression and the absolute conservation of most isotype-specific amino acid sequences between mammalian species is strongly suggestive of selective pressure that could be operating to maintain functional differences among different tubulin subunits. Nonetheless, in spite of the strict conservation of naturally occurring tubulins, transfection experiments using a chimeric isotype synthesized from an N-terminal chicken ß-tubulin sequence fused to the yeast ß-tubulin C-terminal sequence resulted in the incorporation of this chimeric isotype into both spindle and cytoskeletal microtubules in mouse 3T3 cells, with no apparent effect on phenotype or viability (Bond *et al*., 1986).

Are all naturally-occurring mammalian tubulin isotypes indiscriminately scrambled among functionally diverse microtubules? To address this question, we decided to raise tubulin isotype-specific sera that are capable of distinguishing among the various ß-tubulin isotypes, so that their distribution could be examined both qualitatively and quantitatively *in vivo*. These sera, which discriminate among five of the six known mammalian ß-tubulin isotypes, have been used to demonstrate the free intermingling of constitutively expressed isotypes between cytoskeletal and spindle microtubules. In addition, the most divergent naturally occurring ß-tubulin isotype, Mß1, that is normally expressed only in megakaryocytes and fetal nucleated erythroblasts, is indiscriminately incorporated into both interphase and spindle microtubules upon introduction into HeLa cells.

PATTERNS OF ß-TUBULIN ISOTYPE EXPRESSION IN MAMMALS

The pattern of expression of the six known mouse ß-tubulin isotypes among various tissues is shown in Table 1. Insofar as expression data is known for the corresponding isotypes in other mammalian species (i.e., human [Lee *et al*., 1984; Lewis *et al*., 1985b] and rat [Farmer *et al*., 1984]), these patterns are conserved. Three of the isotypes are tissue-specific: Mß1 is restricted in expression to hematopoietic tissue, where it is present in megakaryocytes and platelets (in the adult) (see Fig. 4) and fetal nucleated erythroblasts (in the embryo); Mß4 and Mß6 are restricted in expression to the brain. The remaining ß-tubulin

TABLE 1. Summary of Mammalian ß-Tubulin Isotypes and their Expression in Mouse[a]

	br	he	ki	li	lu	mu	sp	st	te	th	corres- ponding human gene	corres- ponding rat cDNA	% homology (human vs mouse)
Mß1				(-)	(-)		-				Hß1		91
Mß2	+++		+	+	++			+		+	Hß9	RßT.1	100
Mß3	+	+	+	+	+	+	+	+	++++	+	Hß2		100
Mß4	+++										H5ß	RßT.2	100
Mß5	+++	+	+	+	++	+	++	+	+	+++	HM40	RßT.3	100
Mß6	+++										Hß4		99

[a]Tissue abbreviations: br, brain; he, heart; ki, kidney; li, liver; lu, lung; mu, muscle; sp, spleen; st, stomach; te, testis; th, thymus. Based partly on Wang et al., 1986.

```
MB1   MREIVHIQIGQCGNQIGAKFWEVIGEEHGIDCAGSYCGTSALQLERISVYYNEAYGKKYVPRAVLVDLEP
HB1                         M        L   DR A                 R

MB1   GTMDSIRSSRLGVLFQPDSFVHGNSGAGNNWAKGHYTEGAELIENVMDVVRRESESCDCLQGFQIVHSLG
HB1            K  A                                 L    H

MB1   GGTGSGMGTLLMNKIREEYPDRILNSFSVMPSPKVSDTVVEPYNAVLSAHQLIENTDACFCIDNEALYDI
HB1                          M                        I      A

MB1   CFRTLRLTTPTYGDLNHLVSLTMSGITTSLRFPGQLNADLRKLAVNMVPFPRLHFFMPGFAPLTAQGSQQ
HB1        K                                                          G

MB1   YRALSVAELTQQMFDARNIMAACDPRRGRYLTVACIFRGKMSTKEVDQQLLSIQTRNSNCFVEWIPNNVK
HB1                               L                           V     S

MB1   VAVCDIPPRGLNMAATFLGNNTAIQELFTRVSEHFSAMFRRRAFVHWYTSEGMDISEFGEAESDIHDLVS
HB1              S     I        I N          K K             N       NN

MB1   EYQQFQDVRAGLEDSEEDAEEAEVEAEDKDH
HB1          AK V  ED  VT    M P   G
```

Fig. 1. Comparison of the sequence of hematopoietic ß-tubulin isotype in mouse and man. The sequence of Mß1 (Wang et al., 1986) is shown in the one-letter code, and compared with that of Hß1. Blank spaces denote sequence identity.

isotypes are expressed ubiquitously, though at varying levels that change during develoment (Lewis <u>et</u> <u>al</u>., 1985a; Villasante <u>et</u> <u>al</u>., 1986; Wang <u>et</u> <u>al</u>., 1986).

COMPARISON OF THE SEQUENCE OF THE HEMATOPOIETIC-SPECIFIC ß-TUBULIN ISOTYPE IN MOUSE AND MAN

Comparison of the isotype sequences encoded by four human ß-tubulin genes corresponding to the mouse isotypes Mß2, Mß3, Mß4 and Mß5 shows that conservation is absolute: not a single amino acid substitution has occurred in any of these four isotypes since the divergence of mouse and man approximately 100M years ago. However, a significant exception exists to this otherwise tight interspecies conservation of isotype sequences: the human isotype corresponding to Mß1 differs at 37 position, and is therefore only 91% homologous (Fig. 1). These findings have important implications; they demonstrate the scope for variation in an assembly-competent ß-tubulin on the one hand, but also the absolute lack of variation in four out of six ß-tubulin isotypes since the mammalian radiation. The significance of the absolute interspecies conservation of Mß2, Mß3, Mß4 and Mß5 is discussed below (see Conclusions).

GENERATION OF ß-TUBULIN ISOTYPE-SPECIFIC SERA

The task of raising sera that are specific for particular tubulin polypeptides is particularly challenging, largely because of the extensive homology between the different isotypes and also because the tubulin molecule itself is a poor immunogen. We therefore adapted a method originally described by Matthew and Patterson (1983), which depends on generating immune tolerance to epitopes that are shared among a group of closely related antigens. Once tolerance to these unwanted epitopes has been established, the animal is challenged with the immunogen of choice; thus, only epitopes unique to this challenging immunogen can be recognized. Because it is impossible to resolve different tubulin isotypes biochemically, we applied this procedure using cloned fusion proteins encoding each ß-tubulin isotype (Lewis, Gu and Cowan, 1987). The specificity of the sera thus obtained was tested in three ways: by Western blotting against cloned fusion proteins encoding all other ß-tubulin isotypes; by Western blotting against whole extracts of tissues; and by immunofluorescence experiments on fixed microtubules (Figs. 2 and 5). The data show that sera generated using cloned fusion proteins encoding Mß1, Mß2 and Mß5 are specific for their respective isotypes, while an antibody raised against Mß3 detects both Mß3 and Mß4, which differ by only seven amino acids. In addition, Western blots of whole extracts of cells and tissues show that the sera detect only ß-tubulin (data not shown).

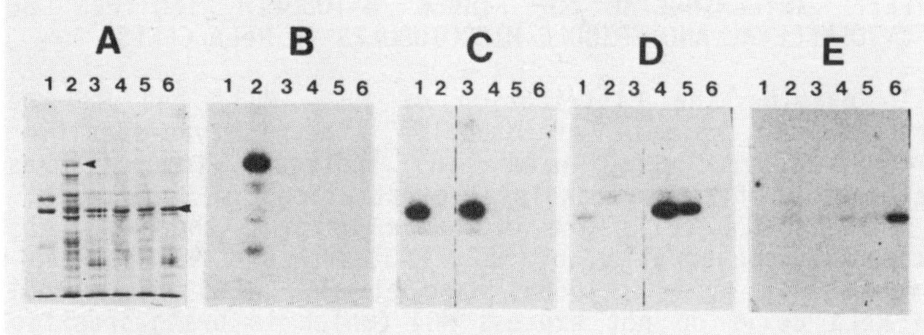

Fig. 2. Specificity of antisera to four mammalian ß-tubulin isotypes. A. Coomassie blue stain of an 8.5% stacking SDS polyacrylamide gel (Laemmli, 1970) loaded with marker brain tubulin (track 1) and whole bacterial cell extracts of cells expressing Mß1 (track 2), Mß2 (track 3), Mß3 (track 4), Mß4 (track 5) or Mß5 (track 6) as fusion proteins (arrowed). B-E. Western blots of gels loaded exactly as that shown in A, reacted with antisera raised against the fusion proteins encoding Mß1 (B), Mß2 (C), Mß3 (D) and Mß5 (E). Note the vastly different amount of each isotype present in adult brain tubulin; these relative amounts are consistent with the known mRNA levels encoding each isotype (Lewis *et al*., 1985a; Wang *et al*., 1986). Data from Lewis *et al*., 1987.

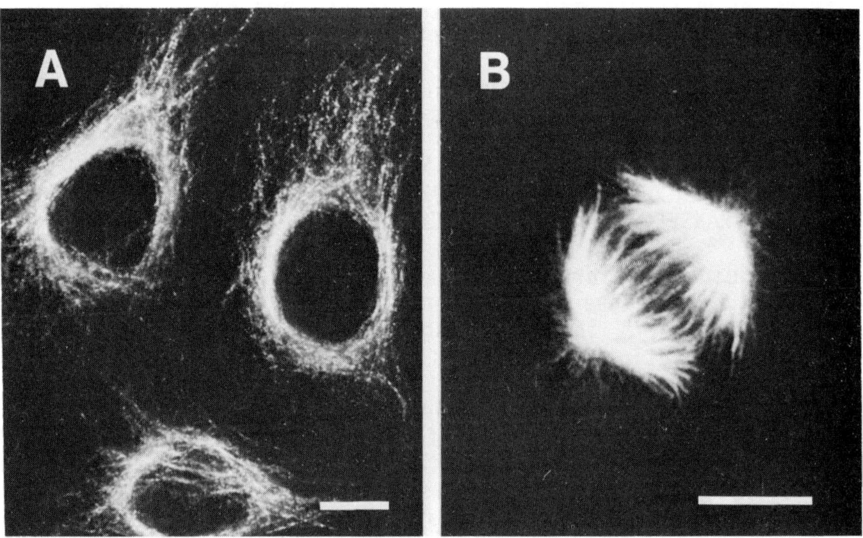

Fig. 3. Intermingling of Mß5 between interphase and spindle microtubules in HeLa cells. Photomicrograph of HeLa interphase (A) and spindle (B) microtubules identified using antibody specific for Mß5 (see Fig. 2). Bar = 10 microns.

FREE INTERMINGLING OF THREE ß-TUBULIN ISOTYPES BETWEEN CYTOSKELETAL AND SPINDLE MICROTUBULES IN HeLa CELLS

To examine the distribution of Mß2, Mß3, and Mß5 between interphase and spindle microtubules, the isotype-specific sera described above were used in indirect immunofluorescence experiments on fixed cytoskeletal preparations of HeLa cells. Note that though HeLa cells are of human origin, there is absolute interspecies conservation of Mß2, Mß3, Mß4 and Mß5 between mouse and man (Lewis et al., 1985a; Wang et al., 1986). In addition, since HeLa cells do not express Mß4 (which is brain-specific) the Mß3/Mß4-specific antibody detects only Mß3 in HeLa cells. An example of the data obtained is presented in Fig. 3, which shows both interphase and spindle microtubules (including aster and pole-to-pole microtubules) stained with the antibody specific for Mß5. Similar data were obtained for Mß2 and Mß3/4. Indeed, double label experiments performed using these sera in conjunction with a monoclonal antibody that detects an epitope common to Mß2, Mß3, Mß4 and Mß5 showed complete coincidence of microtubule labelling (Lewis, Gu and Cowan, 1987). Furthermore, when the soluble tubulin pool and the tubulin in polymerized form were simultaneously assayed for their isotype-specific content of ß-tubulin by Western blotting of extracts made from interphase and mitotic HeLa cells, both preparations contained approximately the same ratios of ß-tubulin isotypes. We conclude, therefore, that interphase and mitotic microtubules are similar, if not identical, in composition with respect to ß-tubulin isotypes (Lewis, Gu and Cowan, 1987).

THE DIVERGENT HEMATOPOIETIC-SPECIFIC MOUSE ß-TUBULIN ISOTYPE, Mß1, IS INCORPORATED INTO SPINDLE AND INTERPHASE MICROTUBULES UPON TRANSFECTION INTO HELA CELLS

An unusual feature of both MB1 and Hß1 (Table 1) is their relatively high degree of divergence compared with the other known mammalian ß-tubulin isotypes (Wang et al., 1986). This finding, plus the observation that Mß1 is restricted in its expression to tissues involved in hematopoiesis in the mouse, suggested that Mß1 (and Hß1) might have become specialized for assembly into the marginal bands of platelets and fetal nucleated erythroblasts. We therefore used the antibody specific for Mß1 (Fig. 2B) to determine the pattern of expression of this isotype at the cellular level. The data show the presence of a strongly fluorescent marginal band in platelets, and a stellar array of punctate spots in the cytoplasm of megakaryocytes (from which platelets are derived by cytoplasmic disintegration) (Fig. 4). In addition, non-mitotic fetal nucleated erythroblasts show a cytoplasmic cytoskeletal array, while dividing cells of the same type contain a strongly fluorescent mitotic spindle (Lewis et al., 1987). These data therefore demonstrate that Mß1 is capable of assembling into functionally distinct kinds of microtubule in vivo.

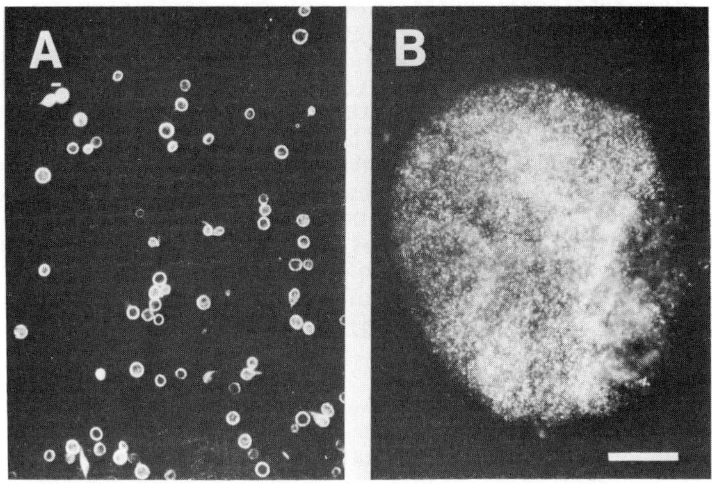

Fig. 4. Immunofluorescent analysis of mouse platelets (A) and megakaryocytes (B) with an antibody specific to Mβ1 (Fig. 2B). Bar = 10 microns.

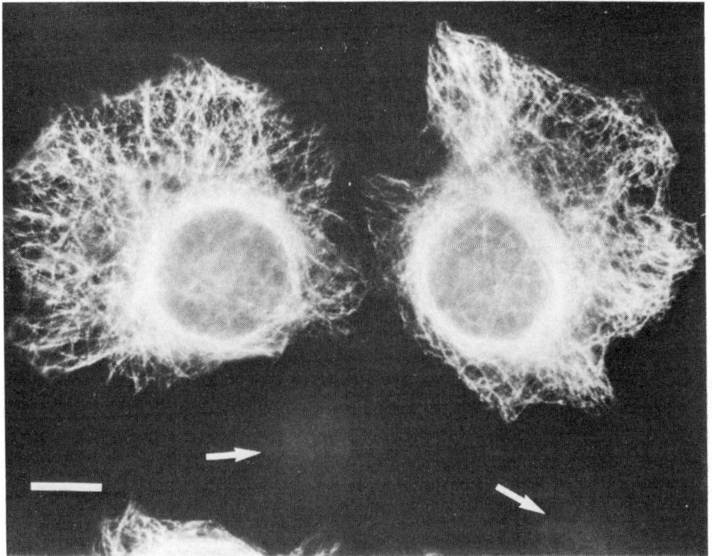

Fig. 5. The Mβ1 isotype is efficiently incorporated into all microtubules upon transfection into HeLa cells. HeLa cells were transiently transfected with a construct consisting of Mβ1 cDNA sequences driven by the SV40 T antigen early region promoter. Transfected cells were analyzed with the Mβ1-specific antibody (Fig. 2B). Note the presence of surrounding untransfected HeLa cells that do not express Mβ1 (arrowed). Bar = 10 microns.

Is the functional versatility of Mß1 restricted to cells in the hematopoietic lineage? To address this question, the cDNA encoding Mß1 (Wang et al., 1986) was coupled to the SV40 T-antigen promotor and terminator. The construct was introduced into HeLa cells, and the transfected culture assayed by immunofluorescence using the anti-Mß1-specific antibody. The results of this experiment are shown in Fig. 5. The Mß1 isotype is indeed incorporated into all HeLa cell microtubules; moreover, double-labelling experiments with a mouse monoclonal anti ß-tubulin antibody that recognizes all ß-tubulin isotypes with the exception of Mß1 showed a completely coincident pattern of microtubule staining. Thus, Mß1 coassembles with all other isotypes present into all kinds of microtubule structure (Lewis et al., 1987).

CONCLUSIONS

The past few years have seen the identification and characterization of five mammalian α-tubulin isotypes and six mammalian ß-tubulin isotypes whose significance with respect to the diversity of microtubule function is uncertain. The patterns of expression of these isotypes among several tissues have shown that some are expressed ubiquitously (though at varying levels), whereas others are restricted to only one tissue type (i.e., brain, testis and hematopoietic tissue). The existence of specialized kinds of microtubule structure in these tissues (namely axonal and dendritic microtubules in neuronal cells, flagellar microtubules in spermatozoa and marginal bands in platelets) plus the remarkable observation that four out of six ß-tubulin isotype sequences have been absolutely conserved since the mammalian radiation, seemed to argue strongly in favor of selective pressure that might be related to microtubule function. The development of immune sera capable of discriminating among different naturally-expressed isotypes therefore became an important step towards investigating the possiblity that segregation of various tubulin isotypes might contribute to functionally different mammalian microtubule structures.

In the cases we have examined, i.e., HeLa cells (which express Mß2, Mß3 and Mß5 constitutively) and megakaryocytes and dividing fetal erythroblasts (which express Mß1), immunofluorescence experiments with isotype-specific sera show that there is no segregation of these isotypes between interphase and spindle microtubules. Indeed, introduction of the relatively divergent Mß1 isotype into HeLa cells, where it is not normally expressed, results in the efficient assembly of this isotype into both spindle and interphase microtubules. If, as our data suggests, there is no subcellular sorting of different tubulin isotypes between functionally distinct microtubule structures, then the absolute conservation of the sequences that define Mß2, Mß3, Mß4 and Mß5 between mouse and man--two species that diverged about 100M years ago--seems paradoxical. Yet the selective pressure operating to maintain the identity of these sequences is made all the more evident by virtue of the differences observed between Mß1 and Hß1 (Fig. 1); the divergence of the latter two sequences has occurred over precisely the same evolutionary time span during which Mß2, Mß3, Mß4 and Mß5 have remained identical. While it is likely that developmental requirements necessitate the timed expression of tissue-specific isotypes (Raff, 1984), such fluctuations would seem to exert selective pressure on tissue-specific transcriptional regulatory sequences rather than on tubulin amino acid sequences _per se_. It seems an inescapable conclusion, therefore, that while tubulin isotypes have been selected for their differential function, the non-sorting of tubulin isotypes implies that the functional differences among them must be subtle, and manifest themselves in microtubules that are mixed copolymers of all expressed tubulin isotypes. It seems probable that differential post-transcriptional modification and differential binding of MAPs will be important ways in which isotypes contribute to the diversity of microtubule function. With isotype-specific sera and cloned fusion proteins in hand, we now have the necessary tools to investigate these questions.

ACKNOWLEDGEMENTS

This work was supported by a grant from the NIH and from the Muscular Dystrophy Association.

REFERENCES

Adachi, Y., Toda, T., Niwa, O. and Yanagida, M. (1986) Differential expression of essential and non-essential α-tubulin genes in _Schizosaccaromyces pombe_. Mol. Cell. Biol. _6_, 2168-2178.

Bond, J.F., Fridovich-Keil, J.K., Pillus, L., Mulligan, R.C. and Solomon, F. (1986) A chicken-yeast chimeric ß-tubulin protein is incorporated into mouse microtubules _in vivo_. Cell _44_, 500-509.

Cleveland, D.W., Lopata, M.A., MacDonald, R.J., Cowan, N.J., Rutter, W.J. and Kirschner, M.W. (1980) Number and evolutionary conservation of α- and ß-tubulin and cytoplasmic ß- and γ-actin genes using specific cloned cDNA probes. Cell _20_, 537-546.

Farmer, S.R., Bond, J.F., Robinson, G.S., Mbangkollo, D., Fenton, M.J. and Berkowitz, E.M. (1984) Differential expression of the rat ß-tubulin multigene family. In _Molecular Biology of the Cytoskeleton_ (Borisy, G.G., Cleveland, D.W. and Murphy, D.B., eds.). Cold Spring Harbor Laboratory, Cold Spring Harbor, N.Y.

Fulton, C. and Simpson, P.A. (1976) Selective synthesis and utilization of flagellar tubulin: the multitubulin hypothesis. In Cell Motility (Goldman, R., Pollard, T. and Rosenbaum, J., eds.), book C. Cold Spring Harbor Laboratory, Cold Spring Harbor, N.Y.

Kemphues, K.J., Kaufman, T.C., Raff, R.A. and Raff, E.C. (1982) The testis-specific ß-tubulin subunit in Drosophila melanogaster has multiple functions in spermatogenesis. Cell 31, 655-670.

Lee, M.G-S., Lewis, S.A., Wilde, C.D. and Cowan, N.J. (1983) Evolutionary history of a multigene family: an expressed human ß-tubulin gene and three processed pseudogenes. Cell 33, 477-487.

Lewis, S.A., Gwo-Shu Lee, M. and and Cowan, N.J. (1985) Five mouse tubulin isotypes and their regulated expression during development. J. Cell Biol. 101, 852-861.

Lewis, S.A., Gilmartin, M.L., Hall, J.L. and Cowan, N.J. (1985) Three expressed sequences within the human ß-tubulin multigene family each define a distinct isotype. J. Mol. Biol. 182, 11-20.

Lewis, S.A., Gu, W. and Cowan, N.J. (1987) Free intermingling of mammalian ß-tubulin isotypes among functionally distinct microtubules. Cell, in press.

Matthew, W.D. and Patterson, P.H. (1983) The production of a monoclonal antibody that blocks the action of a neurite outgrowth-promoting factor. Cold Spring Harbor Symp. Quant. Biol. 48, 623-631.

May, G.S., Gambino, J., Weatherbee, J.A. and Morris, N.R. (1985) Identification and functional analysis of ß-tubulin genes by site-specific integration in Aspergillus nidulans. J. Cell. Biol. 101, 712-719.

Raff, E.C. (1984) Genetics of microtubule systems. J. Cell. Biol. 99, 1-10.

Shatz, P.J., Solomon, F. and Botstein, D. (1986) Genetically essential and non-essential α-tubulin genes specify functionally interchangeable proteins. Mol. Cell. Biol. 6, 3722-3733.

Villasante, A., Wang, D., Dobner, P.R., Dolph, P., Lewis, S.A. and Cowan, N.J. (1986) Six mouse α-tubulin mRNAs encode five distinct isotypes; testis-specific expression of two sister genes. Mol. Cell. Biol. 6, 2409-2419.

Wang, D., Villasante, A., Lewis, S.A. and Cowan, N.J. (1986) The mammalian ß-tubulin repertoire: hematopoietic expression of a novel ß-tubulin isotype. J. Cell Biol. 103, 1903-1910.

FORMATION AND FUNCTION OF THE MARGINAL BAND OF MICROTUBULES IN NUCLEATED ERYTHROCYTES

W.D. Cohen

Department of Biological Sciences, Hunter College of the City University of New York, New York 10021, USA

ABSTRACT

Evidence that the marginal band functions in the response of mature erythrocytes to deformation and as an effector of mature cell shape is reviewed. The observations are then considered in relation to data on marginal band assembly and function during erythrocyte morphogenesis.

INTRODUCTION

The theme of this symposium volume encompasses a wide range of studies, from mechanisms of genetic regulation to developmental changes in cell form and function. For the latter, the nucleated erythrocytes of non-mammalian vertebrates are an excellent model system, undergoing transformation from spherical erythroblasts to flattened, elliptical, mature erythrocytes. In order to understand morphogenetic mechanisms in these cells, it is first necessary to understand cytoskeletal structure, composition, and particularly function in the mature erythrocytes.

BACKGROUND

The size and shape of erythrocytes in different vertebrate species is illustrated in Fig. 1 (note: all Figures follow the text). Here the flattened, elliptical nucleated morphology of non-mammalian vertebrate erythrocytes (upper row) is contrasted with the discoid anucleate morphology of mammalian erythrocytes (lower row), with the exception of those of the camel family (llama), which are anucleate but elliptical. The cytoskeletal system of the mature non-mammalian vertebrate erythrocyte consists of a marginal band (MB) of microtubules (MTs) underlying a cell surface-associated cytoskeletal network (SAC) in the plane of flattening. There are also intermediate filaments connecting nucleus and SAC, and possibly connecting the inner SAC surfaces on opposite sides of the cell as well (Granger and Lazarides, 1982; Centonze et al, 1986). The MB is a continuous circumferential bundle of MTs present in the plane of flattening of the cell, close to (but not touching) the bilayer of the plasma membrane. Mature mammalian erythrocytes, lacking MBs (in addition to nuclei), are clearly the cytoskeletal exception among the vertebrates.

The occurrence of MBs is not restricted either to erythrocytes or to vertebrates. They are present in both erythrocytes and clotting cells of a wide range of vertebrates and invertebrates, including the erythrocytes of all non-mammalian vertebrates (fish, amphibians, reptiles, birds), mammalian erythrocytes of the "primitive" (yolk-sac) series, bone marrow erythroblasts of mammals of the camel family (e.g. camel, llama, guanaco), blood platelets of all mammals, non-mammalian vertebrate thrombocytes, hemoglobin- or hemerythrin-containing erythrocytes of various invertebrates (e.g., "blood clams", certain sea cucumbers, Sipunculans), and the clotting cells of various invertebrates (e.g. *Limulus*, lobsters, certain crab spp.; Cohen and Nemhauser, 1985). MBs are thus characteristic of cells adapted to individual existence in a fluid, flowing environment, and such distribution provides insight into MB function.

THE CYTOSKELETON OF DOGFISH AND BLOOD CLAM ERYTHROCYTES

Living erythrocytes of the smooth dogfish (*Mustelus canis*), on which my students and I have worked extensively at the Marine Biological Laboratory in Woods Hole, Mass., are illustrated in Fig. 2a. These cells have the typical morphology of non-mammalian vertebrate erythrocytes. As shown in Fig. 2b,c, the cytoskeleton of a mature dogfish erythrocyte consists principally of the MB and SAC. The major MB component is tubulin, while the SAC consists of fodrin-like (or spectrin-like) proteins as well as actin. With respect to its calmodulin-binding and antibody-binding properties, the 240,000 Mr subunit of the dogfish erythrocyte SAC bears much greater resemblance to non-erythroid α-fodrin than to mammalian erythrocyte α-spectrin (Bartelt et al, 1982, 1984). This has also been found in other species (Repasky et al, 1982; Glenny et al, 1982), and is in agreement with recent work showing that chicken erythrocyte "α-spectrin" has much higher amino acid homology with human α-fodrin than with human erythrocyte α-spectrin (Moon et al, 1986). The actin-spectrin network ("membrane skeleton") of the mammalian erythrocyte, freed from constraints imposed by MB, nucleus, and perhaps intermediate filaments, is therefore quite likely an evolutionarily modified version of the non-mammalian vertebrate erythrocyte SAC.

Another erythrocyte type which has been a favorite of this laboratory is that of the blood clam (*Noetia ponderosa* and other spp.), shown in Fig. 3a. The cytoskeletal system of these hemoglobin-containing invertebrate erythrocytes is structurally much like that of vertebrates, with the exception that a pair of centrioles is associated with the MB in each cell (Fig. 3b,c). Some blood clam erythrocytes are pointed, and the centrioles are located at the tip of pointed cytoskeletons (Fig. 3d,e). This may be a natural stage of MB biogenesis, but unfortunately this cannot be verified at present because the site of erythropoiesis in blood clams is unknown. In most vertebrates, centrioles are not regularly associated with the MBs of mature cells (skates are an exception; Cohen, 1986). In a relatively simple model of the cytoskeletal system for MB-containing erythrocytes of both vertebrates and invertebrates, the MB is visualized as a flexible frame completely enclosed within an elastic SAC (Fig. 4).

MB FUNCTION IN MATURE CELLS

What is the function of the MB in mature cells? To answer this question one must prepare living cells with and without MBs and compare their properties. Protocols to achieve this with both dogfish and blood clam erythrocytes are based on MB disassembly at low temperature and reassembly upon rewarming, utilizing taxol to inhibit disassembly and colchicine or nocodazole to inhibit reassembly (Cohen et al, 1982; Joseph-Silverstein and Cohen, 1984, 1985). Regardless of how they are prepared, or the temperature at which they are examined, cells lacking MBs retain essentially normal shape (Behnke, 1970; Barrett and Dawson, 1974,; Cohen et al, 1982). Does that mean that the MB has no continuing function with respect to the shape of mature erythrocytes? The answer is no! As noted above, these are cells adapted to existence in a flowing medium; their normal environment is not the droplet on a microscope slide in which morphology is casually examined. To test function, one must perturb cells containing or lacking MBs by mechanical or other means, and look for differences in their responses. When dogfish erythrocytes containing MBs are fluxed through capillary tubes as a means of applying mechanical stress of a type which they might encounter *in vivo*, they tend to retain their normal shape. However, when dogfish erythrocytes lacking MBs are similarly treated, large percentages of such cells become deformed, as illustrated in Fig. 5a,b (Joseph-Silverstein and Cohen, 1984). This result is independent of both the temperature at which the cells are tested, and the particular inhibitor used to produce cells lacking MBs. The same result is obtained with the phylogenetically distant blood clam erythrocytes (Fig. 5c,d; Joseph-Silverstein and Cohen, 1985).

Therefore, the universal function of MBs in mature erythrocytes appears to be resistance to deformation and/or rapid recovery from deformation caused by external forces. The experiments of Waugh et al (1986), in which isolated MBs are subjected to micromanipulation and mechanical measurement, show that MBs have the flexibility necessary for such function, and that MBs released from extension into an elongated ellipse will recover their normal geometry.

How might the MB accomplish this? The mechanical properties of isolated MBs may be summarized as follows: (a) circularization of whole MBs upon release from the cytoskeleton (Bertolini and Monaco, 1976; Cohen, 1978); (b) linearization of regions after MB transection and of segments which "fray" from the MB surface (Cohen, 1978); (c) mechanical integrity permitting micromanipulation with hooks and needles (Cohen, 1978; Waugh et al, 1986);

and (d) resistance to bending measured as hundreds of times greater than that of the erythrocyte membrane (Waugh et al, 1986). Taken together, these properties suggest that the MB has internal bending strain, the release of which produces linearization, and the equalization of which produces circularization and resistance to deformation.

Is the morphology of the mature erythrocyte influenced by the morphology of the mature MB? This question can be approached by making use of pointed dogfish erythrocytes which accumulate in the population when the cells are incubated in Elasmobranch Ringer's solution for several hours at room temperature (Cohen et al, 1982). Although this is a long-term storage artifact of unknown mechanism, it is nonetheless useful. As illustrated in Fig. 6, the pointed cells have cytoskeletons containing pointed MBs. If cells are prepared with and without MBs, using the protocols described above, only cells which contain MBs become pointed. In addition, if pointed cells are incubated at 0°C, their points disappear as their MBs disassemble. Therefore the MB is an effector of cell shape. The SAC of the mature cell can conform to pointed MB morphology, but apparently has an elastic memory for the original differentiated shape if the MB disassembles.

MB FORMATION AND ERYTHROCYTE DIFFERENTIATION

What is the relationship between MB formation and erythrocyte differentiation? In blood clam erythrocytes incubated at 0°C, the MB disassembles, but the pair of centrioles as well as the SAC remains (Nemhauser et al, 1983). MB reassembly during rewarming can be followed in time course samples by means of anti-tubulin immunofluorescence, and the centrioles are observed to function as part of an MB-organizing center during reassembly as illustrated in Fig. 7. However, during MB reassembly in vertebrate erythrocytes, MB-associated centrioles have not been observed (Miller and Solomon, 1984). Cell morphology remains virtually unchanged throughout MB disassembly and reassembly in all cases tested thus far, with the exception of immature bone marrow erythrocytes (Behnke, 1970; Barrett and Scheinberg, 1972; Cohen et al, 1982). Thus, experimentally-induced MB disassembly and reassembly in mature erythrocytes of blood clams and other species is not a good model for the role of MB during cell differentiation. The SAC of the mature cell apparently retains its memory for mature cell morphology independent of the MB.

It is known that MB biogenesis is concomitant with erythrocyte morphogenesis, the stages of which are believed to be sphere to disk to ellipse (Barrett and Scheinberg, 1972). Recent work on differentiating chicken erythrocytes indicates that a centrosome is involved early in MB microtubule initiation (Kim et al, 1987; Murphy et al, 1986), and that MB microtubules contain a distinctive beta tubulin which is the product of a gene specifically activated in erythrocytes and thrombocytes (Murphy et al, 1986). However, the mechanistic relationship between MB assembly and changes in cell morphology is unknown. The observations of Barrett and Scheinberg (1972) suggest that the SAC of spherical erythroblasts is in a more fluid or dynamic state than that of the mature erythrocyte. If so, one might expect the forming MB to produce transitional stages of cell morphology which conform to its structural stage.

During the past year, much effort has been expended in my laboratory in a search for a convenient experimental system with which to study MB formation and morphogenetic function. Since we wished to study relatively large cells containing thick MBs easily seen in phase contrast, our search focused primarily on the amphibian literature and ultimately on the amphibian larval spleen (Ginsburg et al, 1986). The larval spleen develops initially as a closed bag of differentiating erythroblasts, visible as a red spot through the transparent wall of the living larva. It can be removed intact and then opened on a microscope slide to release its content of cells. The population of developing erythrocytes contains cells of varying morphology, including spheres, disks, and ellipsoids. Particularly striking, however, is the large percentage of pointed cells (both singly- and doubly-pointed; Fig. 8a). In contrast to the larval circulation, which typically contains 1-2% of such cells, pointed cells normally constitute 7-10% of the developing splenic population (Ginsburg et al, 1986). The possibility that these are intermediate morphogenetic stages is currently being explored in our laboratory, and is supported by our initial observations on their cytoskeletal system. In doubly-pointed cytoskeletons, the MB consists of two relatively thick linear bundles of MTs, which meet at a point at both ends. In this respect they are similar to the discontinuous MBs of experimentally-produced doubly-pointed dogfish erythrocytes (Fig. 6). As illustrated in Fig. 8b, pairs of phase-dense dots are frequently observed in one (and only one) point of doubly-pointed splenic erythroblast cytoskeletons. Singly-pointed cytoskeletons frequently contain a similar pair of dots at the pointed end, reminiscent of certain naturally-occurring blood clam erythrocytes (Fig. 3d,e), but we have not found them in amphibian splenic cytoskeletons with complete elliptical MBs. Although we believe these to be centrioles based on their size, density, and pairing, their identity remains to be confirmed by means of electron microscopy. Our initial observations have been sufficient to construct of a working hypothesis for stages of MB assembly as correlated with stages in the transition from sphere to flattened ellipsoid (Fig. 9; Cohen and Ginsburg, 1986). Obviously there is much work now to be done, not only with respect to the presence of centrioles at critical stages and locations, but also in testing the proposed morphological sequence by following individual cells as they differentiate in culture. Details of the cytoskeletal changes also remain to be studied. However, it seems clear that this will be an excellent system with which to correlate stages of MB assembly with the functional effect of the

MB on cell morphology during differentiation.

ANALYSIS OF CYTOSKELETAL FUNCTION: A GENERAL NOTE.

One lesson brought home by our work on MBs is that cytoskeletal function can be a subtle thing. It was not immediately obvious that the appearance of normal morphology in erythrocytes lacking MBs was illusory, and that function could only be tested by perturbing such cells. The general point is that cytoskeletal function must be considered in the context of the normal environment of a given cell type. For example, while experimental disruption of cytoskeletal elements such as intermediate filaments (e.g., see Klymkowsky in this monograph) might have little effect on the general properties of cultured cells, such disruption might have considerable effect on the properties of similar cells in tissues of an organism, or indirectly on their neighboring cells. A developmental cytoskeletal function might even be detectable only in the context of the whole organism in its normal environment (an embryo developing in a pond or ocean rather than a dish, for example). In addition, the potential subtlety of cytoskeletal function, involving mechanical parameters not easily observed or measured (Waugh et al, 1986), can make analysis difficult. For example, elegant genetic manipulation can now be used to produce cultured cells with microtubules or intermediate filaments incorporating subunit isotypes normally specific for other cell types, and yet there seems to be little effect on the pattern of cytoskeletal structure or on cell functions such as mitosis (e.g., Cowan et al, and Blessing et al, in this monograph). However, if in a developing embryo such alteration of microtubules or intermediate filaments caused only one chromosome to be lost in every 100 mitoses, that might be sufficient to make the presence of an incorrect isotype lethal to the embryo. Similarly, even if incorporation of an incorrect tubulin or intermediate filament protein isotype into the cytoskeleton of a nucleated erythrocyte produced a normal-looking cytoskeletal system, such incorporation could have lethal or at least selectively disadvantageous mechanical consequences for cells in a flowing bloodstream.

ACKNOWLEDGEMENTS

I am indebted to my former students who carried out much of the critical research cited above, particularly Drs. Iris Nemhauser, Diana Bartelt, and Jacquelyn Joseph-Silverstein. I wish also to thank Mary Ginsburg and Dr. Laura Twersky for their important contributions in current studies, and Dr. Richard Waugh (U. Rochester) for helpful consultations. This work was supported by PSC-CUNY grant 6-63177 and NSF PCM-8409159.

REFERENCES

Barrett, L.A. and Dawson, R.P. (1974). Avian erythrocyte development-microtubules and the formation of disc shape. Devel. of Biol. 36, 72-81.

Barrett, L.A. and Scheinberg, S.L. (1972). The development of avian red cell shape. J. Exp. Zool. 182, 1-14.

Bartelt, D., Carlin, R., Scheele, G. and Cohen, W.D. (1982). The cytoskeletal system of nucleated erythrocytes. II. Presence of a high molecular weight calmodulin-binding protein. J. Cell Biol. 95, 278-284.

Bartelt, D., Carlin, R., Scheele, G. and Cohen, W.D. (1984). Similarities between the Mr 245,000 calmodulin-binding protein of the dogfish erythrocyte cytoskeleton and α-fodrin. Arch. Biochem. Biophys. 230, 13-20.

Behnke, O. (1970). A comparative study of microtubules of disc-shaped blood cells. J. Ultrastr. Res. 31, 61-75.

Bertolini, B. and Monaco, G. (1976). The microtubule marginal bundle of the newt erythrocyte. Observations on the isolated band. J. Ultrastruct. Res. 54, 57-67.

Centonze, V.E., Ruben, G.C. and Sloboda, R.D. (1986). Structure and composition of the cytoskeleton of nucleated erythrocytes. III. Organization of the cytoskeleton of _Bufo marinus_ erythrocytes as revealed by freeze-dries platinum-carbon replicas and immunofluorescence microscopy. Cell Motil. Cytoskel. 6, 376-388.

Cohen, W.D. (1978). Observations on the marginal band system of nucleated erythrocytes. J. Cell Biol. 78, 260-273.

Cohen, W.D. (1986). Association of centrioles with the marginal band in skate erythrocytes. Biol. Bull. 171, 338-349.

Cohen, W.D., Bartelt, D., Jaeger, R., Langford, G. and Nemhauser, I. (1982). The cytoskeletal system of nucleated erythrocytes I. Composition and function of major elements. J. Cell. Biol. 93, 828-838.

Cohen, W.D. and Ginsburg, M.F. (1986). A model for marginal band formation in amphibian erythrocytes. J. Cell Biol. 103, 547a.

Cohen, W.D. and Nemhauser, I. (1985). Marginal bands and the cytoskeleton in blood cells of marine invertebrates. In Blood Cells of Marine Invertebrates, W.D. Cohen, ed., pp. 1-49, Alan R. Liss, Inc., NY.

Ginsburg, M.F., Twersky, L.H. and Cohen W:D.(1986). Studies on the differentiation of splenic erythroblasts in _Ambystoma mexicanum_. J. Cell Biol. 103, 546a.

Glenny, J.R., Glenny, P. and Weber, K. (1972). Erythroid spectrin, brain fodrin, and intestinal brush border protein (TW-260/240) are related molecules containing a common calmodulin-binding subunit bound to a variant cell-type-specific subunit. Proc. Natl. Acad. Sci. 79, 4002-4005.

Granger, B.L. and Lazarides, E. (1982). Structural associations of synemin and vimentin filaments in avian erythrocytes revealed by immunoelectron microscopy. Cell 30: 263-275.

Joseph-Silverstein, J. and Cohen, W.D. (1984). The cytoskeletal system of nucleated erythrocytes. III. Marginal band function in mature cells. J. Cell Biol. 98, 2118-2125.

Joseph-Silverstein, J. and Cohen, W.D. (1985). Role of the marginal band in an invertebrate erythrocyte: evidence for a universal mechanical function. Can. J. Biochem. Cell Biol. 63, 621-630.

Kim, S., Magendantz, M., Katz, W. and Solomon, F. (1987). Development of a differentiated microtubule structure: formation of the chicken erythrocyte marginal band in vivo. J. Cell Biol. 104, 51-59.

Miller, M. and Solomon, F. (1984). Kinetics and intermediates of marginal band formation: evidence for peripheral determinants of microtubule organization. J. Cell Biol. 99 (Suppl.), 70s-75s.

Moon, R.T., Giebelhaus, D.H. and McMahon, A.P., (1986). Structure and evolution of α-fodrin. J. Cell Biol. 103,539a.

Murphy, D.B., Grasser, W.A., Wallis, K.T. (1986). Immunofluorescence examination of beta tubulin expression and marginal band formation in developing chicken erythroblasts. J. Cell Biol. 102, 628-635.

Nemhauser, I., Joseph-Silverstein, J. and Cohen, W.D. (1983). Centrioles as microtubule-organizing centers for marginal bands of molluscan erythrocytes. J. Cell Biol. 96, 979-989.

Repasky, E.A., Granger, B.L. and Lazarides, E. (1982). Widespread occurrence of avian spectrin in nonerythroid cells. Cell 29, 821-833

Waugh, R.E., Erwin, G. and Bouzid, A. (1986). Measurement of the extensional and flexural rigidities of a subcellular structure: marginal bands isolated from erythrocytes of the newt. J. Biomech. Eng. 108, 201-207.

FIGURES

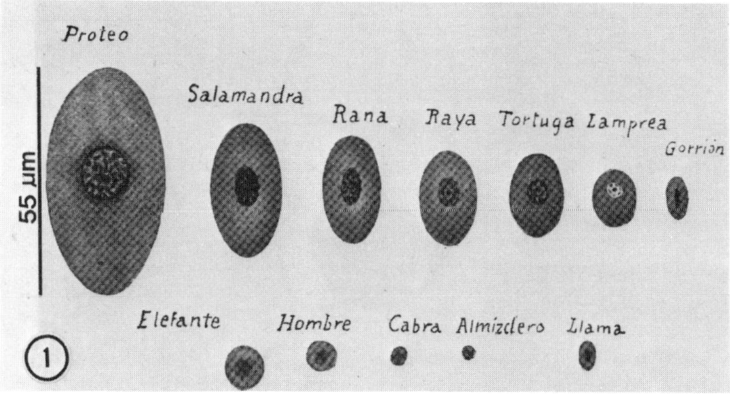

Fig.1. Erythrocyte size and shape: flattened ellipsoidal nucleated erythrocytes of non-mammalian vertebrates (upper row), which contain marginal bands (MBs), vs. anucleate discoid erythrocytes of mammals, which lack MBs (lower row; llama an exception). (Reproduced with slight modification from "Histology" by Ramon Y Cajal, William Wood & Co., 1933, with permission of the Williams and Wilkins Co.).

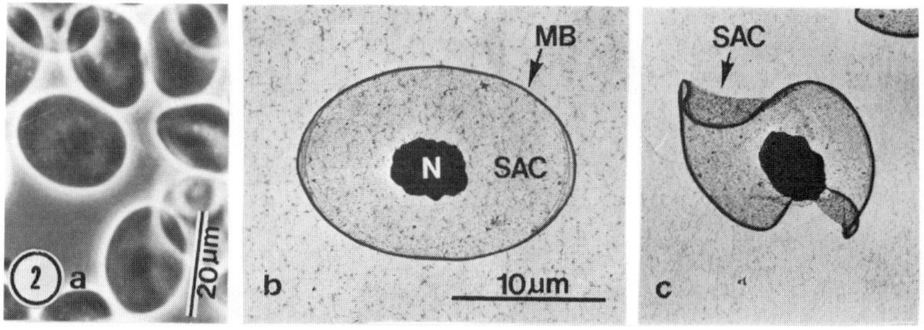

Fig.2. Dogfish erythrocytes and their cytoskeletal system. (a) Living cells, phase contrast; (b) Uranyl acetate-stained cytoskeleton whole mount, with nucleus (N) and MB heavily stained. The cell surface-associated cytoskeletal network (SAC) is more easily seen in twisted cytoskeletons, as in (c). Cytoskeletons are produced by Triton lysis under microtubule-stabilizing conditions. (Reproduced with slight modification from Cohen et al. 1982, with permission of the Rockefeller University Press).

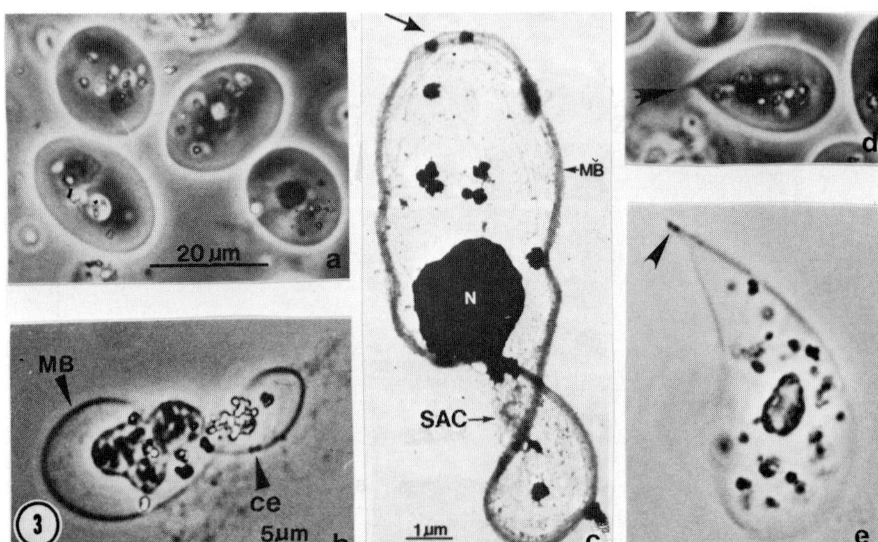

Fig.3. Blood clam erythrocytes and their cytoskeletal system. (a) Living cells in phase contrast, showing morphology similar to non-mammalian vertebrate erythrocytes; (b) Cytoskeleton in phase contrast, with pair of MB-associated centrioles appearing as phase-dense dots (ce); (c) Uranyl acetate-stained cytoskeleton whole mount, TEM. The centrioles (arrow, upper left) are more heavily stained than the MB. Naturally-occurring singly-pointed erythrocytes and similarly pointed cytoskeletons, with the centrioles at the tip, are sometimes observed (d,e; arrowheads). (Reproduced with some modification from Nemhauser et al., 1983, permission of Rockefeller University Press)

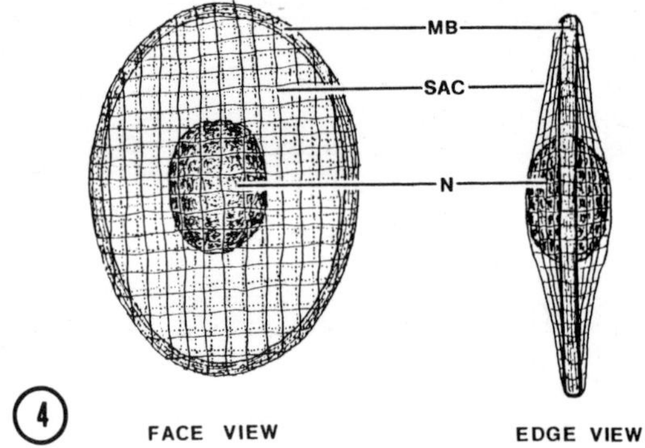

Fig.4. Model of the nucleated erythrocyte cytoskeleton, in which the MB is in contact with the inner face of the SAC. Cell flatness may be achieved by tension within the SAC network, applied across the MB acting as a flexible frame. (Reproduced with slight modification from Cohen and Nemhauser, 1985, with permission of Alan R. Liss Inc.)

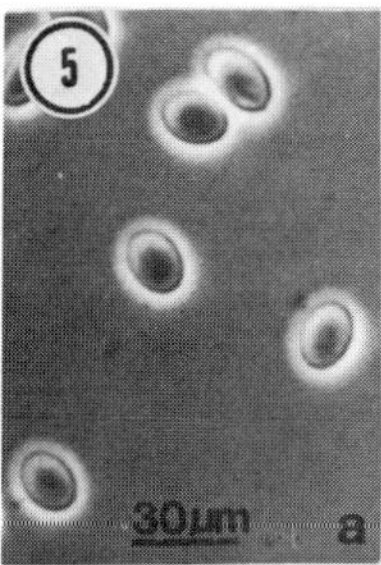

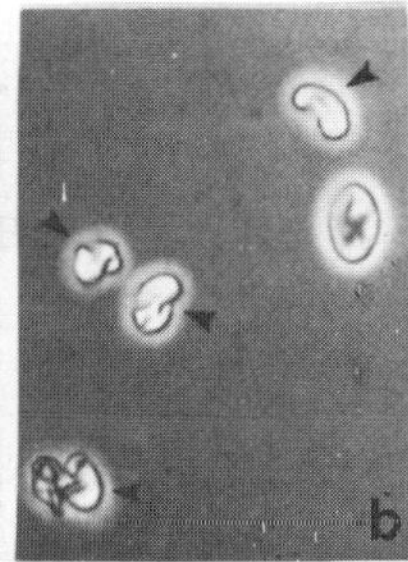

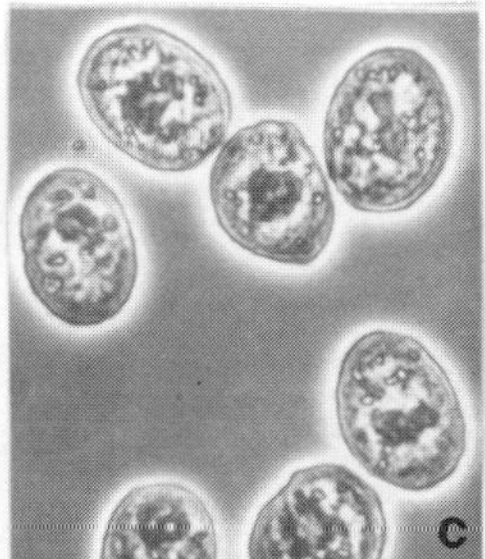

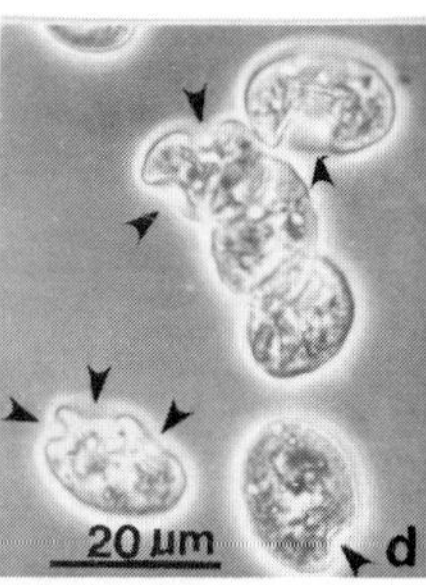

Fig.5. Effect of mechanical stress (fluxing through capillary tubes) on erythrocytes prepared so as to contain or lack MBs at the same temp. (a) dogfish, room temp. + MB; (b) dogfish, - MB; (c) blood clam, 0^oC + MB; (d) blood clam, - MB. Large percentages of cells lacking MBs are deformed (arrowheads). Phase contrast. (Reproduced from Joseph-Silverstein and Cohen, 1984, 1985, permission of Rockefeller University Press and National Research Council of Canada).

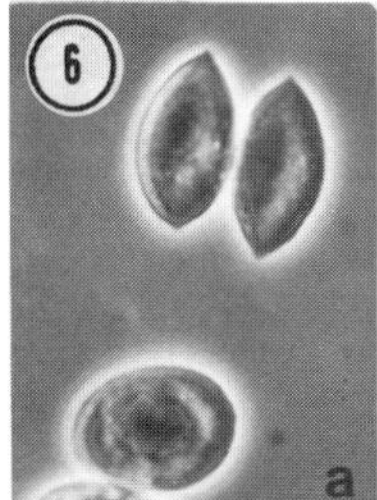

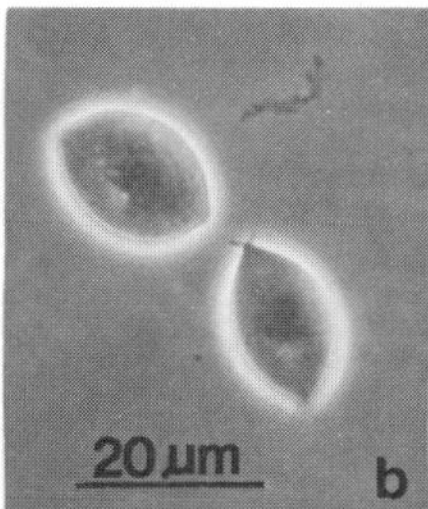

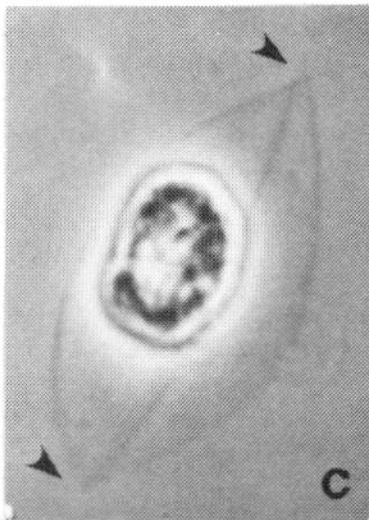

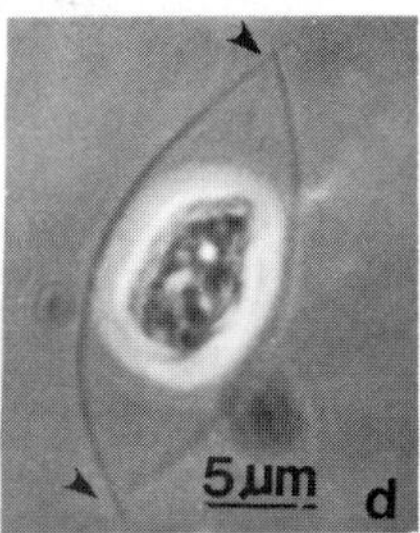

Fig.6. (a,b) Doubly-pointed dogfish erythrocytes, produced by incubation of cells at room temp. for 5 or more hours. (c,d) The cytoskeletons of such cells contain discontinuous MBs which are pointed at both ends (arrowheads), where the two major bundles of microtubules meet. (Reproduced from Cohen et al., 1982, with permission of the Rockefeller University Press).

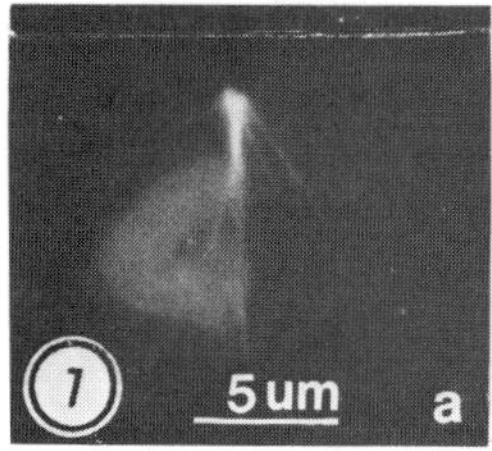

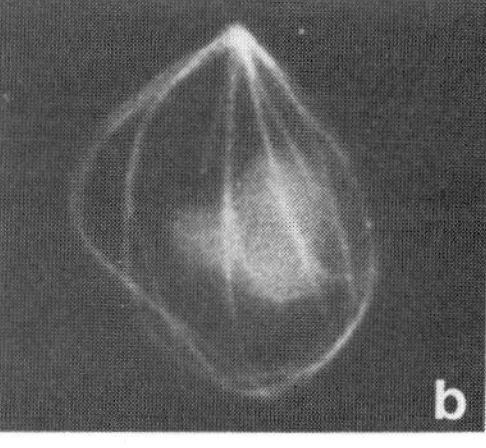

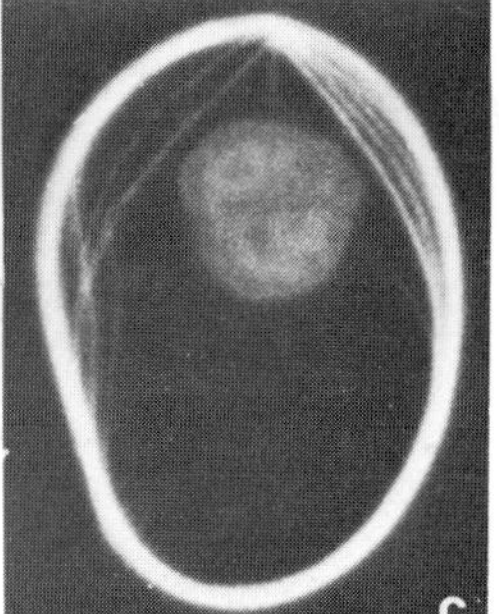

Fig.7. Temperature-induced MB reassembly in blood clam erythrocytes, observed by anti-tubulin immunofluorescence.
(a,b) Early stages of rewarming; centrioles act as part of organizing center generating pointed "polar" arrangement of MTs. (c) Later stage, with nearly complete MB. (Reproduced from Nemhauser et al., 1983, with permission of the Rockefeller University Press).

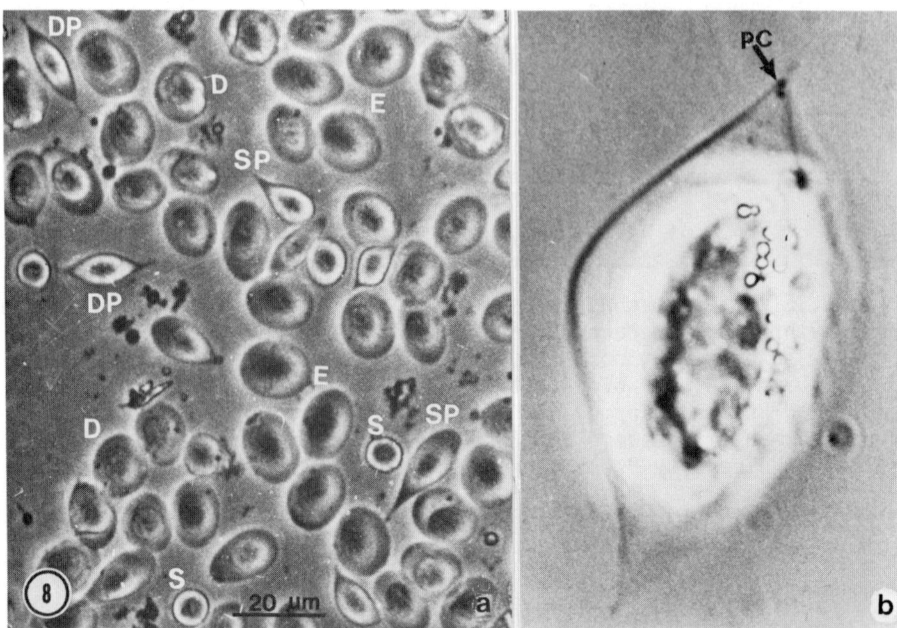

Fig.8. (a) Population of developing erythrocytes from amphibian larval spleen (Ambystoma mexicanum, the axolotl). In addition to flattened elliptical (E) and discoidal (D) erythrocytes, there are spherical (S) erythroblasts and relatively large numbers of singly- and doubly-pointed cells present (SP, DP). (b) Cytoskeleton of a doubly-pointed cell, consisting of two thick MT bundles meeting at each end (lower end is out of plane of focus). The upper point contains a pair of phase-dense dots which may be centrioles (putative centrioles = PC). Phase contrast (Ginsburg et al., 1986).

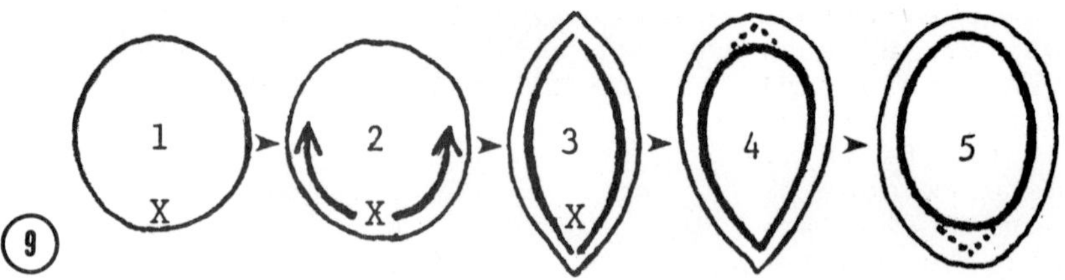

Fig.9. Working hypothesis correlating stages of MB assembly with possible stages of cellular morphogenesis. MTs are nucleated at organizing center X (stages 1,2), growing into two bundles which meet at pointed ends (3). The distal pointed end is then closed to produce a singly-pointed cell (4), and subsequent closure of the proximal end produces a discoidal or elliptical cell with a continuous MB (5). (Cohen and Ginsburg, 1986).

MONOCLONAL ANTIBODY TO PNEUMOCYSTIS DETECTS MICROTUBULUS-ASSOCIATED ANTIGENS IN MAMMALIAN CELLS AND GIARDIA

Linder, E.
The National Bacteriological Laboratory (NBL), S-105 21 Stockholm, Sweden

Monoclonal antibodies of two specificities were developed against Pneumocystis (PC) antigens by immunization of mice with urea-extract of PC-infected human lung tissue (1). Screening tests included an ELISA using urea extract of isolated parasites and paraffin sections of PC-infected lung tissue. The antibodies were of two different specificities, both reacting with an 82 kD parasite component in immunoblotting. Antibody 3F6 of the first specificity reacted only with the 82 kD component and was PC-specific and has been used for diagnostic purposes (2). Antibody 4B8 detected additional parasite polypeptides in the 70-82 kD range, 45 kD and some minor components.

The present report deals with the localization of 4B8-reactive antigenic determinants by immmunological and cytological staining of various parasites, rat tissues and cultured cells.

Cross reactivity of 4B8 was studied by immunofluorescence microscopy using sections of Bouin fixed, paraffin embedded human lung and various rat tissues as antigen. Vero cells were grown on coverslips in Eagles minimal essential medium and fixed in aceton before IFL staining.

The results (see Figure) of immunofluorescent staining show strong reactivity with tracheal and bronchial (br) epithelium, nerve fibres of peripheral nerves, neurons of the central nervous system (ne), epididymal epithelium, spermatocytes (sp) and epithelium of the fallopian tube. In Vero cells a cytoplasmic network consisting of fibres originating from the perinuclear area was observed; in mitotic cells, the spindle showed a distinct reaction (Vero).

Interestingly, only Giardia lamblia ventral flagellae reacted but not the anterolateral and posterolateral flagellae (Giardia). There was no fluorescence in the area of the ventral disc. No reactivity was seen with Trypanosomes, Crithidia, Leishmania.

A common feature of the reactive structures was the association of staining with microtubules. The association between a positive staining reaction and presence of microtubules was consistent, but microtubules, e.g. in mature sperm tails, reacted only weakly and in Giardia selective staining of only the ventral flagella was seen, whereas other flagellae and the microtubulus-containing ventral striated disk of this parasite failed to react with 4B8.

The target antigen recognized by 4B8 may be microtubulus associated proteins or the monoclonal antibody may detect some phylogenetically well preserved determinants present in some but not all tubulins.

REFERENCES

(1) Linder, E., Lundin, L., Vorma, H. (1987) Detection of Pneumocystis carinii in lung-derived samples using monoclonal antibodies to an 82 kD parasite component. J. Imm. Meth. 91:57-62.

(2) Linder, E., Elvin, K., Björkman, A., Bergdahl, S., Morfeldt-Månsson, L., Moberg, L., Sönnerborg, A. (1986) Monoclonal antibody to detect Pneumocystis carinii. Lancet ii, 634.

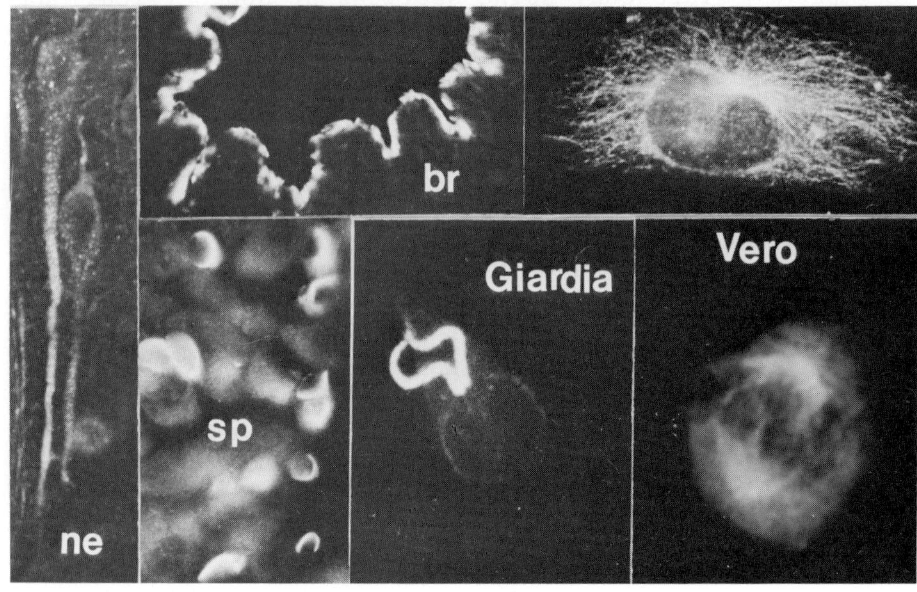

ERYTHROCYTE PROTEIN 4.1 BINDS TO TUBULIN

I. Correas and J. Avila. Centro de Biología Molecular. (CSIC-UAM). Canto Blanco. 28049 Madrid. Spain

The red blood cell membrane skeleton is a network of proteins of which spectrin is the major component. Spectrin binds to actin filaments and this lattice is connected to the membrane via two linker proteins : ankyrin and protein 4.1 (for reviews, see Ref. 1 and 2).

Recent studies have shown that spectrin, 4.1 and ankyrin are not only present in erythrocytes but in every cell type analyzed (1,2). As a result, the interest in these proteins and in their role in non-erythroid cells have notably increased in the last few years.

It has been reported that immunoreactive forms of protein 4.1 are associated with the nucleus and mitotic apparatus in several different cell types (Correas, I, Anderson, R.A., Mazzucco, C.E. and Marchesi, V.T., submitted for publication). A possible explanation is that 4.1-related proteins bind to the spindle microtubules. To analyze this possibility we have studied whether protein 4.1 interacts with tubulin. In vitro experiments suggest that 4.1 copolymerizes with tubulin in the presence of taxol. This result is illustrated in Fig. 1. As expected, when protein 4.1 was incubated with taxol in the absence of tubulin, protein 4.1 remained in the supernatant fraction (lane A). In the presence of tubulin both protein 4.1 and tubulin were seen in the pellet fraction (lane B). The molar ratio of protein 4.1 bound to tubulin oscilates between the same ranges as those described for MAPs. In the experiment shown this value was 6.

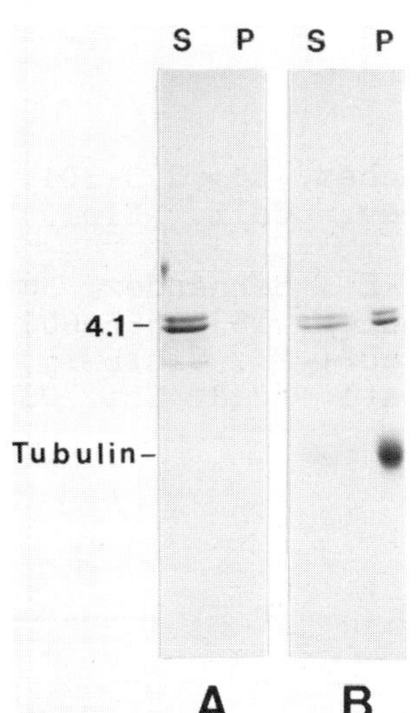

Fig. 1. Cosedimentation of protein 4.1 with intact tubulin. Pig brain tubulin (30 µg) was polymerized in the presence of 10 µM taxol for 20 min at 37ºC. The polymers were incubated with pig erythrocyte protein 4.1 (7 µg) in the presence of taxol for 20 min at 37ºC followed by sedimentation. Equivalent samples of supernatant (S) an pellet (P) were electrophoresed on 7% acrylamide SDS gels. A, protein 4.1 alone. B, protein 4.1 in the presence of intact tubulin.

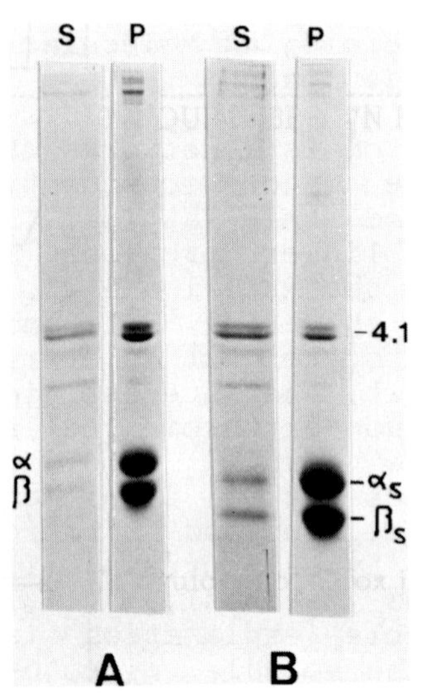

Fig. 2. Cosedimentation of protein 4.1 with S-tubulin. The conditions for this experiment are as described in Fig. 1. A, High M_r MAPs and protein 4.1 in the presence of intact tubulin. B, High M_r MAPs and protein 4.1 in the presence of tubulin which lacks the 4 Kd C-terminal peptide.

MAPs bind to the C-terminal region of tubulin (3) but some of them can also bind to another region localized in the amino-terminal domain of tubulin (4). We have analyzed whether protein 4.1 has the same behaviour as MAPs (Fig. 2). When protein 4.1, high M_r MAPs and intact tubulin were incubated together under the conditions described for Fig. 1, both 4.1 and MAPs cosedimented with tubulin with similar molar ratios (lane A). When subtilisin-digested tubulin was added to 4.1 and MAPs, the amount of protein 4.1 cosedimenting with tubulin notably decreased (lane B). This suggests that the tubulin binding site for 4.1 is localized at the carboxy-terminal region of the molecule where MAPs do interact. However, this does not exclude a lower affinity binding of protein 4.1 for S-tubulin. This possibility is now under investigation.

REFERENCES

(1) Bennet, V. (1985) Annu. Rev. Biochem. 54, 273-304.
(2) Marchesi, V.T. (1985) Annu. Rev. Cell. Biol. 1, 531-561.
(3) Serrano, L., Montejo de Garcini, E., Hernández, M.A. and Avila, J. (1985) Eur. J. Biochem. 153, 595-600.
(4) Littauer, U.Z. Giveon, D., Thierauf, M., Ginzburg, I. and Ponstingl, H. (1986) Proc. Natl. Acad. Sci. USA.

Molecular Genetic Analysis of the Murine Dilute Locus: A Gene Important for Cell Shape or Vesicle Transport? P.K. Seperack, M.C. Strobel, K. J. Moore, N.G. Copeland and N.A. Jenkins. Mammalian Genetics Laboratory, BRI-Basic Research Program, NCI-FCRF, P.O. Box B, Frederick, Md. 21701

Proteins that play important roles either in maintaining cell shape or in transporting intracellular vesicles are being identified continually and the dilute gene product may represent one of these proteins. The dilute, d, mutation is a recessive coat color mutation located on mouse chromosome 9 (1). Mice homozygous for d alleles have a lightened ("diluted") coat color relative to wild-type mice. This variant coat color is correlated with an altered melanocyte morphology (1). Wild-type melanocytes have long thick dendritic processes while dilute melanocytes are essentially adendritic and the pigment granules are clumped around the nucleus. The lesion, therefore, appears to be in a gene whose product is required either to maintain cell shape or to transport intracellular vesicles. Chromosome mapping studies suggest that the dilute gene does not represent any well characterized cytoskeletal protein or transport element (1). One allele of d, d^V, was caused by the integration of a retrovirus, Emv-3, at d providing molecular access to this locus (2,3,4). Unlike d^V, many other mutant d alleles also exhibit a neurological phenotype. We have used a unique sequence fragment, p0.3, isolated from d^V to characterize a new mutation at d and to isolate clones for dilute from a cDNA library.

Molecular analysis of the dilute locus. The structure of the dilute locus surrounding the viral integration site is shown in Figure 1 and the location of the fragment p0.3 flanking the viral integration site is indicated. Wild-type and d^V alleles generate 9 and 18 kb EcoRI fragments, respectively, in a Southern blot analysis (Figure 2A). Recently a spontaneous forward mutation to d was identified, termed d^{120J}. Molecular analysis showed that 3.5 kb of DNA was deleted in this new mutant allele relative to the wild type gene (region deleted spans the bold line in Figure 1). This deletion is 2 kb 3′ to the viral integration site in the d^V allele. No other alterations have been detected in this new mutant allele. Mice homozygous for this new mutation have a lightened coat color and also die by 21 days postpartum from a neurological impairment. Most of the p0.3 probe, which we know contains coding sequence (see below), is deleted in this new mutation. We conclude that the neurological phenotype previously reported to be

Figure 1. d Gene Structure

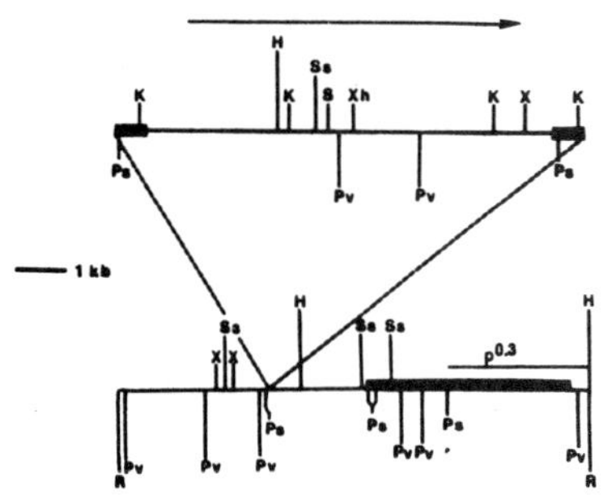

associated with the coat color phenotype is most likely produced by mutations in a single gene (1,4).

The p0.3 probe detected transcripts in Northern analysis using RNA isolated from the B16 melanoma cell line (+/+ at d). Therefore, a cDNA library was made using B16 RNA and several cDNA clones were isolated. One clone contained a 2.5 kb insert and this insert has been used as a probe in additional Southern and Northern analysis.

Figure 2. Southern Analysis

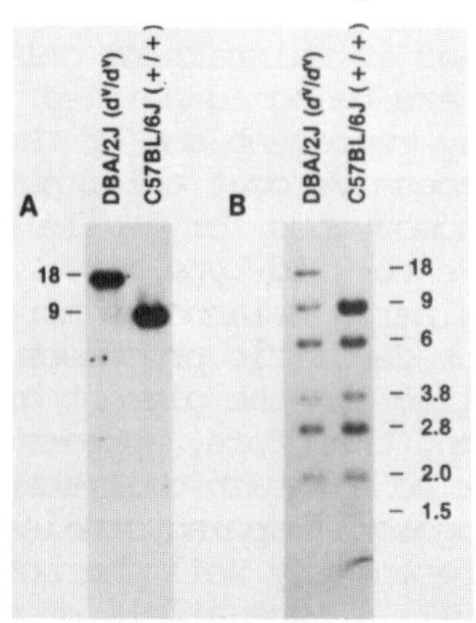

Figure 3. Northern Analysis

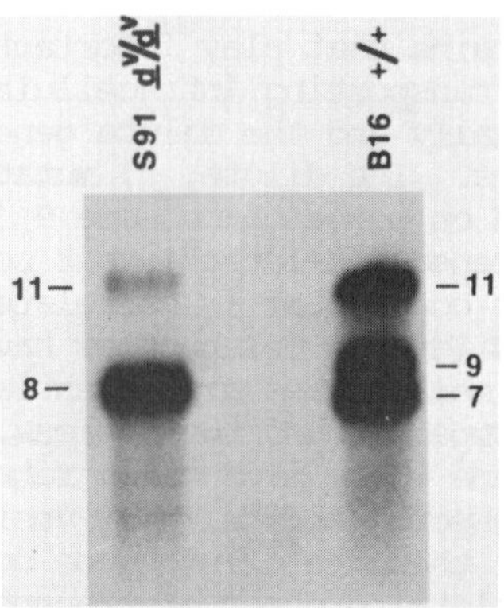

Southern analysis using a dilute cDNA probe. The 2.5 kb cDNA probe detects the same 18 and 9 kb fragments as p0.3 in animals which are d^v/d^v or +/+ respectively and detects additional fragments as well (Figure 2B). The cDNA probe, therefore, recognizes sequences spanning an additional 25 kb of genomic DNA which can be analyzed for d mutations. These data, together with the large transcripts detected by this probe (see below), suggest that the dilute gene is very large, perhaps 150-200 kb.

Detection of dilute transcripts in RNAs isolated from tissues and cell lines using the cDNA probe. Dilute transcripts are detected in A^+ RNA isolated from two melanoma cell lines. In the B16 cell line (+/+ at d) 3 major transcripts of 11, 9 and 7 Kb are detected. However, in the S91 cell line (d^v/d^v) only two major transcripts of 11 and approximately 8 Kb are detected. This indicates that the integrated provirus effects both the number of transcripts and their size in d^v/d^v animals. This same pattern of transcripts is also seen in RNAs isolated from brain, kidney, spleen and thymus but not liver of animals which are either +/+ or d^v/d^v but at reduced levels (data not shown). By continued analysis we will identify the product of this gene and determine its interactions with other cytoskeletal components. Research sponsored by the NCI under contract N01-CO-23909.

REFERENCES

1. Silvers, W.K. 1979. The coat colors of mice. Springer-Verlag.
2. Jenkins, N.A., et al., 1981. Nature 293: 370.
3. Copeland, N.G. et al., 1983. Cell 33: 379.
4. Rinchick, E.M., et al., 1986. Genetics 112: 321.

CYTOSKELETON IN DEVELOPMENTAL NEUROBIOLOGY

CHARACTERIZATION OF A MICROTUBULE ASSEMBLY PROMOTING FACTOR IN DIFFERENTIATED NEUROBLASTOMA CELLS.

Nicholas W. Seeds, Kristin R. Hinds and Ricardo B. Maccioni

Department of Biochemistry, Biophysics & Genetics, University of Colorado Medical School, Denver, CO 80262 USA

ABSTRACT

Clonal cells (N18) of the mouse neuroblastoma C-1300 can be induced to undergo morphological differentiation characterized by the outgrowth of very long neurites containing many microtubules. This differentiation event does not require a major change in the amount of tubulin or its synthesis. However, high speed supernates from differentiated cell extracts can promote the *in vitro* polymerization of purified brain tubulin; in contrast supernates from undifferentiated neuroblastoma cells failed to promote assembly. This neuroblastoma microtubule assembly promoting factor (NbMAPF) is heat and trypsin labile and has a native Mr=160,000. NbMAPF activity is seen at 1 hr. after the induction of differentiation and reaches maximal levels by 4-8 hr., while neurite outgrowth is still minimal; long neurites are not seen until 1 day or more. The appearance of NbMAPF is insensitive to cycloheximide, suggesting that the activity results from a posttranslational modification of a pro-factor molecule. Induction of NbMAPF requires that the cells be attached to a suitable substratum where neurite growth can occur. Furthermore, cytochalasin B blocks the induction of the factor. Interestingly, the NbMAPF does not cycle with microtubules during *in vitro* assembly, and additional cycles of assembly require the readdition of the supernatant fraction from the previous assembly step; suggesting the NbMAPF may play a catalytic role or be inactivated during the assembly process. A rabbit antibody to native NbMAPF has been prepared that blocks NbMAPF activity in a concentration dependent fashion, but does not affect mouse MAP-2 or tau promoted microtubule assembly. These findings suggest that NbMAPF is a unique protein involved in microtubule assembly associated with neurite outgrowth.

INTRODUCTION

Nerve cell differentiation is characterized by the outgrowth of axons and dendrites, composed of many long and well-organized microtubules. These microtubules are thought to function primarily as a cytoskeleton but have been also implicated in axonal transport and sensory transduction in neural cells. To understand the regulation of microtubule assembly during neural differentiation we have used a homogeneous cell system of clonal cell line N18 from the mouse neuroblastoma C1300, which can be induced to undergo a rapid morphological differentiation of neurite outgrowth.

The study of microtubule assembly in this model system has provided a better understanding of the process, as well as some very interesting findings concerning the regulation of microtubule assembly.

NEURITE OUTGROWTH

In the presence of fetal calf serum (10%) clone N18 cells grow as spherical loosely attached and morphologically "undifferentiated" cells characteristic of neuroblasts. Removal of the serum or reducing the concentration to below 1% leads to a greater adherence to the substratum and the initiation of neurite outgrowth (Seeds, et al. 1970). Neurite formation begins within minutes and some long neurites are seen by 1 day, although a true "differentiated" morphology is not attained until 3-4 days in the low serum environment (Figure 1). Removal of serum from the culture medium promotes an immediate increased adhesion of the cell to the substratum and after several days (3-5) a marked reduction in DNA synthesis. However, the initiation of neurite outgrowth is not simply a matter of cell adhesion, since the N18 cells adhere very well to poly-D-lysine coated plastic but they will not extend neurites in the presence of serum. Some serum components in the alpha-globulin fraction appear to inhibit neurite outgrowth and promote retraction of neurites when added back to differentiated cultures.

Neurite outgrowth by N18 cells is temperature dependent and sensitive to inhibition by colchicine and vinblastine (Seeds et al. 1970), indicating its dependence on microtubule assembly. However, neurite formation is insensitive to cycloheximide and apparently independent of de novo synthesis of tubulin subunits for the initial phases of neurite outgrowth.

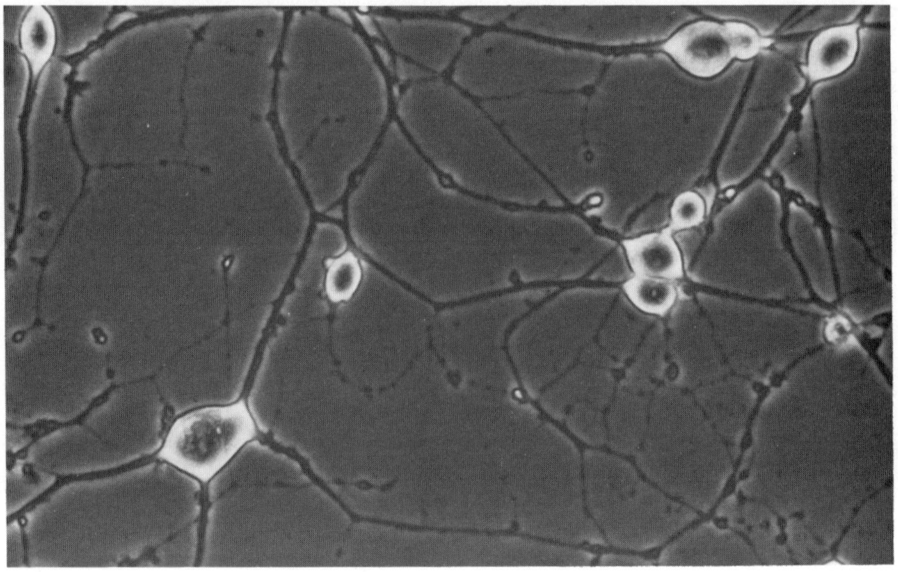

Fig. 1. Differentiated N18 mouse neuroblastoma cells in 0.1% FCS.

TUBULIN CONSTANCY

A more detailed analysis of tubulin concentrations in clone N18 using a colchicine binding assay and quantitative densitrometric SDS-PAGE shows that neuroblastoma cells possess a high concentration of tubulin subunits (140 pmol/10^6cells) compared to other cell lines such as the rat glioma C6(30 pmol/10^6cells). However, both the "differentiated" and "undifferentiated" N18 cells possess similar amounts of tubulin (Morgan and Seeds, 1975). Similar findings have been made by Olmsted (1981) using antibody binding to quantify tubulin pools in a different neuroblastoma clone. Analysis of neuroblastoma clone N103 that fails to differentiate under any experimental conditions also shows a large tubulin pool (Morgan and Seeds, 1975). These studies suggest that while it is probably necessary to have large tubulin pools to support neurite outgrowth during neuronal differentiation, the regulation of microtubule formation is not solely controlled by the size of the tubulin pool.

ASSEMBLY PROMOTING FACTOR

The above findings suggest a possible role for additional macromolecules in the regulation of microtubule assembly during nerve fiber outgrowth. Therefore, cell extracts of both morphologically differentiated and undifferentiated neuroblastoma N18 and several other cell lines were compared for their ability to promote the assembly of purified brain tubulin. Only supernates from differentiated N18 cells are able to promote microtubule assembly (Seeds and Maccioni, 1978). Extracts of the undifferentiated N18 and N103 cells as well as several gliomas are unable to promote assembly (Table 1). The possibility that these inactive supernates contain an inhibitory factor is ruled out in mixing experiments with limiting brain MAPs. In addition, neuroblastoma hybrid NG108 which differentiates in the presence of dibutyryl cAMP shows high levels of assembly promoting activity after growth in db-cAMP for 5 days.

TABLE 1. Microtubule assembly promoting activity of cell extracts.

Cell extract*	Microtubules (ug protein)
None (tubulin only)	0
Undifferentiated N18	3
Differentiated N18	30
Non-differentiating N103	2
Glioma G26	4

*High speed (250,000 X g) supernates were incubated with purified brain tubulin (2 mg/ml) in polymerization buffer 37°C for 20 min. and microtubules isolated by the glass fiber filter assay of Maccioni and Seeds (1978).

Electron microscopy reveals that these active cell extracts do not contain microtubule fragments or other nucleating structures, and that the microtubules formed with the neuroblastoma microtubule assembly promoting factor (NbMAPF) are similar to those seen in cell sections. (Seeds and Maccioni, 1978).

The NbMAPF is insensitive to both RNase and DNase, but is both heat (100°C for 5 min.) and trypsin labile; suggesting that the factor is a protein. Molecular sieve chromotography indicates that the NbMAPF has a native Mr=160,000 (Seeds and Maccioni, 1978).

Although neurite outgrowth begins within 30 min. after serum withdrawal, long neurites are not seen until 1 day or more. Surprisingly, NbMAPF activity is readily detected by 1 hr. and reaches maximal levels by 4-8 hrs. after induction (Table 2). These findings suggest that the NbMAPF is involved in the early initiation steps of MT assembly and neurite outgrowth, rather than being a stabilizing factor generated after neurite formation and elongation.

The appearance of the NbMAPF is insensitive to high concentrations of cycloheximide which inhibit initial rates of protein synthesis in N18 cells by 96%. Thus NbMAPF does not require _de novo_ protein synthesis and the activity probably results from a post-translational modification of an inactive molecule in the undifferentiated cells. Incubation with ^{32}P-phosphate during the induction process has not shown a unique protein phosphorylation in the high speed supernate of differentiated cells.

TABLE 2. Appearance of NbMAPF with differentiation.

Time	Growth Condition	Morphology	Relative Assembly
0	10% FCS	undifferentiated round cells	7%
1 hr.	0% FCS	microspikes	71%
4 hr.	0% FCS	short processes	78%
4 hr.	0% FCS w 10^{-4}M CH*	short processes	73%
24 hr.	0% FCS	long neurites	61%

Relative MT assembly compared to a saturating concentration of brain HMW-MAPs. All samples contained purified brain tubulin and 200 ug supernate. *CH = cycloheximide.

Furthermore, incubation of both NbMAPF from differentiated N18 cells and identical (but inactive) fractions from undifferentiated N18 cells with cAMP and ATP produce no change in either fractions ability to promote assembly.

Induction of NbMAPF requires that the cells be attached to a suitable substratum, where neurite growth can occur; N18 cells in suspension culture in low FCSs do not activate the factor. Similarly, cytochalasin B blocks the induction of NbMAPF and also leads to the loss of NbMAPF activity when added 8 hrs. after induction and factor activity is maximal. These findings suggest that a transmembrane signaling event, possibly related to initial microspike activity preceeding neurite formation is important in the induction of NbMAPF activity.

The NbMAPF appears to be unique when compared to the well studied microtubule associated proteins (MAPs) from brain or neural tissue. These neural MAPs (MAP1, MAP2, MAP3, MAP4 and tau) are by definition associated with microtubules during cycles of assembly - disassembly (Olmsted, 1986). The NbMAPF does not cycle with MTs during assembly, and additional cycles of MT assembly require the readdition of the supernatant fraction from the previous assembly step (Table 3). Thus the NbMAPF appears to act in a catalytical role and may only transiently associate with the MTs, or it may associate with the MTs but be converted to an inactive state in the process of assembly.

TABLE 3. Ability of NbMAPF to promote cycles of microtubule assembly - disassembly.

Assembly Conditions	Cycles	MTs pelleted
1. Purified tubulin & brain MAPs	1	540 ug
Recycle pellet	2	300 ug
2. Purified tubulin & Inactive NbMAPF (undifferentiated cells)	1	30 ug
3. Purified tubulin & Active NbMAPF (differentiated cells)	1	260 ug
Recycle pellet	2	10 ug
Recycle pellet & Cycle 1 Supernate	2	400 ug

ANTIBODY TO NbMAPF

Purification of the NbMAPF from differentiated N18 cells has been difficult with only a 70-fold increase in specific activity attained by ammonium sulphate precipitation, DEAE-Sephadex, Sephadex chromatography; presumably the factor is inactivated during the purification process. Native acrylamide gel electrophoresis of this enriched factor preparation was used to isolate active NbMAPF for injection into rabbits. Following several booster injections an antiserum that specifically blocked NbMAPF activity in a concentration dependent fashion was obtained. Immunoprecipitation of radiolabeled supernates from undifferentiated and differentiated N18 cells shows a similar pattern; thus supporting the view that the NbMAPF is derived by activation of pro-factor molecule. The rabbit antibody to NbMAPF does not inhibit DMSO, mouse tau or MAP-2 promoted microtubule assembly. Furthermore, neither Olmsteds' antibody to mouse MAP-4 (Parysek et al. 1984) nor a monoclonal antibody to the 160 kDa neurofilament protein blocked the NbMAPF activity from N18 cells. These findings suggest that NbMAPF is distinct from tau, MAP-2, MAP-4 and neurofilament protein.

ACKNOWLEGEMENTS

These investigations have been supported in part by USPHS research grants NS-10709 and NS-09818 and a Milheim Foundation for Cancer Research grant to N.W.S. and Grant #1913 from the Council for Tobacco Research U.S.A. to R.B.M.

REFERENCES

Maccioni, R.B. and Seeds, N.W. (1978). Enhancement of tubulin assembly as monitored by a rapid filtration assay. Arch. Biochem. Biophys. 185, 262-271.

Morgan, J.L. and Seeds, N.W. (1975). Tubulin constancy during morphological differentiation of mouse neuroblastoma cells. J. Cell Biol. 67, 136-145.

Olmsted, J.B. (1981). Tubulin pools in differentiating neuroblastoma cells. J. Cell Biol. 89,418-423.

Olmsted, J.B. (1986). Microtubule associated proteins. Ann. Rev. Cell Biol. 2, 421-457.

Parysek, L.M., Asner, C. and Olmsted, J.B. (1984). MAP-4: Occurence in mouse tissues. J. Cell Biol. 99, 1309-1315.

Seeds, N.W., Gilman, A., Amano, T. and Nirenberg, M. (1970). Regulation of axon formation by clonal lines of a neural tumor. Proc. Nat. Acad. Sci. USA 66, 160-167.

Seeds, N.W. and Maccioni, R.B. (1978). Proteins from morphologically differentiated neuroblastoma cells promote tubulin polymerization. J. Cell Biol. 76, 547-555.

EXPRESSION OF JUVENILE AND ADULT TAU PROTEINS DURING NEURONAL DIFFERENTIATION.

J. Nunez[1], D. Couchie[1], J.P. Brion[2] and M. Tardy[1]

[1]INSERM U282, Hôpital Henri Mondor, 94010 Créteil, France

[2]Laboratoire d'Anatomie Pathologique ULB, 1070 Bruxelles, Belgique

ABSTRACT

Several experimental approaches were used to understand the functional significance of the changes in TAU composition and activity seen during brain development: a) only one entity (of 48 kDa) which is present at immature stages of development is immunologically related to adult TAU. The immature 65 kDa entity is immunologically related to MAP_2 (small MAP_2). b) Juvenile TAU is neuron specific; its concentration increases in primary cultures of fetal brain neurons in parallel with neurite outgrowth. c) Cultures of astrocytes contain two entities immunologically related to TAU but with different molecular weights (62-83 kDa). d) Immature TAU is, as adult TAU, axon specific. It is present in the growing axons and its expression reflects a certain sign of maturity of the axonal fibers in the developing cerebellum.

INTRODUCTION

Most of the major microtubule-associated proteins (MAPs) i.e. MAP_1 (350 kDa) , MAP_2 (300 kDa) (Murphy et al, 1977; Sloboda et al., 1975; Kuznetsov et al., 1981) and TAU (52-65 kDa) (Cleveland et al., 1977a,b) undergo quantitative and/or qualitative changes in expression during brain development (see Nunez, 1986 for a review). Changes in composition of the TAU proteins were the first to be documented: it was observed that two MAPs of 48 and 65 kDa are present in the immature rat brain and that they are replaced by a complex of 4-5 proteins (52-65 kDa) in adulthood (Mareck et al., 1980). Peptide mapping analysis suggested that the immature TAU protein of 48 kDa differs from the various adult TAU isoforms not only in its molecular weight but also in its sequence (Francon et al., 1982). In contrast the peptide maps of the 4-5 adult TAU isoforms are very similar (Cleveland et al., 1977). Cloning experiments also suggest that the mRNAs coding for the TAU proteins are heterogenous and that their expression is developmentally regulated (Ginzburg et al., 1982). Recently it has been reported (Lee et al., 1982) that the cDNA of two TAU clones contain common sequences but differ in their carboxy terminal and in the 3' non coding region. It may be assumed therefore that the heterogeneity of the TAU proteins is generated at least partially at the transcriptional level but this conclusion will be proved only when the sequence of all the putative TAU immature and mature gene products will be known. Moreover, both immature and adult TAU being multisite phosphorylated proteins an additional degree of heterogeneity could be generated post transcriptionally (Pierre and Nunez, 1983; Butler and Shelanski, 1985).

The evidence suggesting that the two immature MAPs of 48 and 65 kDa belong to the TAU family being based only on their apparent molecular weight, immunological experiments were performed recently to better characterize these two proteins (Couchie and Nunez, 1985). We will also report in this paper a number of experiments which were deviced in order to establish whether 1) both immature and mature TAU are of neuronal or glial origin (Couchie et al., 1985,1986) 2) immature TAU is axon-specific in the developing brain (Brion et al., unpublished results) as it has been reported to be the case for adult TAU in the mature brain (Binder et al., 1985).

IMMUNOLOGICAL CHARACTERIZATION OF THE 48 kDa AND 65 kDa IMMATURE MAPs

The immature protein of 48 kDa is clearly detected by Western blot analysis both by polyclonal and monoclonal antibodies raised against adult TAU. Thus, although differing in its molecular weight, peptide map and polymerizing activity when tested with purified tubulin (Francon et al., 1982) the 48 kDa protein belongs to the TAU family i.e. probably shares a number of common sequences with adult TAU proteins. The 48 kDa protein was also purified and used to raise an antiserum which, surprisingly, did not react significantly with adult TAU. This suggests that the anti-immature TAU serum we obtained reacts only with a number of epitopes specific of this protein.

Immunoblot experiments also showed that the 65 kDa immature protein did not react with any of the polyclonal or monoclonal anti- TAU antibodies available. In contrast anti-MAP_2 polyclonal antibodies revealed this protein but only at early post-natal stages of brain development. Similarly antibodies raised against the purified 65 kDa immature entity not only reacted with the corresponding antigen but also revealed on Western blots MAP_2 and several others high molecular weight MAPs. This means that the 65 kDa entity, both belongs to the MAP_2 family and probably shares common epitopes with other minor high molecular weight MAPs. An immature MAP of similar molecular weight has also been detected with monoclonal anti-MAP_2 antibodies (Riederer and Matus, 1985).

It is not clear whether "small MAP_2" of 65 kDa is a fragment of MAP_2 produced by a specific protease which would be present and/or active only at early postnatal stages of brain development or a gene product sharing common sequences with MAP_2.

EXPRESSION OF MAPs IMMUNOLOGICALLY RELATED TO TAU DURING GLIAL AND NEURONAL DIFFERENTIATION IN CELL CULTURE

Primary cultures of neurons (Couchie et al., 1986) and astroglial cells (Couchie et al., 1985) have been used to know whether the TAU proteins identified in total brain are neuron specific. Immunoblot experiments showed that the astrocyte contains two components immunologically related to TAU. However the molecular weight of these components (62 and 83 kDa) was different from those of both adult and immature TAU. In addition the 83 kDa cross-reacted both with anti-TAU and anti-MAP_2 antibodies suggesting either that it contains epitopes specific for each one of these MAP_s or that two antigenically distinct MAPs of the same molecular weight are expressed in the astrocyte. Furthermore the concentration of these MAPs was seen to increase markedly during morphological differentiation of the cultured astrocyte which was induced by cAMP derivatives.

In contrast the MAPs present in cultured fetal neurons were identical to those detected in whole brain extracts. Immature TAU was present in fetal brain neurons prior their seeding in the culture dish and its concentration increased several fold during the time course of the culture. Such an increase closely paralleled neuronal morphological differentiation i.e. neuritic outgrowth and branching. No change in TAU composition was seen after 8 days of culture suggesting that immature TAU is sufficient to produce the microtubules required for neurite outgrowth and branching. Adult TAU is therefore probably required to specify later steps of neuronal differentiation.

IMMATURE AND ADULT TAU ARE BOTH AXON-SPECIFIC IN THE DEVELOPING AND MATURE BRAIN.

Previous results have shown that MAP_2 is concentrated in dendrites (Bernhardt and Matus, 1982) whereas adult TAU seems to be specific of axonal microtubules (Binder et al., 1985). Most of the studies leading to these conclusions have been performed in the cerebellar cortex because its development is well known. Since antibodies raised against adult TAU cross-react with immature TAU it has been possible to study the distribution of TAU during early development in this brain region by immunohistochemical techniques (Brion et al., unpublished results). The data showed that whatever the stage of development, anti-TAU antibodies stained several types of axonal fibers. The Purkinje cell body and their dendrites were never labeled. This means that immature TAU is, as adult TAU, localized essentially in axons. Axonal labeling seems to follow the cerebellar developmental pattern. For instance the climbing fibers and the mossy fibers which reach the cerebellum during the embryonic life and make transient contacts with the Purkinje cell perikarya were stained soon after birth by the anti-TAU antibodies. In contrast the parallel fibers, that begin to develop perinatally, do not express TAU at early (5 days) postnatal stages; a clear labeling of the more mature parallel fibers was seen at day 10 after birth in the vicinity of the developing dendrites of the Purkinje cells. This suggests that 1) the appearance of TAU immunoreactivity reflects a certain sign of maturity of the parallel fiber 2) both immature and mature TAU microtubule-associated proteins are axon specific in the developing rat cerebellum.

CONCLUSION

Juvenile TAU, which is expressed only during the early stages of brain development, is composed with one component of 48 kDa whereas at adulthood it is replaced by a complex of several isoforms differing in molecular weight (52-65 kDa). Immature and mature TAU proteins probably share common sequences since they both cross-react with anti-adult TAU antibodies. Whatever the stage adult (52-65 kDa) and immature (48 kDa) TAU proteins seem to be neuron specific. Astroglial cells express two MAPs immunologically related to TAU but displaying different molecular weights (62 and 83 kDa). Both in neurons and astroglial cells *in vitro* morphological differentiation correlates with a marked increase in the concentration of antigenically TAU related components. Neurite outgrowth for instance is accompanied by a marked increase in the expression of immature TAU (48 kDa). This suggests that immature TAU is required to build up growing neurites whereas adult TAU may contribute to the definition of the cytostructure of stable adult axons. Such a conclusion is consistent with previous results (Francon *et al*., 1982) showing that immature TAU is much less effective in promoting the assembly of pure tubulin than the corresponding adult species. This might signify that the microtubules produced in the developing brain are less stable than those polymerized at adulthood.

Nevertheless, immature TAU is, as adult TAU, axon specific. Moreover the expression of immature TAU in the axons seems to follow the developmental pattern i.e. the appearance of TAU immunoreactivity seems to reflect a certain sign of maturity of the growing axonal fibers in the developing rat cerebellum.

These data support therefore the assumption that the microtubules present in the growing axons are different in composition from their counterpart in mature axons. This might be related to requirement for axonal plasticity during axonal outgrowth and for axonal stability in adulthood.

REFERENCES

Bernhardt, R.R. and Matus, A. (1982). Initial phase of dendrite growth: evidence for the involvement of high molecular weight microtubule-associated proteins (HMWP) before the appearance of tubulin. J. Cell Biol. 92, 589-593.

Binder, L.L., Frankfurter, A. and Rebhun, L.L. (1985). The distribution of TAU in the mammalian central nervous system. J. Cell Biol. 101, 1371-1378.

Butler, M. and Shelanski, M.L. (1985). Microheterogeneity of TAU proteins is dependent on age and phosphorylation. J. Cell Biol. 101, 29.

Cleveland, D.W., Hwo, S.Y. and Kirshchner, M.W. (1977a). Purification of TAU, a microtubule-associated protein that induces assembly of tubulin from purified tubulin. J. Mol. Biol. 116, 207-226.

Cleveland, D.W., Hwo, S.Y. and Kirschner, M.W. (1977b). Physical and chemical properties of purified TAU factor and the role of TAU in microtubule assembly. J. Mol. Biol. 116, 227-247.

Couchie, D., Fages, C., Bridoux, A.M., Rolland, B., Tardy, M. and Nunez, J. (1985). Microtubule-associated proteins (MAPs) and in vitro astrocyte differentiation. J. Cell Biol. 101, 2095-2103.

Couchie, D. and Nunez, J. (1985). Immunological characterization of microtubule-associated proteins specific for the immature brain. FEBS Lett. 188, 331-335.

Couchie, D., Faivre-Bauman, A., Puimyrat, J., Guilleminot, J., Tixier-Vidal, A. and Nunez, J. (1986). Expression of microtubule-associated proteins during the early stages of neurite extension by brain neurons cultured in a defined medium. J. Neurochem. 47, 1255-1261.

Francon, J., Lennon, A.M., Fellous, A., Mareck, A., Pierre, M. and Nunez, J. (1982). Heterogeneity of microtubule-associated proteins and brain development. Eur. J. Biochem. 129, 465-471.

Ginzburg, I., Scherson, T., Giveon, D., Behar, L. and Littauer, U.Z. (1982). Modulation of mRNA for microtubule associated proteins during brain development. Proc. Natl. Acad. Sci. USA 79, 4892-4896.

Kuznetsov, S.A., Rodionov, V.I., Gelfand, V.I. and Rosenblat, V.A. (1981). Microtubule associated protein MAP_1 promotes microtubule assembly in vitro. FEBS Lett. 135, 241-244.

Lee, G., Seeburg, P., Cowan, N. and Kirschner, M. (1985). Primary structure of TAU proteins. J. Cell Biol. 101, 135.

Mareck, A., Fellous, A., Francon, J. and Nunez, J. (1980). Changes in composition and activity of microtubule-associated proteins during brain development. Nature 284, 353-355.

Murphy, D.B., Johnson, K.A. and Borisy, G.G. (1977). Role of microtubule-associated proteins in microtubule nucleation and elongation. J. Mol. Biol. 117, 33-52.

Pierre, M. and Nunez, J. (1983). Multisite phosphorylation of TAU proteins from rat brain. Biochem. Biophys. Res. Commun. 115, 212-219.

Riederer, B. and Matus, A. (1985). Differential expression of distinct microtubule-associated proteins during brain development. Proc. Natl. Acad. Sci. USA 82, 6006-6009.

Sloboda, R.D., Rudolph, S.A., Rosenbaum, J.L. and Greengard, P. (1975). Cyclic AMP dependent endogeneous phosphorylation of a microtubule-associated protein. Proc. Natl. Acad. Sci. USA 72, 177-181.

HORMONE INDUCTION OF NEURITE-LIKE PROCESSES IN CULTURED MELANOMA CELLS

S.F. Preston, M. Volpi, C. Pearson and R.D. Berlin

University of Connecticut Health Center, Farmington, Connecticut 06032, USA

ABSTRACT

We describe a new model system for study of the regulation of cell shape. We further suggest that this system will be a valuable tool in analyzing the determinants of neuronal shape. In response to a potent analogue of α-melanocyte stimulating hormone (α-MSH), Cloudman S91 mouse melanoma cells compress into 1-2 hours morphological events and changes which require many days for expression in PC12 cells and other neuronal cell lines. Based on the melanoma system, we also present a novel approach for examination of penetration and diffusion of macromolecules through complex cell processes and neurites.

INTRODUCTION

Perhaps the most studied system for the regulation of cell shape is the PC-12 cell, a well known clonal line derived from a rat pheochromacytoma. PC-12 cells respond to NGF by formation of elaborate neuritic processes within a period of one week (Greene and Tischler, 1976; Gunning _et al_., 1980). In such "primed" cells, mechanically removed neurites can be regenerated, but regeneration requires a 24 hr period in the presence of NGF (Greene _et al_., 1980). In fact, a substantial interval between hormonal signal and morphological change runs through much of the literature on the regulation of cell shape: Puck and colleagues, in their seminal studies, emphasized the role of cAMP in the induction of morphological change during reverse transformation of Chinese hamster ovary cells (Puck, 1977), a process that occurs over 8-24 hr. Westermark and Porter (1982) described thyrotropin-induced arborization and spontaneous reversal to unstimulated morphology, but again, only after 24 hrs had elapsed. It has also been known for some time that in response to MSH, melanoma cells differentiate, synthesize melanin, cease cell division and establish elaborate arborization (Pawelek _et al_., 1975; Johnson and Pastan, 1972). However, all previous reports indicated that arborization in response to MSH required several days in culture.

We have recently reported (Preston _et al_., 1987) that Cloudman S91 mouse melanoma cells undergo extremely rapid and reversible arborization in response to [Nle^4, D-Phe^7]-α-MSH, a potent analogue of α-MSH (Sawyer et al., 1980). We suggested that the rapidity and dramatic character of the morphological change induced by the hormone qualify this melanoma system as an important candidate for studies of cell shape and cytoskeletal reorganization.

In the present report, we briefly summarize salient features of the model, document in greater detail the contribution of membranous elements and intermediate filaments to the neurite-like processes, and discuss the analysis of diffusion within a highly organized cytoplasmic matrix.

THE CLOUDMAN S91 MELANOMA MODEL

In response to [Nle^4, D-Phe^7]-α-MSH, Cloudman S91 mouse melanoma cells cultured in suspension show an immediate increased adherence to tissue culture surfaces. This increased adherence is reminiscent of the response of certain PC12 cell lines to nerve growth factor (NGF) (Goodman et al., 1979). Most importantly, cells grown in monolayers exposed to the hormone undergo extensive, rapid and reversible arborization. Arborization occurs within 20-40 min (Fig. 1) and is similar in structure and complexity to that which occurs in PC12 cells only after days of exposure to NGF. Melanoma cell processes show a configuration characteristic of neurites; they contain a generally parallel arrangement of microtubules, intermediate filaments, mitochondria and other membranous elements. The morphological changes induced by α-MSH are mimicked both by agents that increase intracellular cAMP (e.g., forskolin, dbcAMP) and by inhibition of protein kinase C (with the potent inhibitor H-7), suggesting involvement of multiple signal pathways.

Finally, although α-MSH causes an increase in intracellular cAMP, the hormone produces long-term effects that cannot be mimicked by cAMP. Specifically, even in the continued presence of α-MSH, arborization is followed by morphological reversal to the unstimulated, flattened configuration wtihin 2 hr. (This does not occur with other agents that elevate cAMP, or with H-7). Importantly, whereas MSH-induced arborization occurs in the presence of cycloheximide, actinomycin D, or in enucleated cells, the reversal of arborization does not. Thus, MSH induces a program of rapid shape change that is dependent on new protein synthesis and gene transcription.

ULTRASTRUCTURE OF MELANOMA CELL PROCESSES

In the light microscope, control cells are approximately 20-30 μm long, flat and fibroblast-like. Cells treated with α-MSH for 20-40 min have numerous processes, some the equivalent of several control cell lengths (Fig. 1). Processes induced within minutes of exposure to the hormone resemble PC12 neurites generated only after 3-7 days of NGF treatment (Luckenbill-Edds et al., 1979) (refer to Preston et al., 1987). These processes form mostly by extension from the cell body and not by retraction of cytoplasm. Extended, narrow processes contain microtubules aligned parallel to their long axis. Microfilaments are concentrated at the growing tips and other regions of membrane ruffling. Although intermediate filaments are not a conspicuous feature of PC12 cell neurites (Jacobs and Stevens, 1986), they make a striking contribution to the intracellular matrix of the melanoma cell process (Fig. 2a). Mitochondria and other membranous elements are distributed irregularly throughout the processes occasionally at very high density (Fig. 2b and c, Fig. 3a). It should be noted that despite the large volume taken up by the complex arrangement of membranous elements, granules and vesicles scattered along the processes, ribosomes are not seen in association with any of the process membranes.

The general organization of the neurites formed may be viewed as a highly complex array of membranous elements and mitochondria that are given directional and positional orientation by intermediate filaments and microtubules that run parallel to the long axis of the process.

DIFFUSION OF MACROMOLECULES THROUGH NEURITES

We have discussed previously the implications of parallel microtubule arrays for the diffusion of macromolecules (Brown and Berlin, 1985). The analysis depended on the spacings between microtubules measured indirectly on purified microtubule preparations. It is particularly relevant, we believe, for highly differentiated neurons in which packed microtubules figure prominently in axons or dendrites. In this analysis microtubules and their associated proteins that extend outward from the microtubule wall are considered to form a fibrous network that impedes diffusion. The degree of impedance is a function of the density of the fibrous network and of the dimensions of the diffusing particle. These restraints should apply in structures in which the organization of filaments is less regular (or obvious). However, it is clear from the description of the melanoma cell processes that the organization of filaments and other structures is so complex that a quantitative description of fiber density, etc. is virtually impossible. This realization has forced us to develop an alternative, practical approach to the description of structure or rather the functional implications of structure for molecular diffusion. We are mindful that there are special mechanisms (e.g., kinesin-based) that are responsible for the translation of molecular complexes or particles. We offer here an approach to differentiating these from simple diffusional processes.

In viewing sectioned material, we were repeatedly struck by the presence of membrane vesicles (and mitochondria) which would not allow the penetration of diffusing species. Physically this means that molecules must go around these impediments increasing the pathlength for their diffusion. But how to quantify this pattern given the complexity of the biological structure? Analogous problems have been considered by geologists, for example, who have been concerned with movement of water through soils containing particulates of assorted sizes and shapes. Since diffusion, heat flow, and electrical conductivity can be described by the same mathematical formalism, a single physical model of the same geometry in principle can be used to study all of these processes. To apply this notion to the diffusion problem at hand, we traced the patterns of membranes and organelles and the neurite process outline captured in electron micrographs onto electrically conducting paper (Recorder paper; Datascope, Newark, N.J.). This paper has a thin layer of conductive graphite sandwiched between insulating paper. The resistance of the paper as used for our purposes is measured by clamping its narrow ends between two flat copper electrodes, applying a battery-powered DC voltage and measuring resistance (end-end) with a digital ohm meter. Resistance is measured first on the paper with the pattern traced (paper shaped to the neurite section). The traced vesicles and organelles are then cut out and the resistance remeasured. Fig. 3 illustrates the original micrographs of one such section (top) and the cut-out (bottom). The results may be expressed as the percent change in resistance. The actual area removed by cutting is determined gravimetrically. In the example shown, 27% of the 'section' was excised. The resistance change was 273%. Sampling across a neurite gave an average of nearly 200%. We emphasize that this results mostly from changes in pathlength and to a lesser degree from changes in the area available for diffusion. Analogous considerations may largely explain the 2-3 fold decrease in the diffusion of macromolecules in cytoplasm (e.g., Jakobson and Wojcieszyn). Diffusion will also be retarded, of course, by the fibrous network. In this case, unlike the strictly geometrical factors considered in our paper model, retardation will depend on molecular size.

Electrical analogs have been used often in modelling diffusion problems. Barrer _et al_. (1963) used a conductive paper analog to validate their theoretical prediction of diffusion through polymers filled with regular spherical 'crystals', but there is no reason that more complex geometries cannot be similarly modeled. In fact, we consider that the transference of the complex biological geometry to paper allows information of high structural resolution to be reduced to simple functional terms. Percolation theory, largely developed for analysis of electrical conductivity through media containing obstructions, has been discussed by Saxton (1982) in relation to two-dimensional diffusion. It is characteristic of the 2-dimensional case that the conductivity decreases sharply once a critical area is removed. In 3-dimensions, because there are more alternative pathways for conduction, this feature occurs only with much larger removed volumes. We doubt thus that this will be a factor in limiting diffusion in biological systems. However, these considerations underline the fact that the two dimensional analog is only an approximation. At the same time, it should be noted that the area removed from the 2-dimensional projection is (by conventional morphometry) a measure of volume in 3-dimensions. Our analog results signify that a considerable barrier to diffusion can be created by a relatively small volume of obstructive barriers. We hope now to test this possibility by direct measurements of diffusion within structurally defined cytoplasm. Meanwhile the analog method described provides a useful guide to the magnitude of the effects likely to be observed.

CONCLUSION

Implicit in the early studies of PC12 cell neurites was the notion that this cell line could serve as a model for neurite outgrowth. We suggest that the Cloudman S91 melanoma cell line also serves as a model for neurite outgrowth. We further suggest that the compressed time scale for the unfolding of a program of hormone-induced morphological change makes the melanoma cell model especially relevant to studies of the relationship of signal transduction to cytoskeletal reorganization.

Since the neurite processes formed lend themselves to comprehensive structural analysis, we have begun to determine the significance of the structural organization for molecular diffusion within the cytoplasmic matrix. To this end, we present a technique for reducing complex structure to a simple electrical analog that defines partially the geometrical restraints to conductivity/diffusion.

ACKNOWLEDGEMENTS

We are grateful to Drs. Mac E. Hadley and Victor Hruby for their generous gift of [Nle^4, D-Phe^7]-α-MSH. We also thank Mariellen H. Scoopo for maintenance of the melanoma cell culture and Joan Jannace for careful preparation of the manuscript. This work was supported by a grant from the American Cancer Society, BC-518.

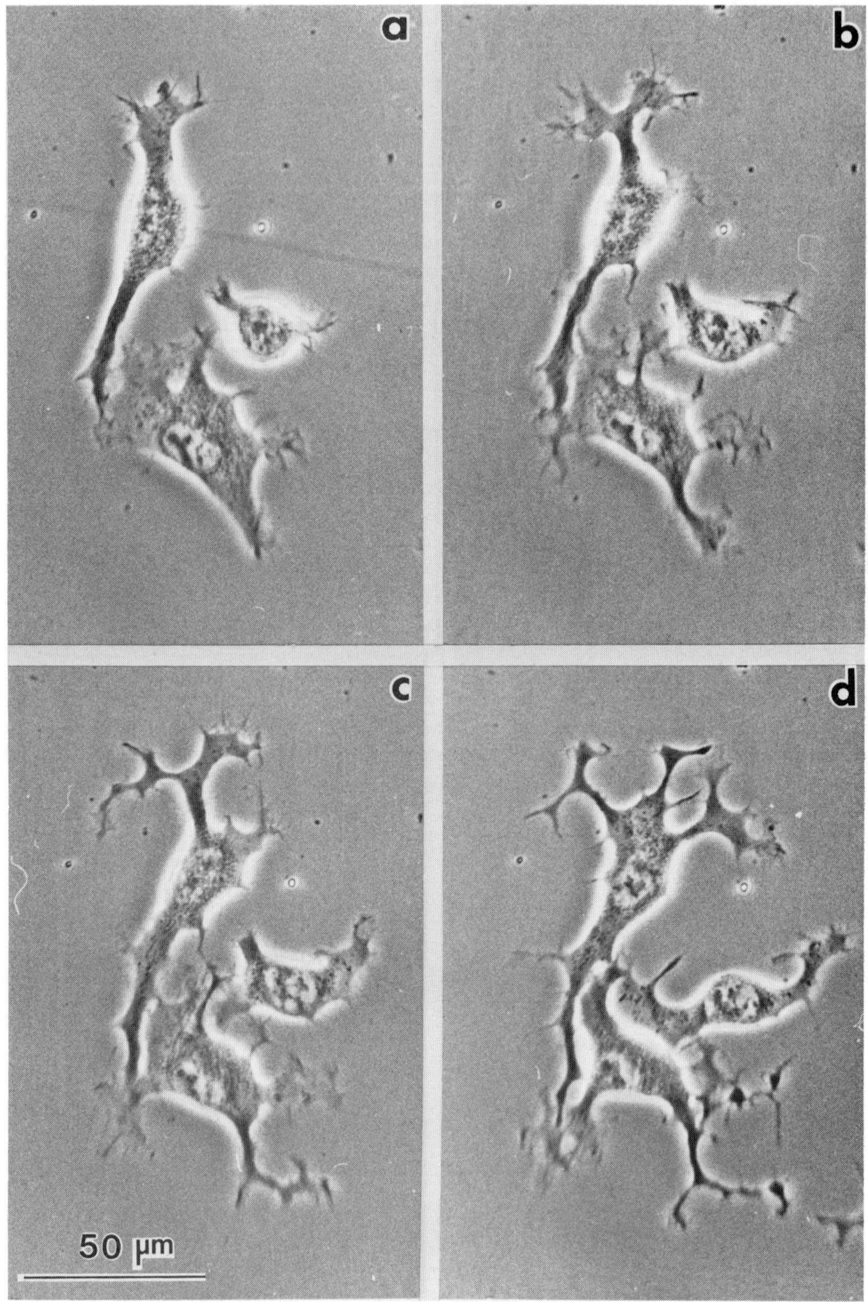

Figure 1.

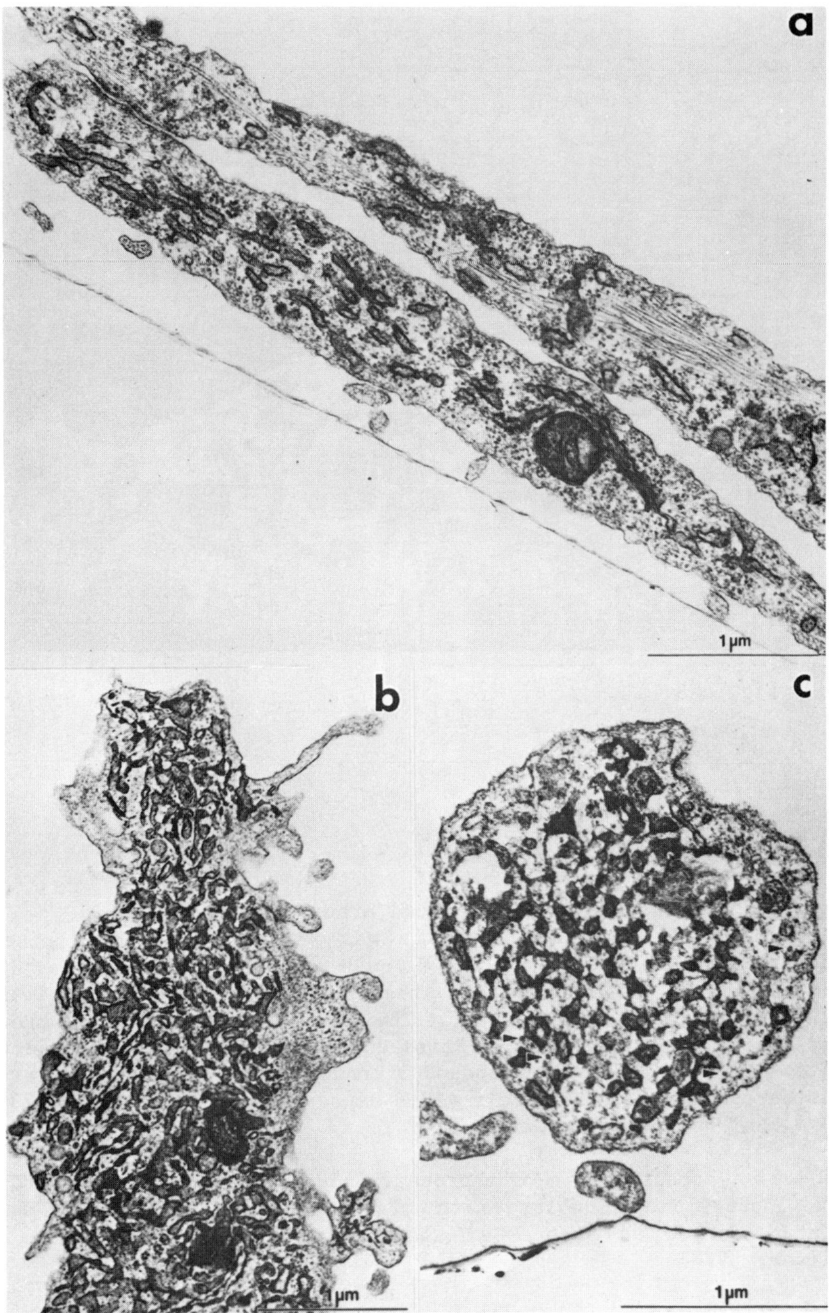

Figure 2.

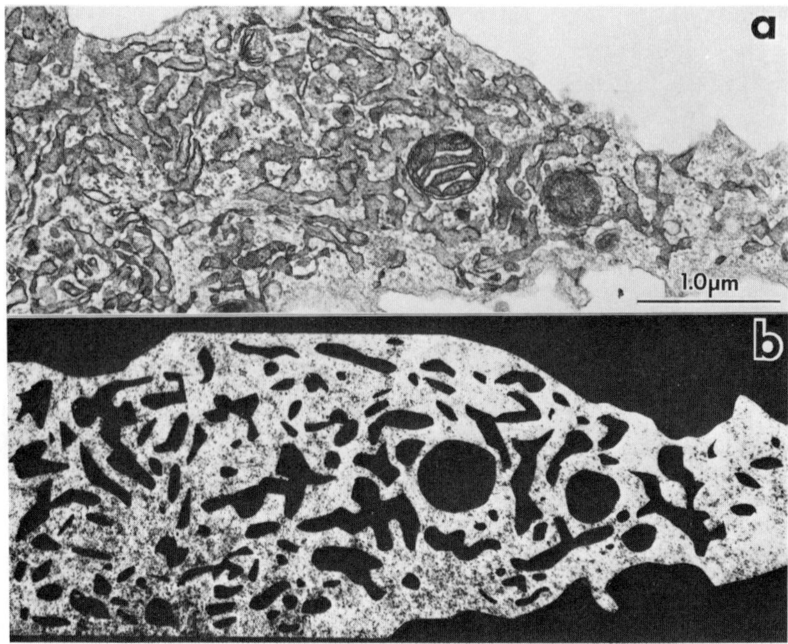

Figure 3.

FIGURE LEGENDS

Fig. 1. Time course of MSH-induced arborization.
A monolayer on a 13 mm coverslip was inverted on a drop of warm complete medium containing 10^{-8} M α-MSH and sealed onto a glass slick with Lubriseal. The microscope stage was maintained at approximately 37° by an air curtain incubator. The field of observation was brought rapidly to focus utilizing a Zeiss 25X oil immersion phase objective (N.A. 0.8), and the cells exposed at regular intervals thereafter and photographed. a) immediately after exposure to α-MSH; b-d) 10,20 and 40 min of hormone treatment.

Fig. 2. Ultrastructure of a process induced by 10^{-8} M α-MSH after 20 min incubation. Monolayers on glass coverslips were fixed in 2% glutaraldehyde and processed as previously described (Berlin and Oliver, 1978).

Fig. 3. Example of electron micrograph of longitudinal section through melanoma process and its excised conductive paper analog. (See text).

REFERENCES

Barrer, R.M., Barrie, J.A. and Rogers, M.C. (1963). Conductivity in heterogeneous media. J. Polym. Sci. A. 1, 2665-2669.

Berlin, R.D. and Oliver, J.M. (1978). Analogous ultrastructure and surface properties during capping and phagocytosis in leucocytes. J. Cell Biol. 77, 789-804.

Brown, P.A. and Berlin, R.D. (1985). Packing volume of sedimented microtubules: regulation and potential relationship to an intracellular matrix. J. Cell Biol. 101, 1492-1500.

Goodman, R., Chandler, C. and Herschman, H.R. (1979). Pheochromocytoma cell lines as models of neuronal development. In Hormones and Cell Culture (ed. G.H. Sato and R. Ross), pp. 653-670. Cold Spring Harbor Laboratory.

Greene, L.A., Burstein, D.E. and Black, M.M. (1980). The priming model for the mechanism of nerve growth factor: evidence derived from clonal PC12 pheochromocytoma cells. In Tissue Culture in Neurobiology (ed. E. Giacobini, A. Vernadakis and A. Zahar), Raven Press, New York

Greene, L.A. and Tischler, A.S. (1976). Establishment of a noradrenergic clonal line of rat adrenal pheochromocytoma cells which respond to nerve growth factor. Proc. Natl. Acad. Sci. USA 73, 2424-2428.

Gunning, P.W., Landreth, G.E., Bothwell, M.A. and Shooter, E.M. (1981). Differential and synergistic actions of nerve growth factor and cyclic AMP in PC12 cells. J. Cell. Biol. 89, 240-245.

Jacobs, J.R. and Stevens, J.K. (1986). Changes in the organization of the neuritic cytoskeleton during nerve growth factor-activated differentiation of PC12 cells: a serial electron microscopic study of the development and control of neurite shape. J. Cell Biol. 103, 895-906.

Jacobson, K. and Wojciezyn, J. (1984). The translational mobility of substances within the cytoplasmic matrix. Proc. Natl. Acad. Sci. USA 81, 6747-6751.

Johnson, G.S. and Pastan, I. (1972). N^6, O^2-Dibutyryl adenosine 3',5'-monophosphate induces pigment production in melanoma cells. Nature 237, 267-268.

Luckenbill-Edds, L., Van Horn, C. and Greene, L.A. (1979). Fine structure of initial outgrowth of processes induced in a pheochromocytoma cell line (PC12) by nerve growth factor. J. Neurocytol. 8, 493-511.

Pawelek, J., Sansone, M., Koch, N., Christie, G., Halaban, R., Hendie, J., Lerner, A.B. and Varga, J.M. (1975). Melanoma cells resistant to inhibition of growth by melanocyte stimulating hormone. Proc. Natl. Acad. Sci. USA 72, 951-955.

Preston, S.F., Volpi, M., Pearson, C.M. and Berlin, R.D. (1987). Regulation of cell shape in the Cloudman melanoma cell line. Proc. Natl. Acad. Sci. USA In press.

Puck, T.T. (1977). Cyclic AMP, the microtubule-microfilament system and cancer. Proc. Natl. Acad. Sci. USA 74, 4491-4495.

Sawyer, T.K., Sanfilippo, P.J., Hruby, V.J., Engel, M.H., Heward, C.B., Burnett, J.B. and Hadley, M.E. (1980). 4-Norleucine, 7-D-phenylalanine-α-melanocyte stimulating hormone: A highly potent α-melanotropin with ultralong biological activity. Proc. Natl. Acad. Sci. USA 77, 5754-5758.

Saxton, M.J. (1982). Lateral diffusion in an archipelago. Effects of impermeable patches on diffusion in a cell membrane. Biophys. J. 39, 165-173.

Westermark, B. and Porter, K.R. (1982). Hormonally induced changes in the cytoskeleton of human thyroid cells in culture. J. Cell Biol. 94, 42-50.

CYTOPLASMIC AND EXTRACELLULAR CONTROL OF NEURITE FORMATION

A. Krystosek

Department of Biochemistry, Biophysics and Genetics, University of Colorado Health Sciences Center, Denver, Colorado 80262, USA

ABSTRACT

Differentiation of the neural hybrid cell line NG108-15 has been studied with the aim of elucidating the mechanisms regulating neurite formation and growth. Organization of tubulin and the number and placement of the microtubule organizing centers (MTOCs) were explored during proliferation and inducible neurite formation using indirect immunofluorescence with an anti-tubulin monoclonal antibody. The results indicated that undifferentiated NG108-15 cells contained a loose network of cytoplasmic microtubules, whereas a dense packing of these polymers was present in the neurites of maturing cells. Under all growth conditions, 94% or greater of the cells possessed a single MTOC. The location of MTOCs in differentiating cells extending one or more neurites was examined in Triton X-100 extracted cytoskeletons. The placement of the MTOC in neurite-bearing cells was not consistent with a determining role in the spatial control of neurite emergence from the cell cortex. The use of a novel silver staining method for visualization of the pericellular matrix has made possible a companion investigation of the orientation of neurite growth in relation to extracellularly deposited materials. The evidence indicated that the neurites of NG108-15 cells respond to such shed cellular products by high precision haptotactic guidance of neurite growth. The detection of the adhesive boundary occurs at the level of the actin-containing filopodia of the advancing neuronal growth cone. Thus, neurite formation by NG108-15 cells is under several levels of control. Microfilament-containing filopodia may participate in the transduction of extracellular information mediating the directional choice for neurite initiation. The intracellular organization of microtubules appears to be a secondary structural response facilitating extension of neurites.

INTRODUCTION

The large soma and long nerve processes are distinctive features of the neuronal lineage. Understanding the genesis of this highly asymmetric form is important not only in regards to neural specific phenomena such as synaptic transmission, but also for an appreciation of the plasticity of cell structure which allows differentiation to diverse tissues and organs. Biochemical examination of the structural components underlying axonal form is a powerful approach to probing cellular specialization. For example, the disintegration of axons in the cold was known to Ramon y Cajal (1960) and earlier workers, but it would be to modern investigators to show that this was the property of a specific molecule and indeed to fashion it into a purification procedure for tubulin (Weisenberg et al., 1968).

A survey of the literature on the regulation of neurite growth reveals, in addition to the cytoskeleton, the following factors as being implicated: genetic predetermination of neuronal shape; hormones, growth factors and their receptors and second messengers; cell surface recognition molecules; competition and cellular interactions in the innervation process; various external forces mediating taxis such as chemical, adhesive and electrical. This is, of course, a selective short list compiled from many experimental preparations and methodologies. Thus, beyond defining the biochemical components of the cytoskeleton, salient questions for current research must include the interrelationship between structural components and other elements of the motile response of living cells.

APPROACH TO THE PROBLEM

It would be advantageous to know which regulatory factors apply to a single cell type and what is the interrelationship between multiple control systems. An approach to this problem is the use in culture of a clonal cell line which can be induced to express the neuronal developmental program. The a priori hope would be to study genetically identical cells in a defined environment. The actual situation is less investigator-controlled but more informative. As will be shown, neural cells in culture create their own trophic microenvironment and exhibit heterogeneous growth patterns based thereon.

The NG108-15 hybrid cell line was used as a model neuron (Hamprecht, 1976) in the present work. Favorable attributes of the cell line include a neuronal pattern of gene expression (Rosenberg et al., 1978) and inducibility of neurite formation (Krystosek, 1985), advanced morphological maturation (Krystosek, 1984), and the membrane components needed for synaptogenesis (Nirenberg et al., 1983). Procedures for the different growth regimens (Krystosek, 1985) and for indirect immunofluorescent demonstration of tubulin in cytoskeletal preparations (Spiegelman et al., 1979) have been described previously.

CYTOSKELETAL INVOLVEMENT IN NEURITOGENESIS

Two striking and related observations that sparked interest in the possible role of the tubulin polymerization system in directed neurite growth were the related neurite morphologies of doublets formed by cell division (Fig. 1) and, also in daughter cells, the apparently directed neurite elongation along the previous parental neuritic pathway, marked by a pinched-off growth cone in the case illustrated in Fig. 2a. The related positioning of sites at the cell cortex from which neurites emerge from daughter cells was initially reported by Solomon (1979), who postulated that the phenomenon could be attributed to an identical spatial distribution, following mitosis, of "endogenous determinants" governing cellular form. The MTOC and microheterogeneity of the cell cortex have been suggested as possible locales of these determinants (Solomon, 1981).

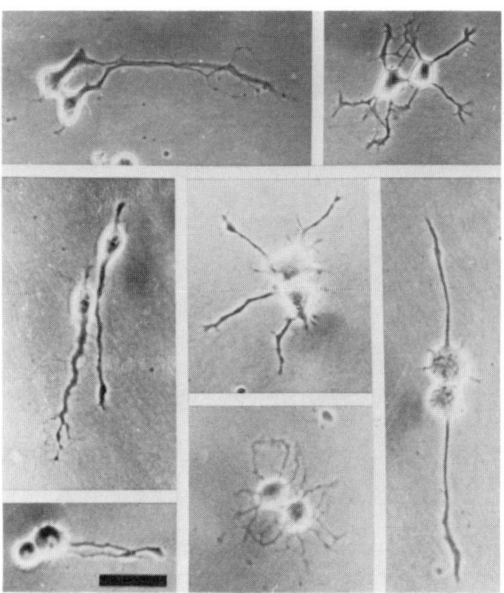

Fig. 1. Related neurite morphologies of daughter cells. Bar = 50 μm. Reproduced from Krystosek, 1985, by permission of Alan R. Liss, Inc.

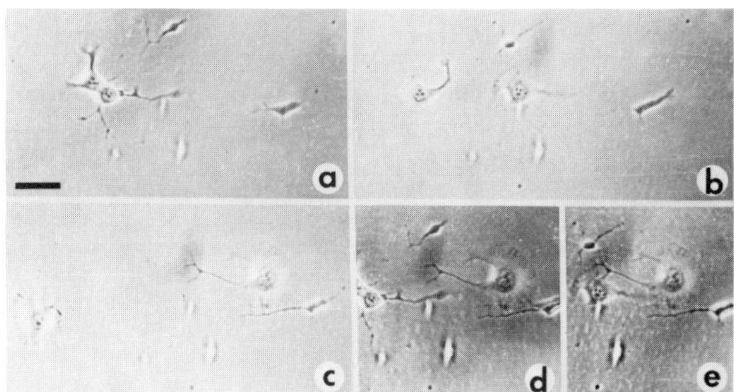

Fig. 2. Directed neurite growth. (a) initial observation of daughter NG108-15 cells and pinched-off growth cones from the parental cell; (b) 1 hr later; (c) 20 hr later; (d) double print of a and c; (e) double print of b and c. Bar = 50 μm.

The following experiments were performed to determine the location of microtubules and the MTOCs during differentiation of NG108-15 cells and to assess whether or not these sites were consistent with a directive role in orienting neurite growth. Cells proliferating in serum containing medium are flattened with large surface areas. A loose array of microtubules spreads from a perinuclear area to the plasma membrane (Fig. 3a). NG108-15 cells induced for differentiation are more rounded and display a thick packing of microtubules in the neurites (Fig. 3b).

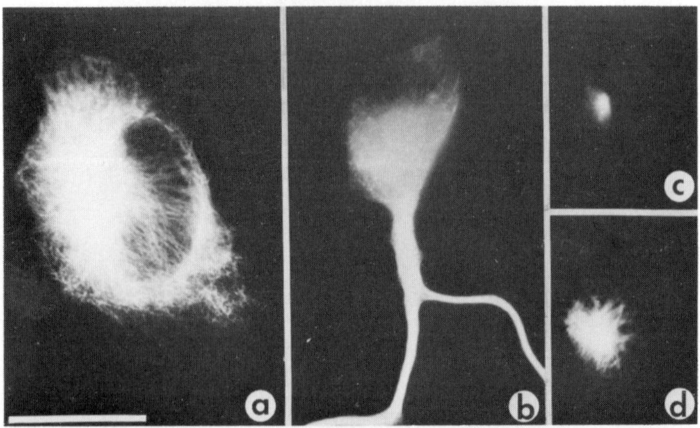

Fig. 3. Distribution of microtubules and the MTOC in NG108-15. Indirect immunofluorescence was performed with a monoclonal antibody (Amersham) to α-tubulin. (a) proliferative cell; (b) differentiation induced with dibutyryl cyclic AMP; (c) and (d) typical intermediates in microtubule regrowth (5 min) following depolymerization with 2 x 10^{-5} M nocodazole. Cytoskeletal preparations according to Spiegelman et al. (1979).

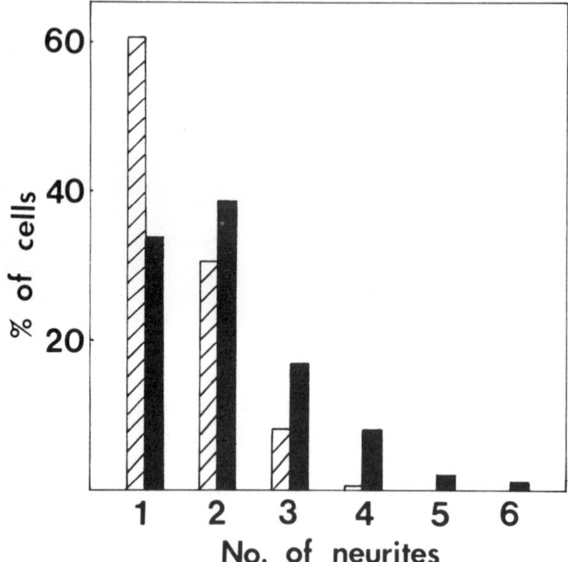

Fig. 4. Histogram showing the percentages of cells having different numbers of neurites. Cross hatch, N2 alone; black bars, N2 supplemented with 0.5 mM dibutyryl cyclic AMP. 500 cells were counted for each condition.

Since a change in tubulin distribution upon neurite formation could be demonstrated, it was next of interest to examine the location and number of organizing centers with regard to the elaboration of neurites. The plasticity of the NG108-15 cell line is especially valuable in this context as the number of neurites expressed may be regulated by the inducing conditions. Spontaneous differentiation in N2 medium yields a majority of unipolar cells with lesser percentages of bi- and tripolar cells (Fig. 4). When N2 is supplemented with dibutyryl cyclic AMP, bi- and multipolar cells form the majority phenotype (Fig. 4). The number of MTOCs was examined during proliferative growth, where greater than 90% of the cells lacked processes, and during the two protocols for neurite induction. The results showed that 94% or greater of the cells in all three cases contained only a single MTOC (Fig. 3c, d).

Three conclusions immediately follow from these data. First, the number of MTOCs is not modulated during the transition from proliferative growth to differentiation. Second, the model of Spiegelman et al. (1979) for the aggregation of multiple MTOCs into a single large initiation site preceding neurite formation does not apply to this cell line. Other work as well has disputed this proposed model (Brinkley et al., 1981; Sharp et al., 1982). Third, the existence of a single MTOC makes it unlikely that its placement alone could control the sites for neurite emergence since the degree of polarity can be varied greatly (Fig. 4).

This last point was subjected to further experimental testing. The location of the MTOC with respect to the origin of the neurite was determined for unipolar cells growing in N2 medium. By using a short (15 min) incubation with nocodazole followed by a 4 min regrowth period, neurite retraction is only partial; both the perinuclear MTOC and disintegrating microtubules in the neurite could be visualized in the absence of organized microtubules in the soma (Fig. 5). For control purposes, phase contrast microscopy verified that neurite retraction was incomplete at 15 min and that cells considered as unipolar did not originate from the complete retraction of other neurites. The relationship between the neurite and MTOC was evaluated in the following way. Cells processed for indirect immunofluorescence were viewed at 630X magnification. Using a protractor reticle in the microscope eyepiece, an axis was aligned through the centers of the MTOC and cell nucleus. The angle of projection of the neurite was read with respect to the original axis. The mean angle of projection was 60° (range 0° - 165°) for 27 observations on NG108-15 cells. A representative example is shown in Fig. 5. This result, which does not support the orientation (0° projection) proposed by Spiegelman et al. (197), showed that there was not an obligatory relationship between the placement of the MTOC and the origin of the neurite. In other words, the neurite need not form at the area of the cell cortex nearest the MTOC.

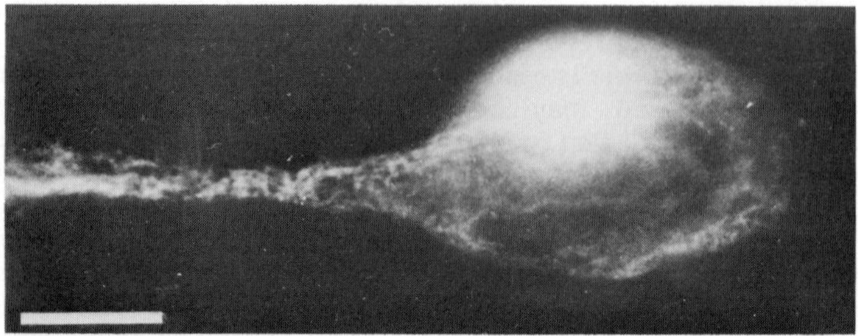

Fig. 5. Location of neurite origin with respect to the MTOC in a unipolar cell. The neurite projects at an angle of 70° from an axis drawn through the centers of the MTOC and nucleus. Bar = 10 µm.

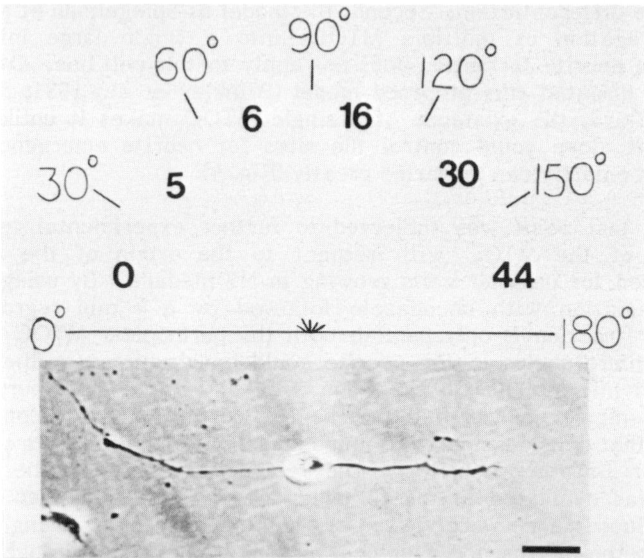

Fig. 6. Neurite form in bipolar cells. Upper panel, the numbers in bold indicate the percentages of cells exhibiting angles of projection between the indicated arcs. Angles were determined using a protractor reticle in the microscope eyepiece. The soma is placed at the origin and one neurite along the 0° axis. The angle of projection of the second neurite is read in the protractor scale. 148 bipolar cells in N2 medium were evaluated. Lower panel: A bipolar cell showing an angle of projection of close to 180°, the most prevalent case. Bar = 25 µm.

The cases of bipolar NG108-15 cells also presented an opportunity to explore these topographical arrangements further. A striking feature of cells extending two neurites was the strong tendency of the processes to exit from opposite sides of the cell (Fig. 6). The location of the MTOC in such cells was examined by the depolymerization-regrowth procedure as indicated above (Fig. 7b) or directly on Triton X-100 extracted cytoskeletons (Fig. 7a) as the MTOC could be inferred from the brightest perinuclear region radiating microtubules. Again the most frequently observed site for the organizing center was not nearest one of the neurites but rather in a more central area of the cell from whence microtubules coursed bidirectionally along the nucleus and into the neurites. This pattern for bipolar cells was also noted in the studies of Spiegelman et al. (1979). This arrangement, which varies somewhat from cell to cell, does not lend itself to an interpretation in which the placement of the MTOC may be viewed as guiding the directionality of neurite growth. A possible explanation for the opposing directionality of neurites might be the need for a balancing of mechanical tension during growth as proposed by Bray (1979). This has not yet been directly tested in the present cell system.

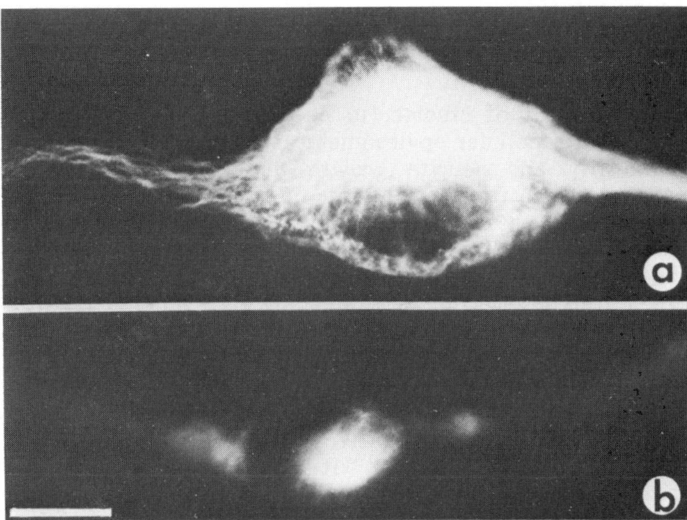

Fig. 7. Location of the MTOC in bipolar cells. (a) Triton X-100 cytoskeletal preparation of untreated cell; (b) nocodazole induced depolymerization and regrowth as described in the text. Collapsing microtubules mark the sites of neurite exit from approximately opposite ends of the cell. Bar = 10 μm.

Finally, it was of interest to learn if the distribution of MTOCs following mitosis could underlie the symmetrical elaboration of neurites in daughter cells. For this purpose, doublets generated by cell division were examined for a possible mirror image relationship in the placement of MTOCs. Preliminary results have indicated that daughter cells need not exhibit a related positioning of MTOCs (Fig. 8). This result as well as the disparity between the single MTOC and neurite relatedness in multipolar cells (Fig. 1) argues against the MTOC as being the sole component of the "endogenous determinants."

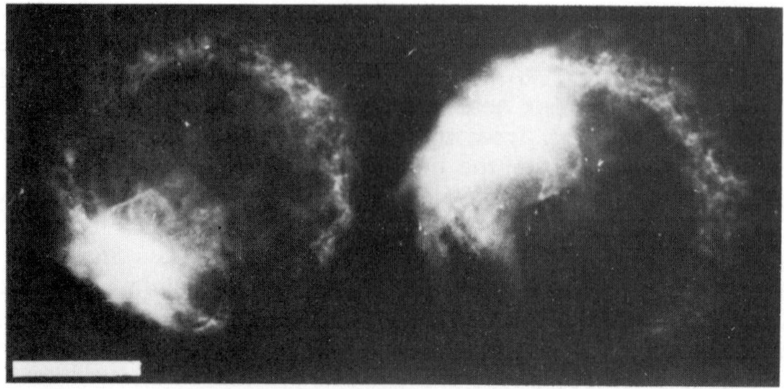

Fig. 8. MTOC location in a postmitotic cell doublet. No symmetry with respect to the plane of division is obvious in the relative placement of the MTOC in the daughter cells. Bar = 10 µm.

GUIDANCE FROM WITHOUT

A concurrent set of studies (to be reported in detail elsewhere) has implicated the extracellular environment as a repository of macromolecules directing motility and neurite growth. A simple but richly detailed experimental approach from which hints of haptotactic or adhesive guidance were first gleaned is the photomicrographic documentation of cell positional changes over time. The re-extension of a daughter cell neurite along the path of the previous parental neurite was already alluded to in Fig. 2a, which presents the initial observation of a randomly viewed cell doublet. Simply by recording cell position at subsequent times it was shown that one cell both continues neurite growth and translocation of the soma towards the pinched-off growth cone (Fig. 2b), achieves contact (Fig. 2c) and surprisingly redirects a neurite back along the terrain it has just traversed. By double-printing of the photographic negatives it can be realized that this new neurite is not randomly deployed but rather impacts upon the previous resting spots of the cell body, neurite and the second (upper) growth cone (Fig. 2d, e). Likewise, the spread membrane of the soma itself is not only in obvious contact with its "target" growth cone but also grazes the previous attachment positions of cell and neurite. Observation of a large number of cells has revealed that this tendency to establish the maximal extent of contact with previously occupied spaces is highly reproducible. This degree of determinism in the elaboration of neuronal form is somewhat surprising in view of the previous demonstration that the actual initiation of neurites has a probabalistic component (Krystosek, 1985).

From this analysis of cell behavior one might suspect the initial orientation of the neurite toward the parental growth cone (Fig. 2a) was mediated by adhesive material along the putative pathway. This appears to be the case as shown by a novel method developed to study deposited cellular material in relation to motility. The procedure, which will be described in detail elsewhere, visualizes the pericellular matrix of cells by its ability to exclude fine grains of nascent elemental silver which otherwise attaches to the culture substratum as a continuous carpet. As shown in Fig. 9a, silver (dark grains in the photograph) is absent from the area immediately around the cell body and the elongating neurite and is partially excluded from an area which represents the pathway of previous translocation by this cell. The border of this path or track is the distance at which filopodia extend and retract from the soma; the sweep is thereby wider than the cell itself. The method is general and also reveals the pericellular matrices of normal neurons and fibroblasts.

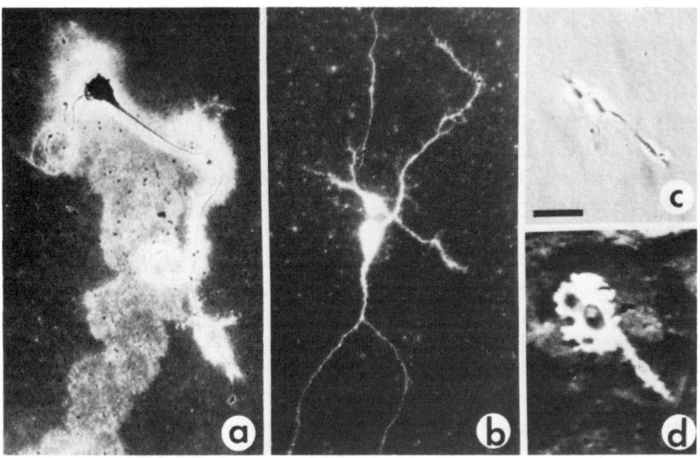

Fig. 9. Visualization of the pericellular matrix in relation to neurite growth. (a) neurite extending backwards along the filopodial border demarcating the path of previous cell movement; (b) silver exclusion marks the former positions of cells and neurites following EGTA treatment, indicating the argyrophobic material remains bound to the substratum; (c) phase contrast micrograph of neurite extending towards an isolated growth cone; (d) silver stain of cells in (c) indicates continuous pathway (derived from parental neurite). Bar = 50 µm.

The chemical nature of the material excluding silver is not known but the following properties have been determined. The material remains active when cells are removed with EGTA (Fig. 9b) while it is destroyed by trypsin treatment. Chondroitinase, heparitinase, and hyaluronidase were without effect. These characteristics would be consistent with a proteinaceous component of the substrate adhesion material.

The ability to obtain a record of all of the areas on the culture substratum that have been conditioned has provided a means to directly test the idea that cells and neurites preferentially associate with previously "touched" regions. The example of Fig. 9a clearly shows that neurite extension is exclusively constrained by the filopodial border from cell transit. Such observations of neurites bearing along conditioned terrain and even branching to maintain maximal contact have been made repeatedly.

When applied to the paradigm of neurite re-extension already discussed in reference to Fig. 2, the technique reveals that a continuous path does indeed exist between the isolated growth cone and the approaching neurite (Fig. 9c, d). Staying on the path presumably reflects a haptotactic guidance similar to growth cone behavior in distinguishing artificially created adhesive boundaries (Letourneau, 1975; Collins and Lee, 1984).

CONCLUSIONS AND PROSPECTS

Neurite extension during differentiation of NG108-15 cells was correlated with a change from a loose meshwork of microtubules to a dense packing within neurites. However, no evidence was obtained that the intracellular organization of microtubule assembly governs the details of neurite form, especially the sites of origin of these processes. Other studies showed that neurite initiation and coursing was oriented in response to the arrangement of cellularly deposited material. This aspect of cellular decision making may mimic the choices made by developing neurons *in vivo* - that is, interpreting the signals for start, stop and which way to go.

The control of neurite growth is multifaceted, yet in each circumstance only one component can be rate-limiting or site-limiting. The rapid and precise directive influence of local substratum-bound guidance cues suggests that other aspects must be secondary or non-directive in this model system. The tubulin assembly apparatus may be such a necessary but non-directive component of exploratory neurite propagation. The role of the endogenous determinants becomes even more mysterious when considered against a known regulatory feature such as guidance from the extracellular milieu. Questions dealing with the hierarchical interrelationship between these two levels of control can be phrased for further research. Such as, can one system override the other?

Lest it be thought that the role of the cytoskeleton is being denigrated, it should be pointed out that the actin containing filopodia are the key elements in both the laying down of and detection of the guidance pathways. An important question to resolve is, therefore, what is the transduction mechanism from extracellular matrix to membrane/microfilaments to directed tubulin polymerization? The answer may hinge on the other major question: What is the molecular nature of the deposited material which initiates this chain of events? The ability to solubilize and functionally evaluate (Krystosek and Seeds, 1986) components of the substrate adhesion material may facilitate a combined biochemical/cell biological evaluation of information transfer at the developing neuronal growth cone.

ACKNOWLEDGEMENTS

This work was supported by a grant from the USPHS, R23-CA32260, awarded by the National Cancer Institute. I thank R.B. Maccioni and N.W. Seeds for many stimulating discussions of the chemistry and biology of tubulin.

REFERENCES

Bray, D. (1979). Mechanical tension produced by nerve cells in tissue culture. J. Cell Sci. 37, 391-410.

Brinkley, B.R., Cox, S.M. and Fistel, S.H. (1981). Organizing centers for cell processes. Neurosci. Res. Prog. Bull. 19, 108-124.

Collins, F. and Lee, M.R. (1984). The spatial control of ganglionic neurite growth by the substrate-associated material from conditioned medium: An experimental model of haptotaxis. J. Neurosci. 4, 2823-2829.

Hamprecht, B. (1976). Neuron models. Angew. Chem. Int. Ed. Engl. 15, 194-206.

Krystosek, A. (1984). Nucleolar maturation accompanies neuroblastoma differentiation (Abstract). J. Cell Biol. 99, 154.

Krystosek, A. (1985). Neurite formation by neuroblastoma-glioma hybrid cells (NG108-15) in defined medium: Stochastic initiation with persistence of differentiated functions. J. Cell. Physiol. 125, 319-329.

Krystosek, A. and Seeds, N.W. (1986). Normal and malignant cells, including neurons deposit plasminogen activator on the growth substratum. Exp. Cell Res. 166, 31-46.

Letourneau, P.C. (1975). Cell-to-substratum adhesion and guidance of axonal elongation. Dev. Biol. 44, 92-101.

Nirenberg, M., Wilson, S., Higashida, H.,Rotter, A., Krueger, K., Busis, N., Ray, R., Kenimer, J.G. and Adler, M. (1983). Modulation of synapse formation by cyclic adenosine monophosphate. Science 222, 794-799.

Rosenberg, R.N., Vance, C.K., Morrison, M., Prashad, N., Meyne, J. and Baskin, F. (1978). Differentiation of neuroblastoma, glioma, and hybrid cells in culture as measured by the synthesis of specific protein species. Evidence for neuroblast-glioblast reciprocal genetic regulation. J. Neurochem. 30, 1343-1355.

Ramon y Cajal, S. (1960). Studies on Vertebrate Neurogenesis (transl. L. Guth), pp. 57-58. Charles C. Thomas, Springfield, Illinois.

Sharp, G.A., Weber, K. and Osborn, M. (1982). Centriole number and process formation in established neuroblastoma cells and primary dorsal root ganglion neurones. Eur. J. Cell Biol. 29, 97-103.

Solomon, F. (1979). Detailed neurite morphologies of sister neuroblastoma cells are related. Cell 16, 165-169.

Solomon, F. (1981). Specification of cell morphology by endogenous determinants. J. Cell Biol. 90, 547-553.

Spiegelman, B.M., Lopata, M.A. and Kirschner, M.W. (1979). Aggregation of microtubule initiation sites preceding neurite outgrowth in mouse neuroblastoma cells. Cell 16, 253-263.

Weisenberg, R.C., Borisy, G.G. and Taylor, E.W. (1968). The colchicine-binding protein of mammalian brain and its relation to microtubules. Biochemistry 7, 4466-4479.

ISOTUBULIN EXPRESSION DURING MOUSE NEURONAL DIFFERENTIATION

Ph. Denoulet, B. Eddé, A. Koulakoff, Y. Netter and F. Gros

Biochimie Cellulaire, Collège de France, 11 place Marcelin Berthelot 75231 PARIS Cedex 05 (FRANCE)

We have previously reported that α and β tubulins, the major constituents of microtubules, display an extensive heterogeneity in nervous tissues and especially in neurons, due to the expression of numerous specific isotubulins. Moreover, we showed that these diverse neurospecific isotubulins progressively appear and accumulate through out brain development (1, 2). The use of cellular models of neuronal differentiation such as mouse teratocarcinoma or neuroblastoma cell lines brought further evidence that some neurospecific isotubulins are expressed during crucial phases of early neural differentiation, i.e. neuronal commitment (β'1) and neurite outgrowth (α2, β'2) (3, 4).

Here, we present additional data concerning (i) the neuronal tubulin polymorphism and (ii) the control levels involved in the expression of the diverse isotubulins.

I) THE NEURONAL TUBULIN PHENOTYPE

Later steps of neuronal differentiation were studied using long term primary cultures of mouse neurons. Cells were isolated from 15-day embryonic brain and cultured for one month or more. Virtually pure neurons were obtained. Tubulin was purified by the taxol method and analyzed by resolutive isoelectric focusing. Immunoblotting and peptide mapping experiments were carried out for a precise identifi cation. At day 2 of culture, α2, α3, β'1 and β'2 neurospecific iso tubulins were already expressed, together with the common α1 and β3 isoforms, and increased until day 7. Later on, acidic forms of β tubulin (β4 to β6) appeared then accumulated up to one month in culture. The most striking observation was that the isotubulin pattern displayed by the well-differentiated neurons was qualitative ly similar to that observed in adult mouse brain. By contrast, glial cells even after one month in culture expressed only a very simple tubulin pattern (α1, α2, β3).

II) CONTROL OF ISOTUBULIN EXPRESSION IN NEURONS

Using different approaches (pulse-chase experiments, in vitro translation of mRNA, Northern blot ...), we showed that the express ion of the diverse isotubulins was controlled at two distinct levels: whereas α1, β'1, β3 and β4 isoforms were directly translated from distinct mRNA, α2, α3, β'2, β5 and β6 were produced by posttransla tional modification.

In the adult brain, up to 6 distinct β tubulin mRNA were identi fied. 4 out of 6 were neurospecific messages, one coding for β'1 and expressed very early, and 3, differring in length, coding for β4 and expressed later in development (5).

Concerning the posttranslational level, the chemical nature of some modifications was determined. Using specific metabolic labeling with $^{32}PO_4$ or ^{3}H-acetate, we found that β'2 derived from β'1 by phosphorylation and that α2, α3 and β5 were acetylated isoforms (6).

III) ISOTUBULIN POLYPORPHISM AND NEURONAL DIFFERENTIATION

Taken altogether, these results can be summarized as follows:

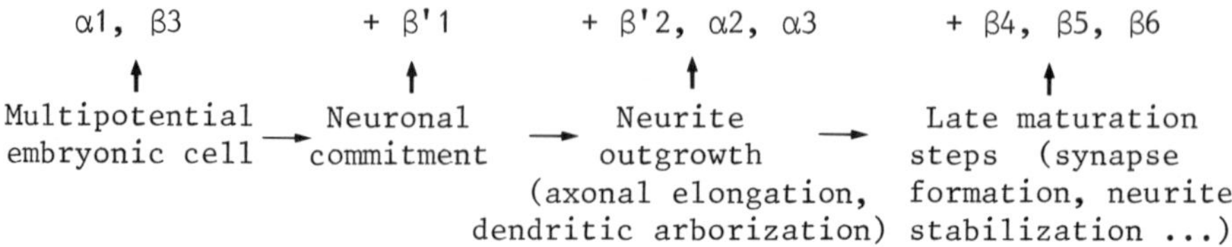

The simple tubulin pattern (α1, β3) expressed by multipotential embryonic cells is greatly and specifically increased in those cells which engage into the neuronal pathway of differentiation. By the time of neuronal commitment, β'1 isotubulin is expressed, presumably after derepression of a specific β tubulin isogene. During the early steps of morphological differentiation, neurons supporting neurite outgrowth express α2 (teratocarcinoma), β'2 (neuroblastoma) or α2, α3 and β'2 (cultured neurons). These three last isotubulins are produced by posttranslational modification. As differentiation proceeds and up to the late steps of neuronal maturation, β4 (messenger-directed) and β5 + β6 (modified isoforms) are progressively produced and accumulated in neurons.

These results support the idea that within the tubulin multigene family, distinct tubulin isogenes could be expressed at crucial steps of neuronal differentiation and might be involved in the important developmental changes of neuronal structures and functions. In addition, posttranslational modifications occurring specifically on these distinct tubulin gene products might play essential roles in conferring specific properties to the diverse isotubulins.

This work was supported by grants from Collège de France, CNRS and INSERM (CRE 85.6008).

REFERENCES

(1) Denoulet, P., Eddé, B., Jeantet, C. and Gros, F. (1982)Biochimie 64, 165-172.
(2) Denoulet, P., Jeantet, C. and Gros, F. (1982) BBRC 105, 806-813.
(3) Eddé, B., Jeantet, C. and Gros, F. (1981) BBRC 103, 1035-1043.
(4) Eddé, B., Jakob, H. and Darmon, M. (1983) EMBO J 2, 1473-1478.
(5) Denoulet, P., Eddé, B. and Gros, F. (1986) Gene 50, 289-297.
(6) Eddé, B., de Néchaud, B., Denoulet, P. and Gros, F. (1987) Devel. Biol. (in press)

CYTOSKELETAL PROTEINS IN NEURITE OUTGROWTH

M.A. Cambray-Deakin, A. Morgan and R.D. Burgoyne
The Physiological Laboratory, University of Liverpool, P.O. Box 147, Brownlow Hill, Liverpool, L69 3BX

The exact mechanisms involved in the development of neuronal connectivity are still in the main unknown. However the neuronal cytoskeleton must play an important role. For example, a major re-organisation of the cerebellar granule cell axonal cytoskeleton is known to occur at the time of synaptogenesis[1]. In contrast little is known of the detailed changes in cytoskeletal structure which accompany the early stages of neurite outgrowth. Here we have used cerebellar granule cell cultures and immunofluorescence to follow the appearance of cytoskeletal components during initial neurite extension.

Trypsin-digested week old rat cerebella were dissociated by mechanical disaggregation and plated onto poly-D-lysine coated glass coverslips in serum-free medium. Formaldehyde-fixed cultures (8' - 2 days in vitro, DIV) were processed for double-label immunolfluorescence using: YOL/34 (anti-α tubulin); YL1/2 (anti-tyrosylated α-tubulin); 6-11B-1 (anti-acetylated α-tubulin); RT97 (anti-200Kd neurofilament sub-unit, NF); Amersham monoclonal antibodies to MAP2, MAP1A, 70Kd, 160 Kd and 200 Kd NF; rhodamine-phalloidin to visualise filamentous actin.

Granule cells rapidly produced highly variable lamellipodal expansions or thin processes within 8' after plating. The processes lengthened and elaborated over 2 DIV and became more uniform, thin and varicose. Rhodamine-phalloidin labelling showed that all cell bodies and extensions contained filamentous actin from the earliest times, showing a uniform pattern of labelling until 60' in vitro when there appeared an enrichment of actin in the growth cones of some neurites. YOL/34 immunofluorescence indicated that α-tubulin containing microtubules (MT's) first appeared in cell processes from 30' - 60'. Labelling with anti-MAP1A + anti-MAP2 appeared concomittantly with or slightly after YOL/34 labelling in all processes over this time course. YOL/34 labelled MT's extended only to the base of the neurite growth cone and never ramified throughout the structure. YL1/2 labelling in neurites appeared before 6-11B-1 labelling suggesting that α-tubulin may be insitially assembled in a non-acetylated form and that acetylation occurs subsequently. Only the 200 Kd NF subunit could be identified in granule cell bodies and processes, first appearing after 1 DIV.

Although an intact MT network is essential for the extension of long neurites[2], initial neurite growth proceeds without MT's, the primary cytoskeletal element involved at this stage being actin microfilaments (MF's). Indeed granule cells treated with 10 µg/ml cytocholasin B do not extend neurites, MF's remain the only cytoskeletal element in the growth cone and may be concentrated there. Growth cones are known to be highly motile structures which may determine the correct orientation of neurite growth[3], responding to extracellular cones. Nascent neurites and growth cones[4,5] contain high levels of Ca^{2+} channels and also $[Ca^{2+}]_i$ which may allow cytoskeletal interactions resulting in neurite growth and growth cone activity e.g. actin filament reorganisation. Recent work in our laboratory suggests that the initiation of granule cell neurite growth may be influenced by glutamate acting via NMDA receptors and voltage-activated Ca^{2+} channels (Pearce, Cambray-Deakin and Burgoyne, unpublished results). The delayed appearance and acetylation of MT's may also be associated with an elevated level of Ca^{2+} in the growing neurite tip as Ca^{2+} inhibits both MT assembly[6] and α-tubulin acetylation[7]. As MT acetylation appears to be correlated with increased MT stability[8] acetylated MT's may appear in neurites after tyrosylated MT's as part of a stabilisation process. Alternatively, acetylated MT's may subserve some function not directly related to process elongation, e.g. organelle transport. The role of the late appearing 200 Kd NF subunit is unclear.

We thank Drs. Piperno, Kilmartin and Anderton for their generous gifts of antisera. This work was funded in part by The Wellcome Trust.

REFERENCES

(1) Burgoyne, R.D. (1986) Comp. Biochem. Physiol. 83B, 108.
(2) Daniels, M. (1975) Ann. N.Y. Acad. Sci. 253, 535-544.
(3) Bentley, D. and Toroian-Raymond, A. (1986) Nature 323, 712-715.
(4) Connor, J.A. (1986) Proc. Nat. Acad. Sci. U.S.A. 83, 6179-6183.
(5) Bolsover, S.R. and Spector, I. (1986) J. Neurosci. 6, 1934-1940.
(6) Weisenberg, R.C. (1972) Science 177, 1104-1105.
(7) Maruta, H., Greer, K. and Rosenbaum, J.L. (1986) J. Cell Biol. 103, 571-579.
(8) Cambray-Deakin, M.A. and Burgoyne, R.D. Cell Motil. Cyto. in press.

COMPARATIVE ASPECTS ON COD AND BOVINE BRAIN MICROTUBULES

E. Strömberg and M. Wallin
Department of Zoophysiology, Comparative Neuroscience Unit,
University of Göteborg, Box 250 59, S-400 31 Göteborg, Sweden

Most investigations on microtubules have been performed on microtubule proteins prepared by a temperature-dependent assembly-disassembly method. The method takes advantage of the fact that these microtubules break down at low temperature and then reassemble with increasing temperature. From a functional aspect it is obvious that poikilothermal animals living at low temperatures also have microtubules that must fulfill their tasks at these temperatures. Since most of the studies on microtubule proteins have been carried out with mammalian material little is known about microtubules from lower animals. In an earlier investigation (1) it was found that microtubules in cod (Gadus morhua) splanchnic nerve axons exist even at $0^{o}C$. In contrast, cold-labile microtubules were found upon isolation of cod brain microtubules. Preliminary results showed that cod brain microtubules differed from bovine brain microtubules in many aspects. The purpose of the present study was therefore to highlight these differences and make comparisons between microtubules from homeo- and poikilothermal vertebrates.

Microtubule proteins were prepared from bovine brain by two cycles of assembly-disassembly in the absence of glycerol. Tubulin was separated from the microtubule-associated proteins (MAPs) by ion-exchange chromatography. Microtubule proteins from cod brain were isolated by an assembly-disassembly method as described in (1). Assembly was performed at $30^{o}C$ and disassembly at $4^{o}C$. Once-cycled preparations were used, although for some experiments twice-cycled preparations were used. Cod brain microtubules were also isolated by a taxol-dependent procedure as described in (2). Cod brain MAPs
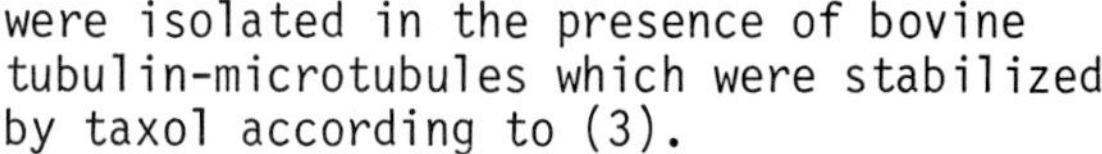
were isolated in the presence of bovine tubulin-microtubules which were stabilized by taxol according to (3).

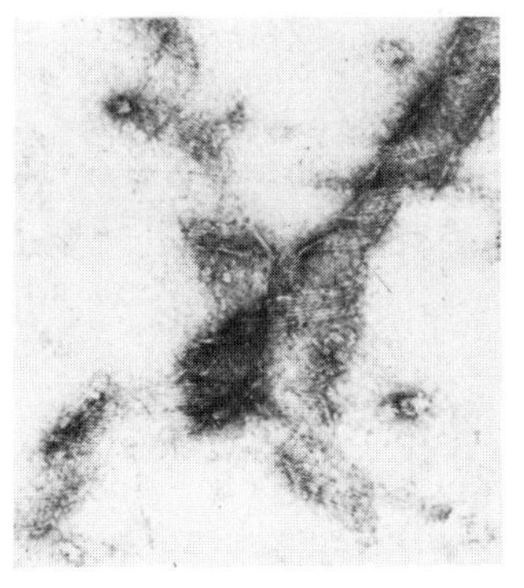

Ca^{2+} had only a slight effect on the extent of assembly at a concentration of 2 mM. At 4 mM a higher turbidity was found when the assembly was monitored at $30^{o}C$ by the change in absorbance at 350 nm. The majority of the microtubules were spiralized (Fig.1). A

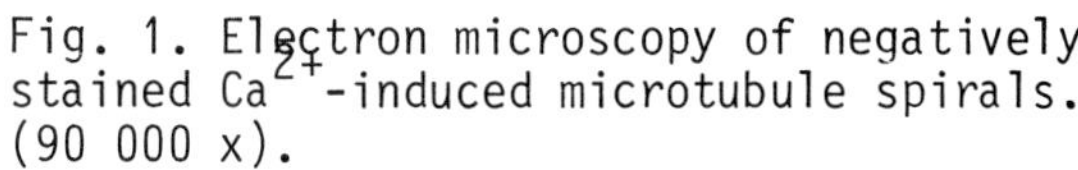
Fig. 1. Electron microscopy of negatively stained Ca^{2+}-induced microtubule spirals. (90 000 x).

very high turbidity was found in the presence of 10 mM Ca^{2+}, and both spirals and macrotubules were present. In the presence of colchicine (0.1-1 mM) the assembly was unaffected.

A low amount of MAPs was found in the assembly-disassembly purified microtubule preparations. However, several MAPs were isolated from taxol-purified cod brain microtubules. High molecular weight proteins were present, of which MAP2 was identified by immunohistochemistry on Western blots. The cod brain MAP2 was not heat-stable in contrast to bovine brain MAP2. No MAP with a similar molecular weight as MAP1 was found, and no MAP1-like proteins were identified by immunohistochemistry. However, a protein with a higher molecular weight than MAP1 was found (400 kDa). Several proteins of lower molecular weight were also found, of which tau was identified, although in low amounts. Cod MAPs were found to co-assemble with bovine tubulin-microtubules assembled in the presence of taxol (fig. 2).

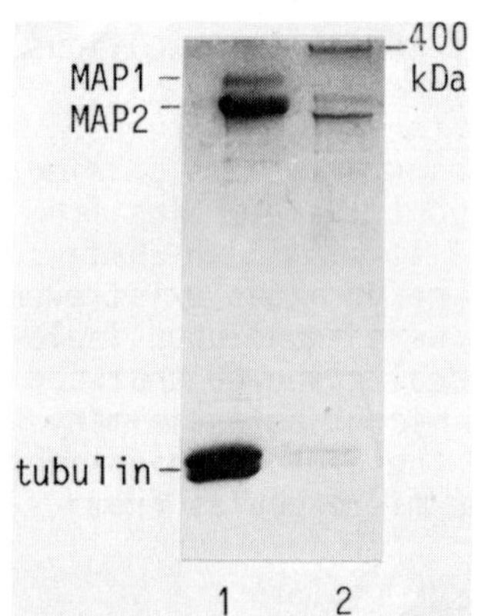

Fig. 2. SDS-polyacrylamide gel electrophoresis of bovine brain microtubules (lane 1) and MAPs from microtubules composed of bovine tubulin and cod MAPs (lane 2).

Microtubules from cod brain were found to consist predominantly of 13 protofilaments as judged from cross-sections of tannic-acid fixed microtubules. Arm-like projections were seen on the surface of the microtubules.

In conclusion, cod brain microtubules were found to differ from mammalian microtubules in many aspects. Both Ca^{2+} and colchicine are known to bind to the C-terminal of mammalian tubulin and to inhibit the assembly of microtubules (for a review see 4). Cod brain microtubule assembly was not inhibited by these drugs, suggesting the presence of microheterogeneities in the C-terminals of cod and mammalian tubulin. The characteristics of the MAPs were also different. MAP2 was not heat-stable and a protein with a high molecular weight (400 kDa) was detected, but no MAP1. The results indicate that species-differences between microtubules exist.

REFERENCES

(1) Strömberg, E., Jönsson, A.-C. and Wallin, M. FEBS Lett. 204, 111-116.
(2) Vallee, R.B. (1982) J. Cell Biol. 92, 435-442.
(3) Bloom, G.S., Luca, F.C. and Vallee, R.B. (1985) Biochemistry 24, 4185-4191.
(4) Maccioni, R.B., Serrano, L. and Avila, J. (1986) BioEssays 4, 165-169.

CYTOPHOTOMETRIC ANALYSIS OF GLIAL FIBRILLARY ACIDIC PROTEIN (GFAP) IN PRIMARY CULTURES OF RAT ASTROCYTES

L. Megias[1], J. Renau-Piqueras, M. Burgal, C. Guerri, R. Báguena Cervellera and M. Sancho-Tello.
Instituto Investigaciones Citológicas, Valencia, and (1) Dept. of Anatomy, Fac. Medicine, Granada, Spain.

Primary culture astrocytes have been widely used as a model for the study of normal astroglial cells. These cultures are a source of large amounts of glial material, and the metabolism of these cells derived from neonatal rat brain resembles that of astrocytes "in vivo" (1). Glial fibrillary acidic protein (GFAP), a type of intermediate filament protein, is now generally considered to be a specific astrocyte marker in tissue sections as well as in cultures. Morphological differentiation of astrocytes "in vitro" occurs in parallel with the biochemical development of the GFAP (2). Quantitative evaluation of GFAP has been carried out by immunoelectrophoresis and immunoradiometric methods. However, these procedures are disruptive and do not permit direct correlation of the amount of GFAP with the astrocyte cytoskeletal morphology. To establish this correlation we have applied cytophotometric methods to analyze micrographs of astrocytes after 1,4,12,21 and 25 days in primary culture.

MATERIALS AND METHODS. Primary cultures of astrocytes were established from 21 day-old rat foetuses (1). Growth curves were obtained by removing cells from petri dishes at various stages of culture with a trypsin solution and counting suspensions in a hemocytometer. DNA and protein were also determined. For indirect immunofluorescence, cultures grown on coverslips were fixed in acetone for 10 min at -20ºC, washed in PBS and incubated with a monoclonal antibody to GFAP (Boehringer) (20 μg/ml) for 60 min at room temperature, rinsed with PBS and incubated with a FITC-conjugated goat antimouse Ig (1:50) for 60 min at room temperature. After washing, the coverslips were mounted with PBS-glycerol (1:1). Immunofluorescent staining was observed with a Zeiss standard microscope equipped with epifluorescent optics and recorded on 24x36 mm negatives. To avoid the problem related with the topological determination of fluorescence intensities, the fluorescent reaction pattern in isolated astrocytes was quantified by means of photographic negatives. These were densitometrically analyzed with the HIDACSYS program of image analysis (3) with a Zeiss Cytoscan SMP-05 scanning cytophotometer interfaced to a PDP 11/24 Computer. A 10 μm Zeiss scanning stage was used, and 120 μm square areas of the negative were quantified with 16 intensity levels.

RESULTS and COMMENTS. Growth curves, DNA and protein content of astrocytes in primary culture were similar to those previously described (1). On the basis of GFAP staining we conclude that our cultures were composed mainly (80%) of astrocytes. Qualitative

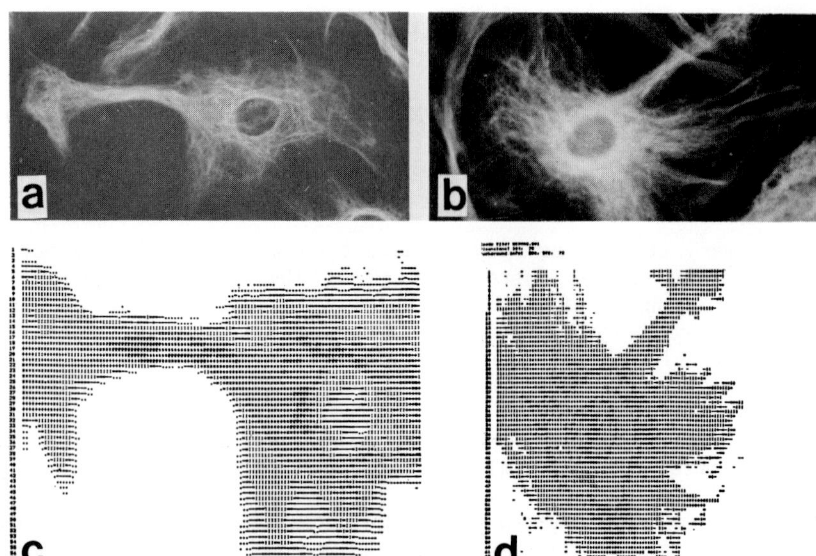

Micrograph showing the GFAP staining pattern: a) 4 days, b) 12 days c) and d) correspond to densitometric maps of these cells.

analysis of astrocytes incubated with anti-GFAP demonstrated that after 1 day in culture immunofluorescence was restricted to a small area surrounding the nucleus. After 4 days in culture the cells showed a reticular cytoskeleton whereas cells from 12,21 and 25 days of culture displayed a filamentous pattern (Fig 1). Cytophotometric measurements showed that the amount (absorbance) of fluorescence in 12-day cultured cells was approximately three times greater than that of 4-day cultured astrocytes (mean absorbance/cell:4 days, 0.1444; 12 days, 0.457). This method also permits, determining quantitatively the distribution pattern of GFAP in the cytoskeleton and comparing this with morphological features (Fig 2). In conclusion, our results using this morphological method show that during the development of the culture the GFAP content increases, in parallel with the morphological of the differentiation astrocyte, together with a redistribution of the cytoskeleton pattern. The cytophotometric procedure used here could therefore be a useful tool to analyze qualitatively and quantitatively the cytoskeletal changes during astrocyte development as well as to evaluate the effects of drugs on this cell component.

REFERENCES

1.- Sensenbrenner, M. (1977) in Cell Tissue and Organ Cultures in Neurobiology (Federoff, S. and Hertz, L. eds.) pp. 191-213, Academic Press, New York.
2.- Goldman, J.E. and Chin, F.C. (1984) J. Neurochem. 42, 175-184.
3.- Van der Ploeg, M., Van den Broek, K., Smeulders, A.W.M., Vosepoel, A.M. and Van Duijn, P. (1977) Histochem 54, 273-288.

INTERMEDIATE AND ACTIN FILAMENTS IN CELL DIFFERENTIATION AND CANCER BIOLOGY

INTERMEDIATE FILAMENTS FOCAL CENTERS AND INTERRELATIONSHIP WITH OTHER CYTOSKELETAL SYSTEMS. CELL CYCLE AND MICROINJECTION STUDIES

P. Madsen and J.E. Celis

Department of Medical Biochemistry, Aarhus University, DK-8000 Aarhus C, Denmark

ABSTRACT

Immunofluorescence analysis with keratin antibodies of the distribution of intermediate filaments focal centers (IFFC) in African green monkey kidney BS-C-1 cells revealed that these were not restricted to any particular stage of the cell cycle. Strong perinuclear staining was also observed in some interphase cells reacted with tubulin antibodies, but these focal arrays of microtubules did not always colocalize with IFFC. Intermediate filaments focal centers did not codistribute with the centrioles in mitosis or with the microtubule organizing centers (MTOC) observed in interphase cells following nocodazole treatment (10 μg/ml, 20 hrs).

Cytoplasmic microinjection of vimentin antibodies into BS-C-1 cells resulted in the perinuclear aggregation of the vimentin and keratin filaments, but had no effect on the distribution of microfilaments, microtubules, mitochondria or the Golgi apparatus. Similar results were observed in BS-C-1 cells injected with keratin antibodies, although in the latter case a significant dispersion of the Golgi was observed. The results are discussed in terms of the function of intermediate filaments.

INTRODUCTION

Intermediate-sized filaments (7-11 A° in diameter) are ubiquitous cytoskeletal elements that can be subdivided in five biochemically distinct classes: keratins present in epithelial cells, neurofilaments present in most but not all neurons, desmin filaments in muscle, glial filaments in glial cells and vimentin in cells of mesenchymal origin (for reviews see Lazarides, 1980, 1982; Osborn and Weber, 1983). The function(s) of these filaments is at present unknown although they have been implicated in various intracellular roles such as nuclear anchorage (Small and Celis, 1978; Letho et al. 1978), mechanical integrators of cytoplasmic space (Lazarides, 1980), organelle interactions (Lee et al. 1979; Goldman et al. 1979, David-Ferreira and David-Ferreira, 1980; Chen et al. 1982; Mose Larsen et al. 1982, 1983; Tokujasu et al. 1983; Traub et al. 1985,1986) and gene expression. (Traub, 1985 and references therein). Two lines of evidence suggest that intermediate filaments may not play a fundamental role in cell growth. First, some cell lines do not have any detectable intermediate filament subunits (Jackson et al. 1980; Paulin et al. 1980; Mose Larsen et al. 1983; Venetianer et al. 1983; Duprey et al. 1985; Giese and Traub, 1986,; Heidelberg and Chen, 1986; Lilienbaum et al. 1986), and second, microinjection of intermediate filaments antibodies results in perinuclear filament aggregation without affecting cell growth (Gawlitta et al. 1981; Lane and Klymkowsky, 1982; Lin and Feramisco, 1982). Also, little is known concerning their synthesis and structural and functional relationships with other cytoskeletal systems.

Recently, we reported the occurrance of intermediate filaments focal centers (IFFC) in African green monkey kidney TC7 cells (a subclone of BS-C-1 cells) and presented evidence for the interaction of intermediate filaments (keratins and vimentin) with actin microfilaments (Celis et al. 1984). Here we studied the occurrance of these focal centers in cells at different stages of the cell cycle and examined the distribution of various cytoskeletal systems and organelles in cells microinjected with vimentin and keratin antibodies.

INTERMEDIATE FILAMENTS FOCAL CENTERS (IFFC) ARE NOT RESTRICTED TO ANY PARTICULAR STAGE OF THE CELL CYCLE

Immunofluorescence staining of methanol fixed BS-C-1 cells with keratin antibodies revealed a discontinous staining of filaments (Fig. 1A) as well as the presence of strongly fluorescent perinuclear centers (Fig. 1B) in about 62% of the interphase cells. A similar staining pattern was observed in cells reacted with vimentin antibodies (results not shown). Double immunofluorescence studies with keratin (Fig. 1C and E) and tubulin antibodies (Figs. 1D and F) showed that the IFFC not always colocalized with the focal array of microtubules in interphase. Furthermore, these centers (Figs. 2A and C) did not codistribute with the centrioles in mitosis (Figs. 2B and D) or with the microtubule organizing centers (MTOC) observed in interphase cells after treatment with nocodazole (10 μg/ml, 20 hrs, Figs. 2E and F).

Indirect immunofluorescence analysis of synchronized G_1 BS-C-1 cells (analyzed a few hours after mitotic shake off) with keratin antibodies showed that only a fraction of these cells, similar to that observed in asynchronous populations, contained focal centers (results not shown). Likewise, asynchronous BS-C-1 cells reacted with keratin (Figs. 3A and C) and PCNA antibodies (Miyachi et al. 1978) specific for cyclin (Figs. 3B and D) failed to reveal a specific distribution of these centers in S-phase (see also Celis and Celis, 1985; Madsen and Celis, 1986). Cells in G_2 were not analyzed due to poor synchrony.

MICROINJECTION OF VIMENTIN AND KERATIN ANTIBODIES

Previously (Celis et al. 1984), we showed that treatment of TC7 cells with cytochalasin B (10 μg/ml, 1 hr) produced a star-like arrangement of the keratin (Fig. 4A) and vimentin filaments that in most cases codistributed with patches of actin (Fig. 4B)(see also Knapp et al. 1983). These results have been confirmed in cytochalasin B treated BS-C-1 cells reacted with vimentin (Fig. 4D; compare with Fig. 4E, phalloidin staining) and keratin antibodies (not shown).

Since the above observations argued strongly for an interaction of intermediate filaments with microfilaments we proceeded to examine the effect of microinjected vimentin and keratin antibodies on the distribution of these filaments. The effect on microtubules, mitochondria and Golgi was also examined. Figs. 5A to F show immunofluorescence photomicrographs of BS-C-1 cells injected into the cytoplasm with vimentin monoclonal antibodies, fixed with methanol or acetone 20-22 hrs later, and stained for vimentin (Fig. 5A), keratin (Fig. 5B), actin (Fig. 5C), tubulin (Fig. 5D), mitochondria (Fig. 5E) and Golgi (Fig. 5F). Clearly, the injected antibody had a marked effect on the distribution of the vimentin (Fig. 5A) and keratin (Fig. 5B) filaments, but had no effect on the distribution of microfilaments (Fig. 5C), microtubules (Fig. 5D), mitochondria (Fig. 5E) or the Golgi apparatus (Fig. 5F). Likewise, microinjection of keratin antibodies caused a redistribution of the keratin (Fig. 6A) and vimentin (Fig. 6B) filaments , but had no effect on the distribution of the microfilaments (not shown), microtubules (not shown) or mitochondria (not shown). A significant effect on the Golgi apparatus was however observed (Fig. 6C). In most cases, the Golgi dispersed or collapsed around the focal centers suggesting a link between this organelle and keratin filaments. Injection of various tubulin antibodies did not result in a redistribution of microtubules (Fig. 6D) and therefore no further studies were carried out with these antibodies.

DISCUSSION

The presence of focal centers in African green monkey kidney cells that contain both keratins and vimentin is of interest as these cytoplasmic structures may correspond to organizing centers (Borenfreund et al. 1980; Eckert et al. 1982; Fey et al. 1983). These centers have not been observed in most cell lines and therefore may not represent typical organizing centers as in the case of microtubules (Pickett-Heaps, 1969). Often, these centers codistributed with focal array of microtubules in interphase (most likely centrosomes), but were distinct from the centrioles in mitosis and in many cases the MTOC reappeared shortly after recovery from nocodazole treatment, suggesting that MTOC and IFFC are independent structures (Hormia et al. 1982; Eckert et al. 1982), although there may be some degree of association. Indeed, an association between centrioles and vimentin filaments has been proposed in various cell types (Goldman et al. 1979; Borenfreund et al. 1980; Wang et al. 1979; Aubin et al. 1980; Blose, 1981; Geuens et al. 1983).

The results concerning microinjection of vimentin and keratin antibodies are interesting as both antibodies induced aggregation of the intermediate filaments (see also Lane and Klymkowski, 1982; Lin and Feramisco, 1982), but failed to affect the distribution of the actin microfilaments which in cytochalasin B treated cells aggregate and codistribute with patches of intermediate filaments (Celis et al. 1984). Thus, it would seem that the putative interaction between these filaments observed in cytochalasin B cells may not be a direct one, but rather the result of a generalized local change induced by the drug.

The observation that intermediate filament aggregation did not cause a redistribution of the mitochondria is suprising as this organelle has been shown to interact with these filaments (Lee et al. 1979; David-Ferreira and David-Ferreira, 1980; Chen et al. 1981; Mose Larsen et al. 1982, 1983). However, since mitochondria may also be linked to microtubules (Raine et al. 1971; Allen, 1975; Smith et al. 1975, 1977; Ball and Singer, 1982), it is likely that disruption of one filamentous system may not be enough to alter its distribution. Interestingly, microinjection of keratin antibodies had a significant effect on the Golgi apparatus which in some cells desintegrated while in others collapsed around the IFFC. This effect was not observed however with vimentin antibodies implying a specific interaction between this organelle and keratin filaments. A similar association of the Golgi apparatus with microtubules has been proposed by Rogalski and Singer (1984).

Taken together, the above observations as well as data referred to in the introduction tend to support the notion that intermediate filaments do not play a fundamental role in some cell types. It is likely that these filaments function in relation to a cell differentiation process (Francke et al. 1987)

ACKNOWLEDGEMENTS

We would like to thank S. Himmelstrup Jørgensen for typing the manuscript and O. Sønderskov for photography. P. Madsen is a recipient of a fellowship from the Medical Research Council. This work was supported by grants from the Danish Medical and Natural Science Research Councils, the Danish Cancer Foundation and NOVO.

REFERENCES

Allen, R.D. (1975). Evidence for firm linkages between microtubules and membrane-bound vesicles. J. Cell Biol. 64, 497-503.

Aubin, J.E., Osborn, M., Franke, W.W. and Weber, K. (1980). Intermediate filaments of the vimentin-type and the cytokeratin-type are distributed differently during mitosis. Exp. Cell Res. 129, 149-165.

Ball, E.H. and Singer, J.S. (1982) Mitochondria are associated with microtubules and not with intermediate filaments in cultured fibroblasts. Proc. Natl. Acad. Sci. USA 79, 123-126.

Blose, S.H. (1979). Ten-nanometer filaments and mitosis: Maintenance of structural continuity in dividing endothelial cells. Proc. Natl. Acad. Sci. USA 76, 3372-3376.

Borenfreund, E. Smith, E., Bendich, A. and Franke, W.W. (1980). Constitutive aggregates of intermediate-sized filaments of the vimentin and cytokeratin type in cultured hepatoma cells and their dispersal by butyrate. Exp. Cell Res. 177, 215-235.

Celis, J.E. and Celis, A. (1985). Cell Cycle dependent variations in the distribution of the nuclear protein cyclin (proliferating cell nuclear antigen) in cultured cells: subdivision of S-phase. Proc. Natl. Acad. Sci. USA 82, 3262-3266.

Celis J.E., Graessmann, A. and Loyter, A. eds. (1986). Microinjection and organelle transplantation techniques. Academic Press, London.

Celis, J.E., Small, J.V.,Mose Larsen, P., Fey, S.J., De Mey, J. and Celis, A. (1984). Intermediate filaments in monkey kidney TC7 cells: Focal centers and interrelationship with other cytoskeletal systems. Proc. Natl. Acad. Sci. USA 81, 1117-1121.

Chen, L.B., Summerhayes, I.C., Johnson, L.V., Walsh, M.L., Bernal, S.D. and Lampidis, T.J. (1982). Probing mitochondria in living cells with rhodamine 123. Cold Spring Harbor Symp. Quant. Biol. 46, 141-151.

David-Ferreira, K.L. and David-Ferreira. J.F. (1980). Association between intermediate-sized filaments and mitochondria in rat Leydig cells. Cell Biol. Int. Rep. 4, 655-662.

Duprey, P., Morello, D., Vasseur, M., Babinet, C., Condamine, H. Brûlet, P. and Jacob, F. (1985). Expression of the cytokeratin Endo A gene during early mouse embryogenesis. Proc. Natl. Acad. Sci. USA 82, 8535-8539.

Eckert, B.S., Daley, R.A. and Parysek, L.M. (1982). Assembly of keratin onto PtK1, cytoskeletons: evidence for an intermediate filament organizing center. J. Cell Biol. 92, 575-578.

Fey, S.J., Mose Larsen, P., Bravo, R., Celis, A. and Celis, J.E. (1983). Differential immunological crossreactivity of HeLa keratin antibodies with human epidermal keratins. Proc. Natl. Acad. Sci. USA. 80, 1905-1909.

Francke, W.W., Hergt, M. and Grund, C. (1987). Rearrangement of the vimentin cytoskeleton during adipose conversion: Formation of an intermediate filament cage around lipid globules. Cell 49, 131-141.

Gawlitta, W., Osborn, M. and Weber, K. (1981). Coiling intermediate filaments induced by microinjection of a vimentin-specific antibody does not interfere with locomotion and mitosis. Eur. J. Cell Biol. 26, 83-90

Geuens, G., de Brabander, M., Nuydens, R. and De Mey, J. (1983) The interaction between microtubules and intermediate filaments in cultured cells treated with taxol and nocodazole. Cell Biol. Int. Rep. 7, 35-47.

Giese, G. and Traub, P. (1986). Induction of vimentin synthesis in mouse myeloma cells MPC-11 by 12-O-tetradecanoylphorbol--13-acetate. Eur. J. Cell Biol. 40, 266-274.

Goldman, R.D., Zackroff, R.V., Starger, J.M. and Whitman, M. (1979). Intermediate filaments. Assembly, disassembly, reorganization and relationship to centrioles. J. Cell Biol. 83, part 2, 343 (abstr.)

Hedberg, K.K. and Chen, L.B. (1986). Absence of intermediate filaments in a human adrenal cortex carcinoma-derived cell line. Exp. Cell Res. 163, 509-517.

Hormia, K., Linder, E., Letho, V.-P., Vartio, T., Badley, R.A. and Virtanen, I. (1982). Vimentin filaments in cultured endothelial cells form butyrate-sensitive juxtanuclear masses after repeated subculture. Exp. Cell Res. 138, 159-166.

Jackson, B.W., Grund, C., Schmid, E., Bürki, K., Franke, W.W. and Illmensee, K. (1980). Formation of cytoskeletal elements during mouse embryogenesis. I. Intermediate filaments of the cytokeratin type and desmosomes in preimplantation embryos. Differentiation 17, 161-179.

Knapp, L.W., O'Guin, W.M. and Sawyer, R.H. (1983) Drug-induced alterations of cytokeratin organization in cultured epithelial cells. Science 219, 501-503.

Lane, E.B. and Klymkowsky, M.W. (1982). Epithelial tonofilaments: investigating their form and function using monoclonal antibodies. Cold Spring Harbor Symp. Quant. Biol. 46, 387-402.

Lazarides, E. (1980). Intermediate filaments as mechanical integrators of cellular space. Nature 283, 249-256.

Lazarides, E. (1982). Intermediate filaments. Annu. Rev. Biochem. 51, 219-250.

Lee, C.S. (1979). Mitochondria and mitochondria-tonofilament -desmosomal associations in the mammary gland secretory epithelium of lactating cows. J. Cell Sci. 38, 125-135.

Letho, V.P., Virtanen, J. and Kurki, P. (1978). Intermediate filaments anchor the nuclei in nuclear monolayers of cultured human fibroblasts. Nature 272, 175-177.

Lilienbaum, A., Legagneux, V., Portier, M.M., Dellagi, K. and Paulin, D. (1986). Vimentin gene: expression in human lymphocytes and in Burkitt's lymphoma cells. EMBO J. 5, 2809-2814.

Lin, J.J.-C. and Feramisco, J.R. (1981). Disruption of the in vivo distribution of the intermediate filaments in living fibroblasts through the microinjection of a specific monoclonal antibody. Cell 24, 185-193.

Madsen, P. and Celis, J.E. (1985). S-phase patterns of cyclin(PCNA) antigen staining resemble topographical patterns of DNA synthesis. A role for cyclin in DNA replication? FEBS Lett. 193, 5-11.

Madsen, P., Nielsen, S. and Celis, J.E. (1986). Monoclonal antibody specific for human proteins IEF 8Z30 and 8Z31 accumulates in the nucleus a few hours after cytoplasmic microinjection of cells expressing these protiens. J. Cell Biol. 103, 2083-2089.

Miyachi, K., Fritzler, M.J. and Tan, E.M. (1978). Autoantibody to a nuclear antigen in proliferating cells. J. Immunol. 121, 2228-2234.

Mose Larsen, P., Bravo, R., Fey, S.J., Small, J.V. and Celis, J.E. (1982). Putative association of mitochondria with a subpopulation of intermediate-sized filaments in cultured human skin fibroblasts. Cell 31, 681-692.

Mose Larsen, P., Fey, S.J., Bravo, R. and Celis, J.E. (1983). Mouse mitochondrial protein IEF 24: Identification and immunohistochemical localization of mitochondria in various tissues. electrophoresis 4, 247-256.

Osborn, M. and Weber, K. (1983). Tumor diagnosis by intermediate filament typing. A novel tool for surgical pathology. Lab. Invest. 48, 372-397.

Paulin, D. Babinet, C., Weber, K. and Osborn, M. (1980). Antibodies as probes of cellular differentiation and cytoskeletal organization in the mouse blastocyt. Exp. Cell Res. 130, 297-304.

Pickett-Heaps, J.D. (1969). The evolution of the mitotic apparatus; an attempt at comparative ultrastructural cytology in dividing cells. Cytobios. 1, 257-280.

Raine, C.S., Ghetti, B. and Shelanski, M.L. (1971). On the association between microtubules and mitochondria within axons. Brain Res. 84, 386-393.

Rogalski, A.A. and Singer, S.J. (1984). Associations of elements of the golgi apparatus with microtubules. J. Cell Biol. 99, 1092-1100.

Small, J.V. and Celis, J.E. (1978). Direct visualization of the 10-nm (100-A)-filament network in whole and enucleated cultured cells. J. Cell Sci. 31, 393-409.

Smith, D.S., Järlfors, U. and Cameron, B.F. (1975). Morphological evidence for the participation of microtubules in axonal transport. Ann. N.Y. Acad. Sci. 253, 472-502.

Smith, D.S., Järlfors, U. and Cayer, M.L. (1977). Structural crossbridges between microtubules and mitochondria in central axons of an insect (periplaneta Americana). J. Cell Sci. 27, 255-272.

Tokuyasu, K.T., Dutton, A.H. and Singer, S.J. (1983) Immunoelectron microscopic studies of desmin (skeletin): localization and intermediate filament organization in chicken skeletal muscle. J. Cell Biol. 96, 1727-1735.

Traub, P. (1985). Intermediate filaments. (Berlin: Springer-Verlag), pp. 1-266.

Traub, P., Perides, G., Scherbarth, A. and Traub, U. (1985). Tenacious binding of lipids to vimentin during its isolation and purification from Ehrlich ascites tumour cells. FEBS Lett. 193, 217-221.

Traub, P., Perides, G., Schimmel, H. and Scherbarth, A. (1986). Interaction in vitro of nonepithelial intermediate filament proteins with total cellular lipids, individual phospholipids, and a phospholipid mixture. J. Biol. Chem. 261, 10558-10568.

Venetianer, A., Schiller, D.L., Magin, T. and Franke, W.W. (1983). Cessation of cytokeratin expression in a rat hepatoma cell line lacking differentiated functions. Nature 305, 730-733.

Wang, E., Connolly, J.A., Kalnins, V.I. and Choppin, P.W. (1979). Relationship between movement and aggregation of centrioles in syncytia and formation of microtubule bundles. Proc. Natl. Acad. Sci. USA 76, 5719-5723.

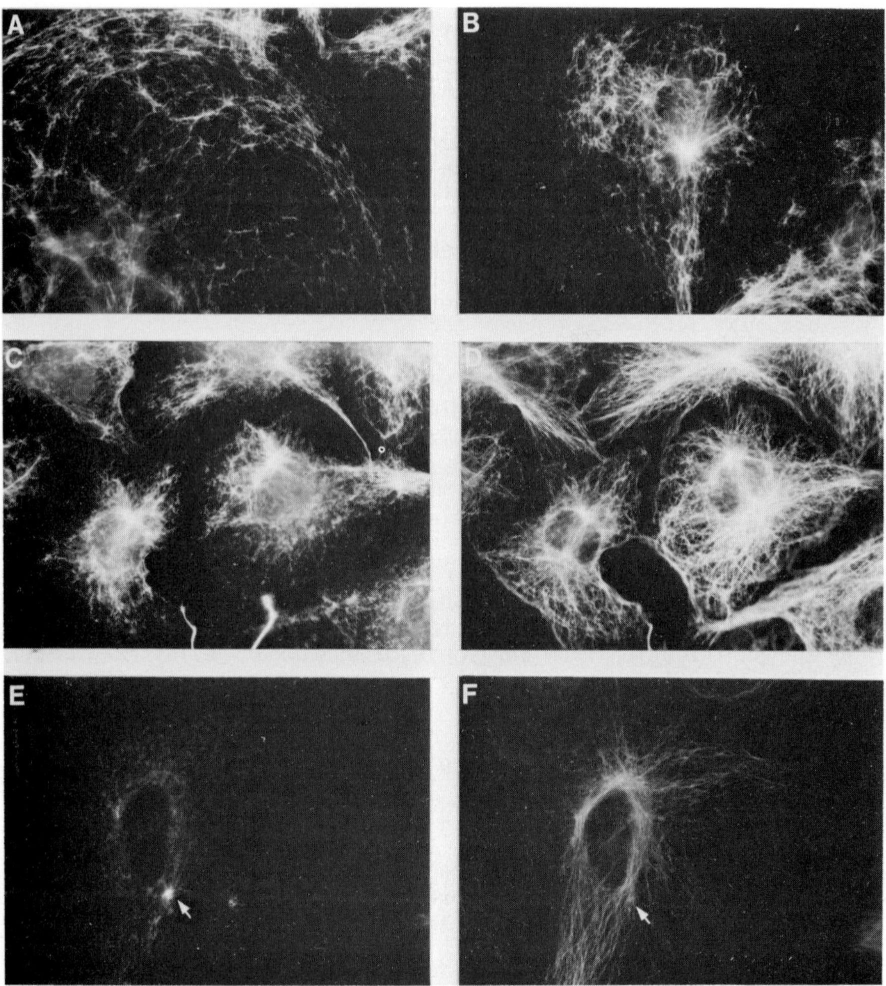

Fig. 1. Indirect immunofluorescence staining of BS-C-1 cells with keratin and tubulin antibodies. (A,B) cells stained with a broad specificity keratin antibody (Fey et al. 1983). (C-F) Double immunofluorescence of BS-C-1 cells reacted with (C,E) keratin IEF 46 antibody and (D,F) tubulin monoclonal antibodies.

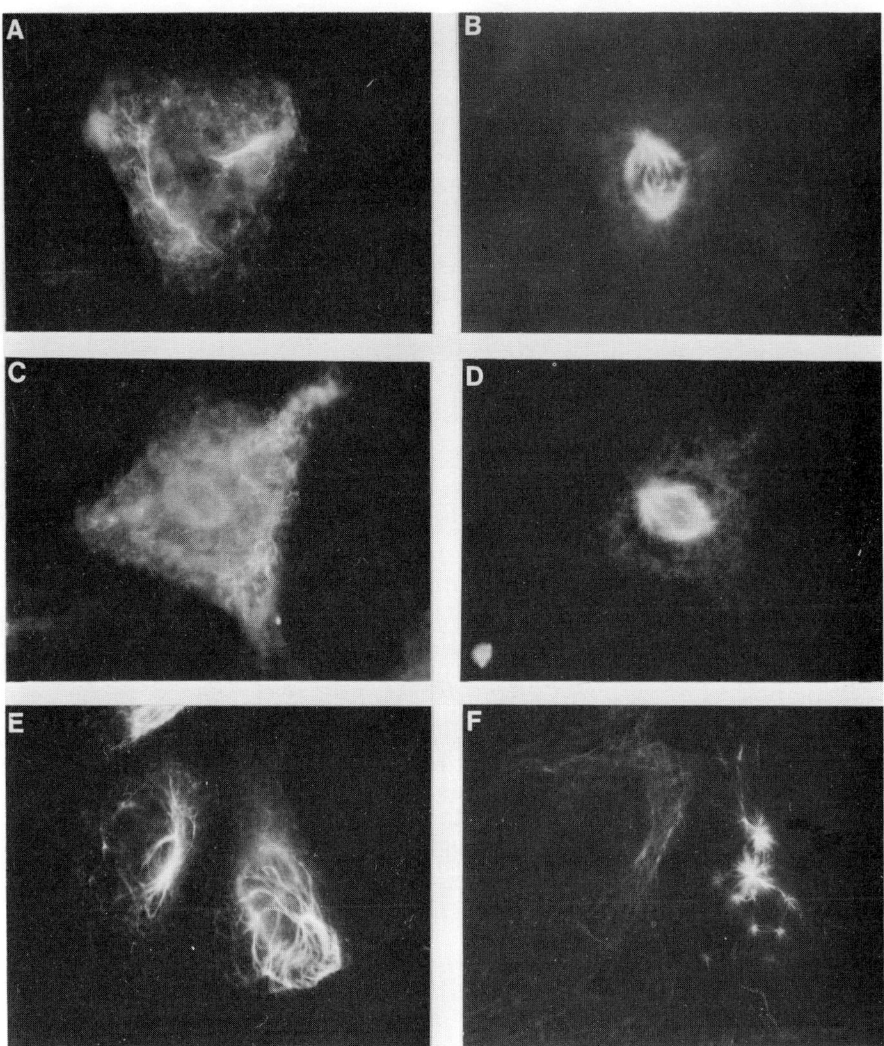

Fig. 2. Double immunofluorescence of BS-C-1 cells reacted with keratin and tubulin antibodies. (A-D) mitotic BS-C-1 cells reacted with (A,C) keratin IEF 46 and (B,D) tubulin antibodies. (E,F) immunofluorescence of nocodazole treated cells (10 µg/ml, 20 hrs, 3 min recovery) reacted with (E) keratin and (F) tubulin antibodies.

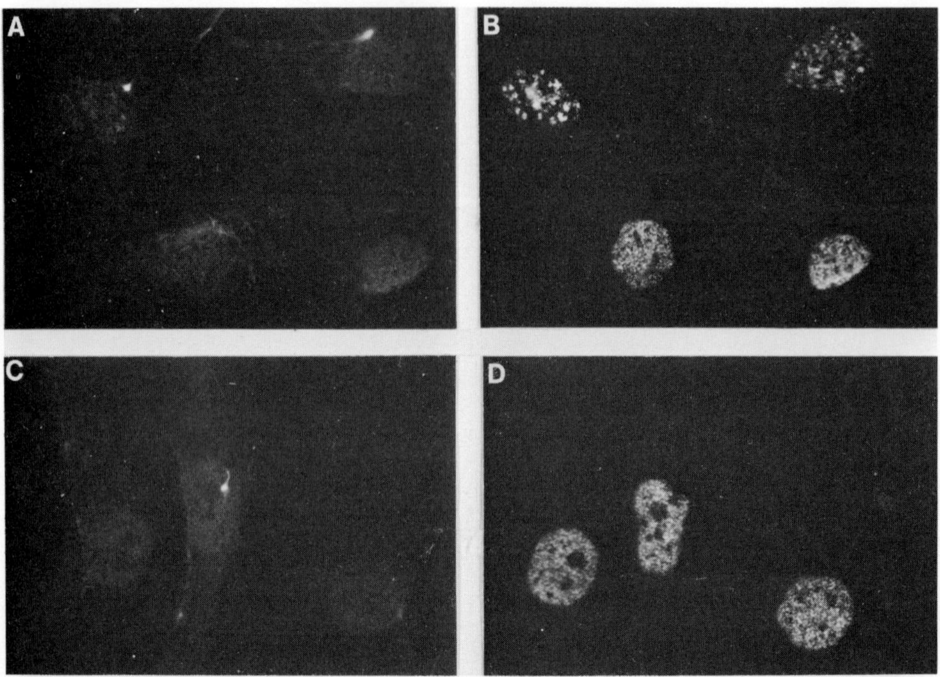

Fig. 3. Double immunofluorescence of BS-C-1 cells reacted with (A,C) keratin IEF 31 and (B,D) PCNA(cyclin) antibodies. Only S-phase cells react with PCNA(cyclin) antibodies (Miyachi et al. 1978; Celis and Celis, 1985). Various nuclear patterns of cyclin(PCNA) antigen distribution are indicated as proposed by Celis and Celis, (1985), and Madsen and Celis (1986).

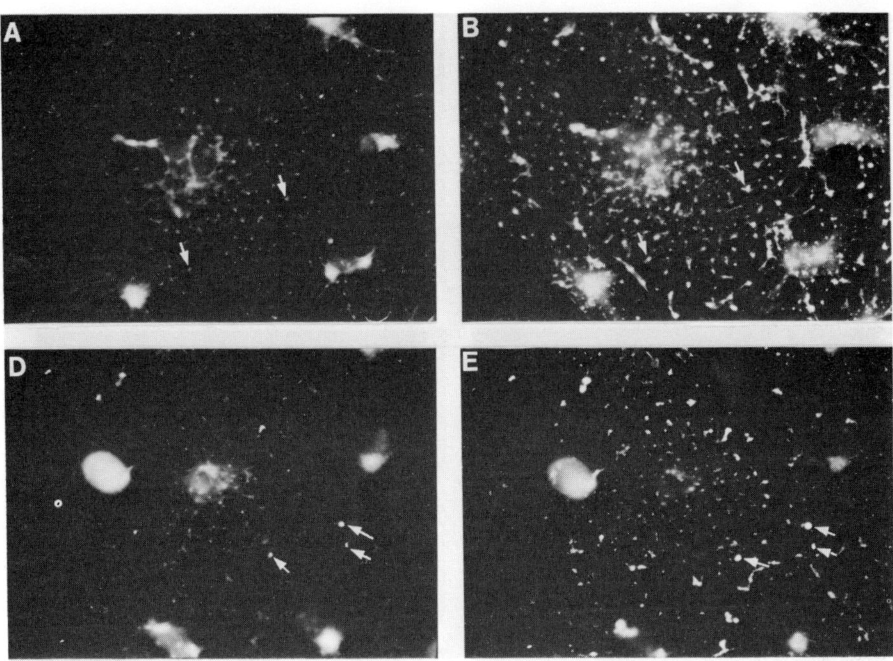

Fig. 4. Double immunofluorescence of TC7 and BS-C-1 cells treated with cytochalasin B (60 min at 37°C). TC7 and BS-C-1 cells fixed in acetone were reacted with keratin (TC7,A), vimentin (BS-C-1,D) and rhodamine labelled phalloidin (B,E). Arrows indicate equivalent points in the micrographs.

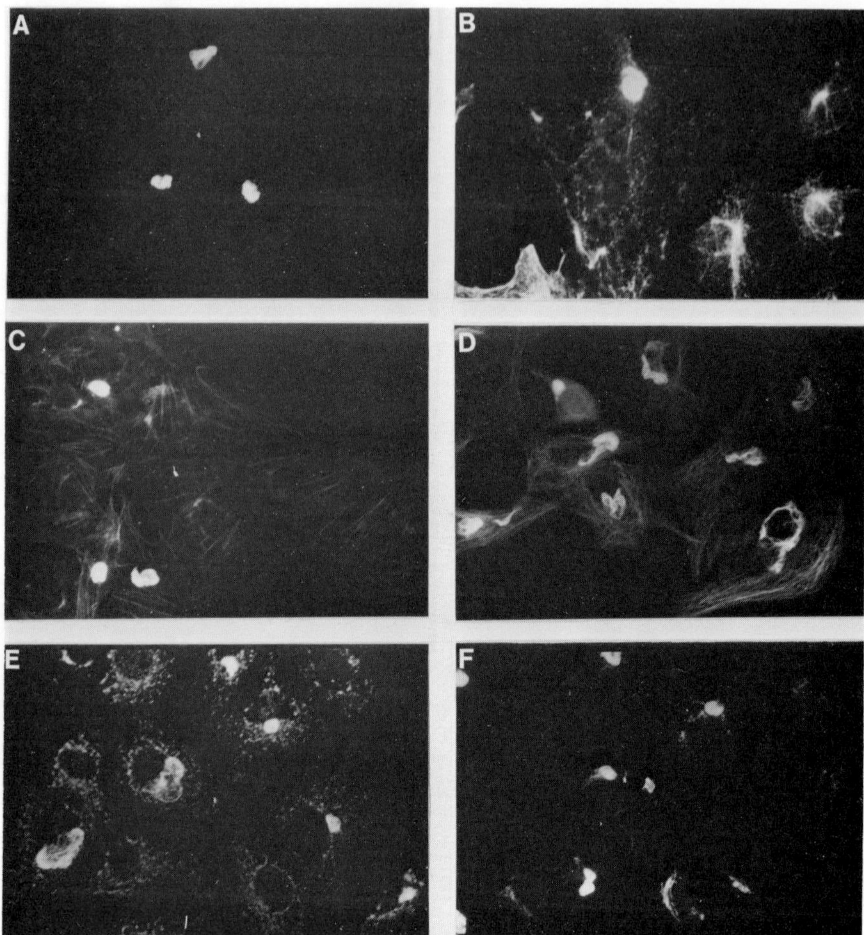

Fig. 5. Indirect immunofluorescence micrographs of BS-C-1 cells microinjeced with vimentin monoclonal antibodies. Cells were microinjected into the cytoplasm and were fixed with methanol or acetone 20-22 hrs after injection (Celis et al. 1986; Madsen et al. 1986). (A) reacted with rhodamine labelled rabbit anti-mouse immunoglobulins. (B) As (A) but reacted with keratin antibodies prior to the rhodamine labelled rabbit antibodies. (C) reacted with rhodamine labelled phalloidin. (D) as (B) but reacted with tubulin antibodies. (E) as (B) but reacted with mitochondria antibodies (Mose Larsen et al. 1982). (F) as (B) but reacted with Golgi antibodies (Celis et al. submitted).

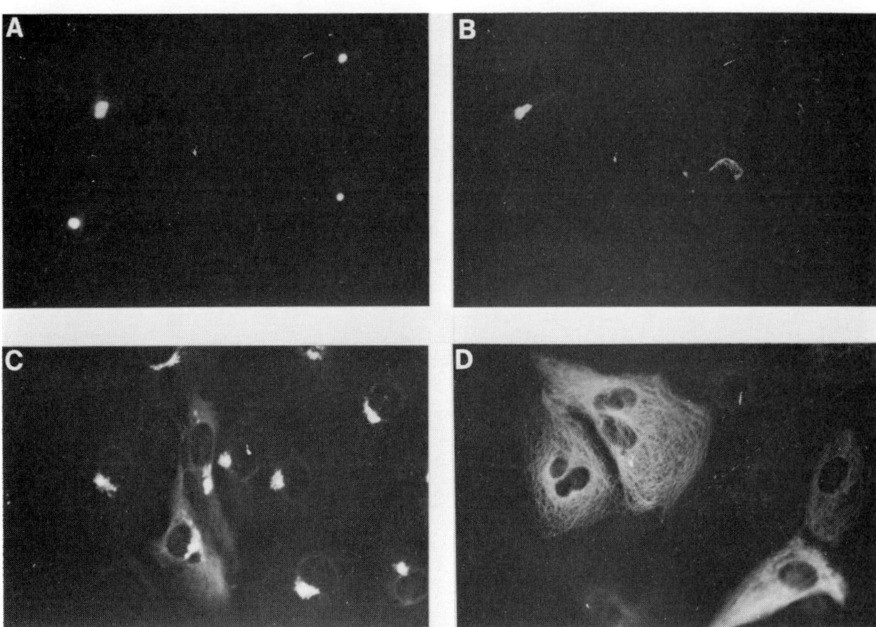

Fig. 6. Indirect immunofluorescence micrographs of BS-C-1 cells microinjected with keratin and tubulin antibodies. Cells were fixed in methanol 20-22 hrs after injection. (A) injected with keratin antibodies and stained with rhodamine labelled rabbit anti-mouse immunoglobulins. (B) as (A) but reacted with vimentin antibodies prior to the rhodamine labelled rabbit antibodies.(C) as (B) but reacted with Golgi antibodies. (D) cells injected with tubulin antibodies and reacted with rhodamine labelled rabbit anti-mouse immunoglobulins.

IN SEARCH OF THE INTERMEDIATE LIKE FILAMENTS IN DROSOPHILA MELANOGASTER

R. Marco, A. Domingo, J. Vinós and M. Cervera.

Instituto de Investigaciones Biomédicas del CSIC and Departamento de Bioquímica UAM Facultad de Medicina. Universidad Autónoma de Madrid. Madrid 28029. SPAIN.

ABSTRACT

Due to the failure in identifying the putative intermediate-like filaments in *Drosophila melanogaster* using antibody crossreactivity at the protein level and hybridization probes at the gene level we have relied on their two major properties, their insolubility and capability of forming filaments under defined *in vitro* polymerization conditions. A set of polypeptides have been purified from extracts of *Drosophila melanogaster* adult homogenates which are capable of polymerizing into filaments with properties reminiscent of these third cytoskeletal elements . Polyclonal antibodies prepared against the major components of this set crossreacted with the whole set of polypeptides even after affinity purification by immunoabsorption. Moreover, the antibodies also recognized myosin, one of the major polypeptides in the Triton X-100 insoluble pellet. This set of polypeptides was present in extracts directly prepared in 4% SDS and in embryonic extracts where myosin is a minor polypeptide, suggesting that they may be present *in vivo*. In addition to nuclear laminas, additional candidates for intermediate-like filaments in *Drosophila* should be found in the high salt insoluble components present in the Triton X-100 extracted sediment.

INTRODUCTION

Intermediate filaments constitute the third element of the celullar cytoskeleton, with distinct morphological and biochemical properties which clearly differentiates them from microtubules and microfilaments (Lazarides, 1980, Traub, 1985). In contrast to these two components of the cytoskeleton, intermediate filaments present a very high diversification in different cell types, being used as differentiation markers and cell typing by medical pathologists (Osborn et al, 1985). Although they were initially described in mammalian cell types, the availability of specific antibodies against them has been used to progressively extend their identification to the majority of vertebrates. In invertebrates, with the exception of neurofilaments, intermediate filaments which were identified taking advantage of the huge nerves present in many invertebrates(Huneeus and Davidson, 1970, Lasek et al, 1979, Eagles et al, 1981), only recently, Bartnick, Osborn and Weber (1985 and 1986) have produced clear evidence of their presence in moluscs and worms. Biessman's group (Falkner et al, 1981, Walter and Biessman, 1984) has described the presence in *Drosophila* Kc cells of a 46,000 polypeptide recognized by a monoclonal antibody which crossreacted with mammalian vimentin, but it was not purified and therefore, never shown to polymerize *in vitro* into electron microscopic intermediate filament structures. Such evidence has not been published yet in *Drosophila melanogaster* or for that matter in the whole range of high invertebrates orders, insects and other arthropods, in spite of the importance of their identification in *Drosophila*, in view of the possibilities of exploiting its genetical and developmental manipulability. In fact Bartnick et al (1985, 1986) conclude that it is even possible that intermediate filaments at least as we know them, may not exist in certain invertebrate orders including *Drosophila*. This is even more a hindrance since the genetic aproach would be very useful in clarifying the function of these cytoskeletal components which remains still a mistery in the whole range of the animal kingdom (Franke, 1987, Fraser et al, 1987). As emphasized by Franke in a recent review (1987), this conclusion cast doubts about assigning a key housekeeping role to a highly insoluble scaffolding cellular framework of the type provided by intermediate filaments. This would be so unless an analogous set of proteins with equivalent properties were found in these organisms.In this article we review our work trying to identify their presence in *Drosophila melanogaster* which indicates the difficulties in finding them.

THE DROSOPHILA MELANOGASTER INTERMEDIATE FILAMENTS

A few years ago, interested in the possible role of these differentation specific components in development, we started to work to identify their presence in the fruit fly. The already published evidence indicated that the degree of homology at the level of gene nucleotide sequence was too low to be used to detect the presence of genes equivalent to the ones cloned in mammals, like desmin, vimentin or cytokeratins (Fuchs and Marchuk, 1983, Quax et al, 1984). We tried to use the probable higher homology at the aminoacid sequence level, taking advantage of the availability of a large repertoire of poly and monoclonal antibodies against every type of mammalian intermediate filament to fish out from the set of insoluble polypeptides present in the pellet from *Drosophila* adult, larval and embryos homogenates the potential candidates more likely to be members of the intermediate filament family in the fruit fly. Our results with this approach have been consistently negative, in agreement with published and unpublished work from other groups (Bartnick et al, 1985, Karr and Alberts, 1986).

A SET OF FILAMENT FORMING POLYPEPTIDES FROM DROSOPHILA MELANOGASTER FLIES.

Since there was the possibility that Drosophila intermediate filaments in fact did exist but may have evolved too much to be detected by these approaches, we decided to proceed with the goal of purifying the more likely candidates, relying for their isolation on their two major properties, namely, their insolubility and capability to form intermediate filament structures in the electron microscope upon polymerization. A complication arises from the tiny size of Drosophila melanogaster, which makes impossible in practice to start a purification protocole with a single tissue as Bartnick et al.(1985) did in the case of Helix pomatia esophagus.In Figure 1, a chart flow summarizing the purification procedure is presented. Homogenates from whole adult flies were prepared in a Triton X-100 containing buffer, the insoluble fraction was recovered by centrifugation and treated with an 8 M Urea solution to extract in soluble form the putative intermediate filament-like components. A set of distinct polypeptides remains in the Triton extracted pellet. The majority of them are solubilized by the urea treatment and precipitate back again as soon as they are dialyzed in the conditions described for polymerization of intermediate filaments. Several solubilization cycles mantained the basic set of polypeptides in this fraction. Nevertheless, two arguments opposed the conclusion that these polypeptides were indeed the intermediate filaments components. First, it was conspicuous that the two major polypeptides present in these solubilized or insolubilized fractions were actin and myosin, obviously not likely candidates for the role of intermediate filaments. Second, the electron microscopic inspection of these polymerized samples failed repeatedly to discovered anything more than huge aggregates in which only rarely very small and thin filaments were visible. Using mammalian cytokeratins we tried to detect whether there were intermediate filaments polymerization inhibiting factors in this relatively crude preparation, with unclear results.

In spite of these negative results, since it was possible that intermediate filaments were a minor part of this insoluble fraction, we decided to proceed with our assumption and apply a protein purification scheme to our urea solubilized pellet as summarized in Figure 1. To handle these polypeptides in a soluble form, high concentrations of urea or guanidine HCl were included in the different chromatographic steps. Gel filtration in the presence of urea failed to resolve the different polypeptides, suggesting that the polypeptides present in this fraction were still capable of interactions even in the presence or urea. On the other hand, gel filtration of the proteins solubilized by SDS, gave a good resolution in gel filtration, but our initial trials to obtain fine polymerized filaments from the separated fractions were inconclusive and hampered by the possible polymerization inhibitory effect of the remaining quantities of SDS, although in principle SDS was removed from the preparation.

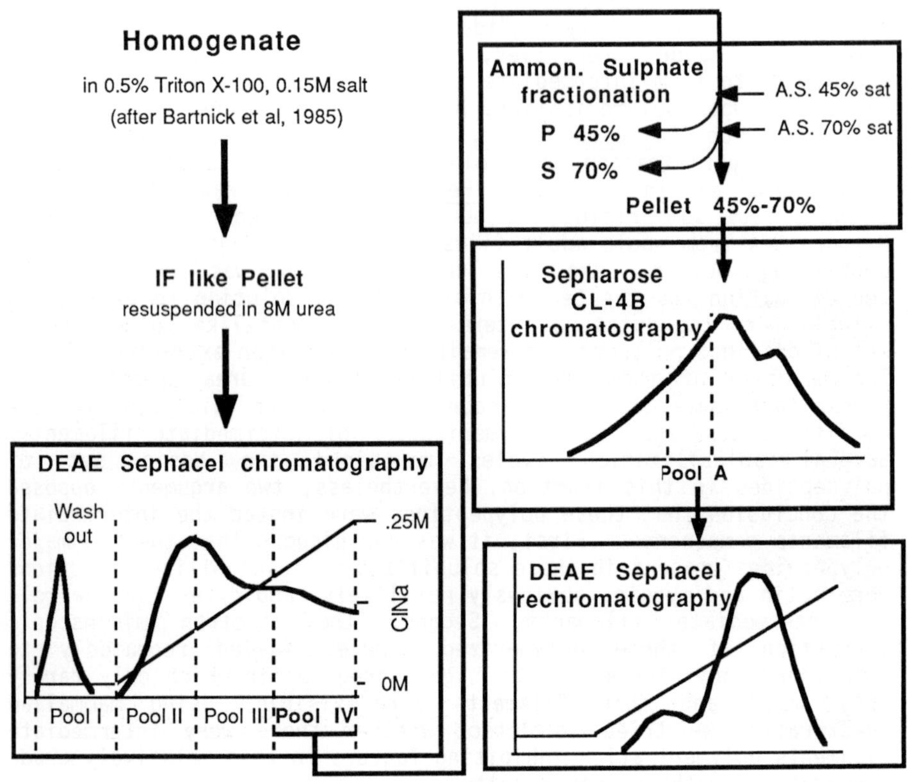

Fig.1.-Purification chart of the Drosophila intermediate filament like set of polypeptides.

On the other hand, the different polypeptides in this fraction could be fractionated in DEAE chromatography in the presence of 8 M urea. Although actin and myosin still tended to impurify every fraction in the column, when different pools from the column were polymerized, fractions marked as pol IV in Figure 1 consistenly gave a higher quantity of thin filaments (Cervera et al.,1987), which tended to arrange themselves in longer fibers. At this stage, it occurred to us that the old technique of salting out could well work even in the presence of urea .

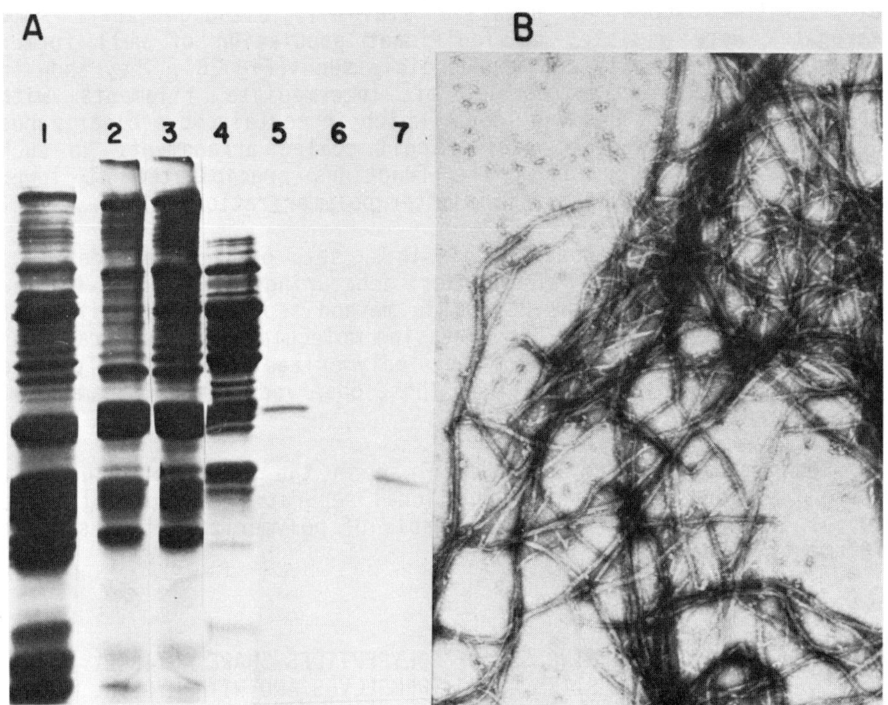

Fig. 2.- Salting out treatment of pool IV from DEAE-Sephacel chromatography. A) Protein profiles. Lane 1 corresponds to pool IV. Fractions precipitated by ammonium sulphate were solubilized in 8 M urea and polymerized by dialysis. 0-35% ammonium sulphate cut: lane 2 (polymerized fraction), lane 5 (soluble fraction): 34-35% ammonium sulphate cut: lane 3 (polymerized fraction), lane 6 (soluble fraction): 45-70% ammonium sulphate cut: lane 4 (polymerized fraction) and lane 7 (soluble fraction). B: Electron microscopy of the 45-70% ammonium sulphate cut polymerized fraction (lane 4).
Bar. 0.5u.

In Figure 2A, we present the results of such a fractionation. Some polypeptides of pool IV, which included the majority of actin and myosin salted out when treated with 35% ammonium sulphate.In the supernatant of 45% ammonium sulphate several polypeptides appear enriched, which salted out when treated by a 70% cut. When this last fraction was solubilized in 8 M Urea, polymerized, and viewed in the electron microscope by negative staining, although still some agregates were visible, a significant population of well formed filaments were clearly and reproducibly seen (Fig 2B). They show an overall structure reminiscent of intermediate filaments with continuous smoothly bending loops, which in certain more clear zones show structures suggestive of a coil coiled arrangment. No such filaments were observed in the fractions precipitated at lower ammonium sulphate concentrations after polymerization.

Additional fractionation of this set of polypeptides was achieved by gel filtration under denaturing conditions in the presence of 7 M guanidine HCl. This method separated the different polypeptides in a ladder of decreasing molecular weights. Fractions from Mr 120,000 to 45,000, when polymerized, consistently gave filaments, cleaner but similar to those observed in the 70% ammonium sulphate cut.

Rechromatography in DEAE cellulose in the presence of urea of these fractions, resulted in additional separation of some of these polypeptides, which were still capable of polymerizing into similar filaments (Figure 3).

THE FILAMENT FORMING SET OF POLYPEPTIDES SHARE ANTIBODY CROSSREACTING SITES AMONG THEMSELVES AND WITH MYOSIN.

The different polypeptides were futher purified by polyacrylamide gel electrophoresis in the presence of SDS, stained with Coomassie Blue, sliced and injected into rabbits after mixing with Freund adjuvant, complete in the first injection and incomplete in three boosts everyother week. After this immunization schedule, the rabbits were bled and the antisera tested against the set of filament forming polypeptides. Interestingly, the different antibodies recognized the whole set of polypeptides, even after immunoabsorption and elution from the different bands indicating that the set of polypeptides shared the same antigenic groups.

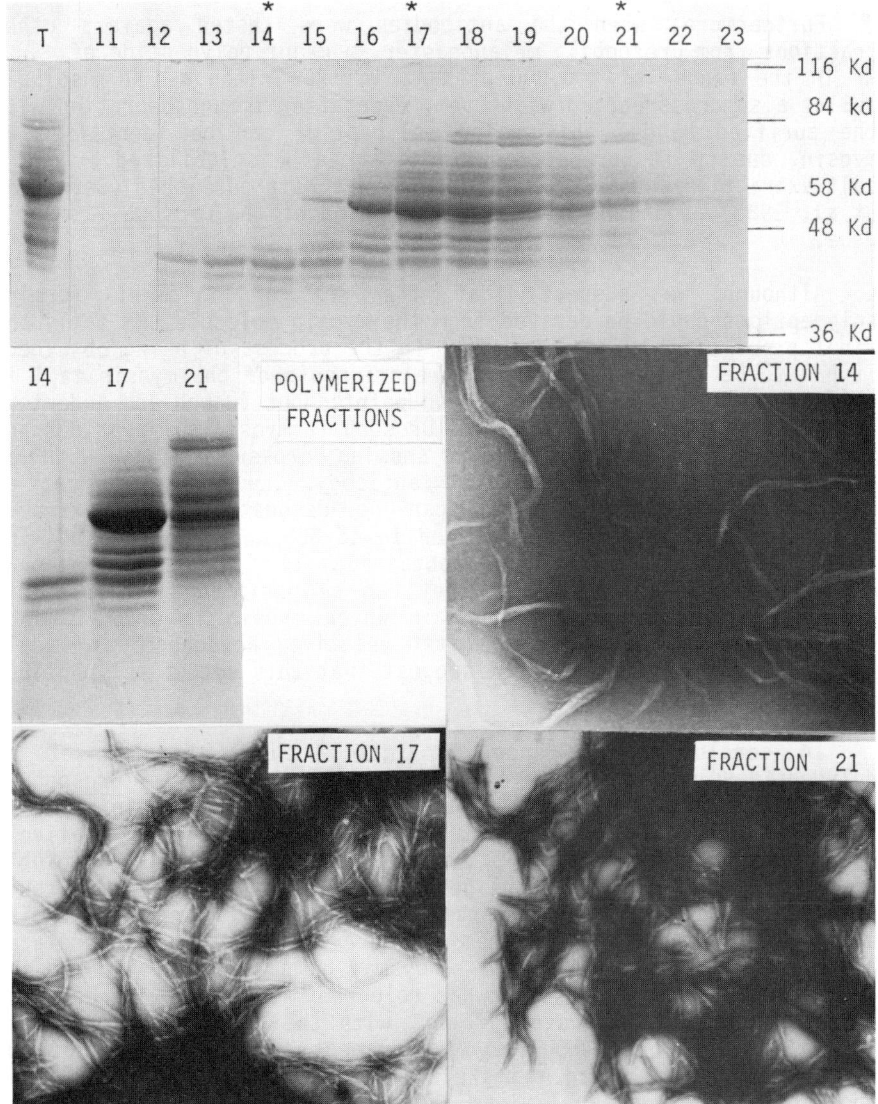

Fig. 3. DEAE- Sephacel rechromatography in the presence of 8 M Urea of the 95,000 to 45,000 Mr fraction purified by gel filtration in the presence of guanidine- HCL. In the upper part of the figure the protein profile and in the lower part the electron microscopy of the fractions indicated is shown.

Furthermore, when the antibodies were tested against other fractions from Drosophila melanogaster, a major polypeptide of Mr in the range of 200,000 present in the Triton X-100 insoluble pellet also crossreacted with them, even after immunoabsorption with the purified bands. This major polypeptide can be identified as myosin, due to its properties,since it can be solubilized by 0.6 M salt extraction and by comparison with the data published (Bernstein et al, 1983) about the cloning and mapping of the Drosophila myosin gene.

Although we suspect that the set of filament forming polypeptides could be derived from the myosin molecule LMM tail, the final proof of this idea is still in the process of being obtained. It has to be remembered that the relationship of the myosin tail to the intermediate filaments has been pointed out (Haugh and Anderton, 1983). In fact, in our hands, the Drosophila myosin polypeptide band has been the main polypeptide showing crossreactivity with an anti-mammalian desmin monoclonal antibody. Moreover, the set of filament forming polypeptides can be demonstrated in extracts obtained by directly homogenization in 4% SDS and immediate boiling in similar quantities to those obtained after a more conventional extraction procedure. In fact they can be purified from embryonic extracts, a developmental stage in which myosin is only a minor polypeptide in contrast to its relative abundance in adult homogenates. All this evidence suggest that this set of polypeptides could well exist in vivo.

In conclusion, we show in this article that a set of polypeptides related to myosin can be purified from Drosophila melanogaster adult homogenates which have properties reminiscent to these described for intermediate filaments. They are a relatively minor component of the fly, even in the Triton X-100 insoluble fraction, and require considerable purification before their polymerization properties can be satisfactorily demonstrated. The antibodies prepared against them indicate that they are crossrelated among themselves and to myosin. They may be present in vivo and the intriguing questions arise of what role they may be playing and what relationship they may have, if any, with the putative intermediate like filaments in Drosophila. With respect to the question of the presence of these third cytoskeletal components in Drosophila, in addition to the nuclear laminas (Fisher, et al 1982) recently shown to be able of polymerizing into intermediate-tipe filaments (Aebi et al 1986), further candidates for this role may be found in the Triton X-100 and salt insoluble sediment, besides the set of polypeptides reported here. Testing this possibility is currently one of our major research goals. In this respect, it is remarkable that the set of polypeptides reported here remains insoluble even in high salt. This makes more difficult their removal from the fractions which may include other candidates for the putative intermediate filaments in Drosophila.

ACKNOWLEDGMENTS

The finantial support of the F.I.S.S. and the C.S.I.C. made possible this work. The suggestions and critical reading of the manuscript by Dr.J. Avila is gratefully acknowdledged.

REFERENCES

Aebi,U., Cohn,J., Buhle,L. and Gerace, L. (1986) The nuclear lamina is a meshwork of intermediate-type filament Nature 323, 560-564.

Bartnik, E., Osborn, M. and Weber, K. (1985) Intermediate filaments in non-neuronal cells of Invertebrates: Isolation and biochemical characterization of intermediate filaments from the esophageal epithelium of the molusc Helix pomatia. J. Cell Biol. 101: 427-440.

Bartnik, E., Osborn, M. and Weber, K. (1986) Intermediate filaments in muscle and epithelial cells of Nematodes. J. Cell. Biol. 102: 2033-2041.

Bernstein, S.I., Mogami, K., Donady, J.J. and Emerson,C.P. (1983). Drosophila muscle myosin heavy chain encoded by a single gene in a cluster of muscle mutations. Nature 302, 393-397.

Cervera, M., Domingo, A., Vinõs, J. and Marco, R. (1987). Drosophila melanogaster contains a set of polypeptides capable of polymerizing into intermediate-like filaments. Biochem. Biophys. Res. Commun. 144, 1043-1048.

Eagles, P.A.M., Gilbert, D.S. and Meggs, A. (1981) The polypeptide composition of axoplasm and neurofilaments from the marine worm Myxicola infundibulum. Biochem J. 199: 89-100.

Falkner, F. -G., Saumweber, H. and Biessmann, H. (1981). Two Drosophila melanogaster proteins related to intermediate filament proteins of vertebrate cells. J.Cell Biol. 91: 175-183.

Fisher, P.A., Berrios, M. and Blobel, G. (1982). Isolation and characterization of a proteinaceous subnuclear fraction composed of nuclear matrix,peripheral lamina and nuclear pore complexes from embryos of Drosophila melanogaster. J. Cell Biol. 92, 674-686.

Franke, W. (1987). Nuclear lamins and cytoplasmic intermediate filaments proteins: A growing multigene family. Cell 48, 3-4.

Fraser, R.D.B., Steinert, P.M. and Steven, A.C. (1987). Focus on intermediate filaments. Trends in Biochem. Sci. 12, 43-45.

Fuchs, E. and Marchuk, D. (1983). Type I and type II keratins have evolved from lower eukaryotes to form the epidermal intermediate filaments in mammalian skin. Proc. Nat. Acad. Sci. U.S.A. 80: 5857-5861.

Haugh, M. and Anderton, B.(1983) Relationships of filaments. Nature 303, 21.

Huneeus, F.C. and Davison, P.F. (1970) Fibrillar proteins from squid axons. I. Neurofilament protein. J.Mol.Biol. 52: 415-428.

Karr, T.L. and Alberts, B.M. (1986). Organization of the cytoskeleton in early Drosophila embryos. J.Cell Biol. 102: 1494-1509.

Laemmli, U.K. (1970) Cleavage of structural proteins during the asembly of the head of bacteriophage T4. Nature, 277, 680-685.

Lasek, R.J., Krishman, N. and Kaiserman-Abramof, I.R. (1979). Identification of the subunit proteins of 10-nm neurofilaments isolated from axoplasma of squid and Myxicola giant axons. J. Cell Biol. 82: 336-346.

Lazarides, E. (1980) Intermediate filaments as mechanical integrators of cellular space. Nature, 283: 249-256.

Osborn, M. Altamannsberger, M., Debus, E. and Weber, K. (1985). Differentiation of the major human tumors groups using conventional and monoclonal antibodies specific for individual intermediate filament proteins. In Intermediate filaments (ed. Wang, E., Fischman D., Liem, R.K. and Sun, T-T) pp 649-668 New York Academy of Sciences, N.Y.

Quax, W. Van der Heuvel, R., Egberts, W.V. Quax-Jeuken Y. and Bloemendal, H. (1984). Intermediate filaments cDNAs from BHK21 cells: Demonstration of distinct genes for desmin and vimentin in all vertebrate classes. Proc. Nat, Acad. Sci. U.S.A. 81: 5970-5974.

Traub, P. (1985) Intermediate filaments, A. review. Springer-Verlag. Berlin-Heidelberg.

Walter, M.F. and Biessmann, H. (1984). Intermediate-sized filaments in Drosophila tissue culture cells. J.Cell Biol. 99: 1468-1477.

EXPRESSION OF GENES CODING FOR INTERMEDIATE FILAMENT PROTEINS IN MOUSE TERATOCARCINOMA CELLS : A MODEL FOR EMBRYONIC DEVELOPMENT

D. Paulin, A. Lilienbaum and H. Jakob

Unité de Génétique cellulaire du Collège de France et de l'Institut Pasteur, Université Paris 7, 25 rue du Dr. Roux, 75724 Paris Cedex 15

ABSTRACT

Embryonal carcinoma cells (EC) and differentiated derivatives grown in tissue culture have been used as a model to study regulation of cytoskeleton proteins. The distribution of different intermediate filament types have been examined with specific antibodies able to distinguish vimentin, desmin, keratin, GFA and neurofilament polypeptides. Twelve EC lines have been shown to express vimentin when cultured as monolayers. One species of vimentin mRNA of 2 Kb can be characterized with a cDNA probe in EC cells (PCC3, PCC4, PCC7 and F9). *In vitro* differentiation of EC cells can be induced by growth in the form of cell aggregates, confluency, or induced with drugs (retinoic acid, HMBA, etc.). In terminal differentiation *in vitro*, the final steps appear to resemble closely the observations during normal embryonic development.

INTRODUCTION

Mouse teratocarcinomas are malignant tumors that are characterized by the presence of a variety of differentiated cell types including derivatives of all three primary germ layers and a distinctive cell type known as "embryonal carcinoma". The latter cells are the stem cells of the tumor. A single embryonal carcinoma cell can give rise to all the differentiated cell types that are observed in the tumors. These multipotent indifferentiated embryonal carcinoma cells are responsible for the malignancy, the progressive growth and transplantability. Teratocarcinomas are readily induced by transplanting a normal young embryo to an extra-uterine site or by grafting parts of a 6-day embryo. Embryo derived tumors can also be obtained spontaneously (Stevens and Little, 1954 ; Pierce et al., 1970). It is not known why teratocarcinoma cells continue to proliferate in an indifferentiated state and in other instances undergo differentiation. Embryo derived cells (EK) can also be obtained by direct culture of the embryo.

Tumors can be transmitted as solid form but ascitic conversion can also be achieved by injecting teratocarcinoma intraperitoneally. The stem cells proliferate in suspension in the ascitic fluid, can differentiate and form structures known as embryoid bodies (Martin and Evans, 1975).

Several groups have now established embryonal carcinoma cell lines (Jakob et al., 1973). Some of them are able to differentiate *in vitro*. The cell types that were identified included keratinizing, epithelium, cartilage, striated muscle, neuronal and endodermal cells. Because the formation of different types of intermediate filaments is related to cell differentiation, the teratocarcinoma cells offer a model of considerable interest to study the molecular aspect of the differentiation processes. The main objective is to assess the extent to which these cells are useful as a model to study normal development. A number of questions arise : (1) what changes occur in embryonic cells to become embryonal carcinoma cells able to grow *in vitro* ? (2) Are the changes reversible or not ? (3) Are the differentiated cell lines established *in vitro* from the tumor identical to the cells of the adult tissues ? (4) When EC cells differentiate either spontaneously or are induced after various treatments, are the different stages valid models for differentiation studies ? (5) Is the inductive treatment influencing the direction of differentiation ? Some of these questions could be addressed by following intermediate filament expression (see references in Traub, 1985).

The cytoskeleton of vertebrate organisms contains various ubiquitous elements such as microfilaments and microtubules. In addition, the same cells may also contain other cytoskeletal elements including intermediate filaments. Intermediate filaments represent a category of structures which are formed in different cell types by different protein constituents. Indistinguishable by electron microscopy and X-ray, these structures are characterized by immunological and biochemical criteria. Today five classes of intermediate filaments are well characterized. Desmin is typical of skeletal, cardiac and visceral smooth muscles as well as certain vascular smooth muscle cells. Cytokeratins include 20 proteins related but not identical to α-keratin from epidermis. They were found in all "true" epithelial cells and their expression is often concomitant with the formation of typical desmosome.

Neurofilaments which take their name from their ubiquitous neuronal distribution consist in a triplet of polypeptides from 65,000 to 70,000 daltons, 140,000 to 160,000 daltons and 200,000 to 210,000 daltons.

GFA, glial fibrillary acidic protein is an intermediate filament of 50,000 to 55,000 daltons present in astrocytes.

Vimentin, a protein of 52,000 to 58,000 daltons has a widespread distribution in mesenchymal derivatives and established cell lines.

INITIAL EXPRESSION OF IF IN THE MOUSE EMBRYO

The formation of different types of intermediate filaments is related to cell differentiation and therefore their order of appearance during embryogenesis is of obvious importance.

Mouse embryos at an early preimplantation stage lack detectable IFs (Jackson et al., 1980 ; Franke et al., 1982). Cytokeratins are the first to appear, concomitantly with cellular differentiation in the outer cells of the blastocyst (Lehtonen and Badley, 1980 ; Paulin et al., 1980 ; Jackson et al., 1981 ; Kemler et al., 1981 ; Lehtonen et al., 1983 ; Duprey et al., 1985). After implantation, co-expression of vimentin and cytokeratin occurs in some cells of the parietal endoderm (Lehtonen et al., 1983). Following the primitive streak stage, vimentin is to be found in primary mesenchymal cells (Franke et al., 1982). Neurofilaments appear early in development and progressively replace vimentin which is expressed before NF in most dividing neuroepithelial cells (Cochard and Paulin, 1984).

VIMENTIN IS PRESENT IN EC CELL LINES

We have characterized twelve distinct EC lines of different origins with specific antibodies against actin, tubulin and for the five intermediate filament classes. EC cells grown in culture are rounded with some but not much cytoplasm. They stick strongly to each other. After fixation with methanol and incubation with antibodies to actin, specific fluorescence is found in surface structures but no filament bundles are revealed. In the same conditions, the inner cells of the blastocyst show diffuse fluorescence and dot-like structures but staining reveals no filament bundles either. These inner cells of the embryo and EC cells do not have actin-cables but microtubules are always present (Paulin et al., 1978, 1979).

We used monoclonal or polyclonal antibodies against each of the five IF classes to determine the type of IF they contain. All EC cell lines tested contained vimentin. Filaments somewhat differently arranged appear in various lines. The number of positive cells varies from 50 to 100 % although all the lines are cloned. Two dimensional gel electrophoresis provides biochemical evidence that the EC lines do express vimentin during growth culture. *In vivo*, however, the earlier cells present in the young embryo do not show expression of any of the intermediate filament classes. Vimentin identification has been obtained only from day 8, in the mesoderm layer. Considering the result obtained, i.e. vimentin expressed, we can conclude that EC cells differ from their progenitors. All the EC cells studied are maintained in culture about 20 generations or more. Thus, it is tempting to conclude that the vimentin expression of EC cells could be due to growth *in vitro*.

IS VIMENTIN GROWTH REGULATED IN EC CELLS ?

However, the question whether there is a relationship between the synthesis of vimentin and the growth condition has to be solved. In this view, EC cells were cultured in different ways : 1) *in vivo* injected into intraperitoneal cavity to give rise to ascites ; 2) *in vitro* in the form of aggregate ; 3) in monolayer.

Ascitic fluid obtained after injection of PCC3 cells contains aggregates which maintain the synthesis of vimentin. This result contrasts with that obtained with embryoid bodies passaged *in vivo* in ascitic fluid, from the original tumor without cultures *in vitro*.

When cells are grown *in vitro* as aggregates, they have the same morphology as the original cells : no actin cables are visible but actin is distributed diffusely over the cells in microvillosities and ruffles. Change in the vimentin synthesis was observed after three days of growth in aggregates, vimentin synthesis was switched off in more than 30 % of the cells. These negative cells are located in the centre of the aggregates (see Fig. 1).

To determine whether vimentin synthesis is also controlled at the transcriptional level, poly A+ mRNA were isolated from the two types of culture. To characterize vimentin mRNA, we use a cDNA clone which includes 90 % of the coding sequence (Lilienbaum et al., 1986).

Typical results are shown on Fig. 1 : 1) there is only one species of vimentin RNA of 2 kb length produced in mouse embryonal EC cells, PCC3, PCC4, PCC7-S, F9 and in fibroblastic cells ; 2) we found a good correlation between the amount of vimentin RNA detected and vimentin polypeptide synthesized ; 3) for the PCC3 cells cultured as aggregate *in vitro* compared with those cultured as monolayer, a fall of 30 % is observed.

Correlation with cellular C-myc, cfos and SV40 T antigen, oncogenes were also studied (Rüther et al., 1985 ; Kellermann and Kelly, 1986). F9 cells with a C-myc gene integrated in the DNA were selected after transfection with a plasmid including the metallothionin promoter. Cultures in presence of cadmium induce the myc transcription. No differences for the amount of vimentin mRNA were found, indicating no direct correlation between the C-myc expression and vimentin synthesis (C. Crémisi and R. Onclercq, personal communication).

DIFFERENTIATION OF EC LINES LEADS TO CELLS EXPRESSING TISSUE-SPECIFIC IF.

The switching off of vimentin could be obtained when EC cells are induced to differentiate in neuronal cells or myotubes. When the teratocarcinoma C175 cells fuse and form myotubes, reaction with the vimentin serum was lost. One week after myotube formation, the myotubes stained positively with the desmin antibody, but did not stain with the vimentin or with the other IF serums. Thus on fusion, an apparent switch occurs from expression of the vimentin type to expression of the desmin type of IF.

Various EC cell lines give rise to neural derivatives (Jones-Villeneuve et al., 1982 ; Liesi et al., 1983 ; Rechardt et al., 1984; Wartiovaara et al., 1984 ; Wartiovaara and Rechardt, 1985).

C175-1-1003 (McBurney, 1976) is a multipotential embryonal carcinoma clonal cell line which can be induced to follow different developmental pathways by altering the composition of the culture medium (Darmon et al., 1982). In serum free medium, cells differentiate into neurons through a stage of preneurons containing both vimentin and 70 k NF. Fully differentiated neurons contain 70 k NF only but no vimentin. These cells display electrically excitability properties.

In vitro differentiation of PCC3/A/1 cells can be induced by growth in the form of cell aggregates (Nicolas et al., 1981). This system consists in growing the cells as aggregate for 4 days and then replating the aggregates under conditions where they can attach to the dish. Processes extended many micrometers in length were detected and stained with tubulin antibody.

PCC7-1009 cell line gives rise *in vitro* either spontaneously or after treatment with retinoic acid to neuronal derivatives (Fellous et al., 1978 ; Pfeiffer et al., 1981 ; Levy et al., 1982). By growth to confluency or by plating cells after 4 days of aggregation culture, outgrowth of neurite-like processes can be observed. Cholinergic transmitters can be characterized in such cultures (Paulin et al., 1982).

When treated with retinoic acid in the presence of dibutyryl cAMP, the 1009 cells give rise to several types of differentiated cells which can be distinguished by the use of the appropriate anti-IF serum. A summary of the results obtained is given in Table I. The proportion of positive cells for neurofilament or GFA or vimentin differ according to the treatment, the time of treatment and the line. Aggregation alone promotes differentiation but retinoic acid treatment accelerates the process. The presence of cell populations containing either NF or GFA indicates that starting with EC cells, completely terminally differentiated and mature neuronal and glial derivatives can be obtained.

Cells expressing more than one IF type have also been noted. The co-expression patterns which always involve vimentin as one member have also been observed *in situ* during embryonic development.

Vimentin positive neuroblasts acquire neurofilament expression and at least transiently both systems co-exist ; however, further neuron maturation leads to a loss of vimentin and only neurofilaments are expressed in most neurons.

CONCLUSIVE REMARKS

Two further points seem of general interest :

1) The inductive treatment to which EC cells are subjected influences not only the type of IF expression, the direction of differentiation, but also the number of cells displaying each IF type.

2) As expected from many previous studies of *in vitro* differentiation of embryonal carcinoma cells, the type of differentiated cell obtained is strongly influenced by the nature of the EC line used.

Our results provide further support for the idea that teratocarcinoma cells are in general valid models of differentiation.

ACKNOWLEDGEMENTS

We thank François Jacob in whose laboratory this work was done for encouragement and discussion. We are indebted to C. Crémisi and R. Onclercq with whom the experiments reported on Fig. 1 were done. We are grateful to G. Houzet and G. Merle for help in the preparation of the manuscript.

This work was supported by grants from the CNES (87.CENS.1224), the CNRS (UA 1148), the Fondation pour la Recherche Médicale, the Ligue Nationale Française contre le Cancer, the MRES (86.C.0945), the Fondation André Meyer and the March of Dimes.

REFERENCES

Cochard, P. and Paulin, D. (1984). Initial expression of neurofilaments and vimentin in the central and peripheral nervous system of the mouse embryo *in vivo*. J.Neurosciences 4, 2080-2094.

Darmon, M., Buc-Caron, M.H., Paulin, D. and Jacob, F. (1982) Control by the extracellular environment of differentiation pathways in 1003 embryonal carcinoma cells : study at the level of specific intermediate filaments. The EMBO J. 1, 901-906.

Duprey, P., Morello, D., Vasseur, M., Babinet, C., Condamine, H., Brûlet, P. and Jacob, F. (1985). Expresssion of the cytokeratin endo A gene during early mouse embryogenesis. Proc.Natl.Acad. Sci.USA 82, 8535-8539.

Fellous, M., Günther, E., Kemler, R., Wiels, J., Berger, R., Guénet,

tron microscopy and X-ray microprobe analysis in detection of acetylcholinesterase in cultured embryonal carcinoma cells. J. Histochem.Cytochem. 32, 1154-1158.

Rüther, U., Wagner, E.F. and Müller, R. (1985). Analysis of the differentiation-promoting potential of inducible c-*fos* genes introduced into embryonal carcinoma cells. The EMBO J. 4, 1775-1781.

Stevens, L.C. and Little, C.C. (1954). Spontaneous testicular teratocarcinomas in an inbred strain of mice. Proc.Natl.Acad.Sci.USA 40, 1080-1087.

Traub, P. (1985). Intermediate filaments. Springer - Verlag Berlin.

Wartiovaara, J., Liesi, P. and Rechardt, L. (1984). Neuronal differentiation in F9 embryonal carcinoma cells. Cell Differentiation 15, 125-128.

Wartiovaara, J. and Rechardt, L. (1985). Neural differentiation in embryonal carcinoma cells. In Developmental Mechanisms : Normal and Abnormal. Alan R. Liss ed., pp. 3-13.

TABLE 1

EC cell lines	Treatment	IF expressed in derivatives	
1003 C17-S1-clone 1003	Serum free medium 4 days	Neuroepithelium	Vim
	7 days	Preneurons 10 %	Vim + 70 K NF
	9-14 days	Neurons 60 %	70 K + 200 K NF
	Serum 2 days (after 2 days SF)	Mesenchym 90 %	Vim
		Endoderm 10 %	Ker
1009 PCC7S-Aza R1-1009	10^{-7}M RA 6 days	Neurons	70 NF
		Mesoderm	vimentin
	10^{-7}M RA + 10^{-3}M db cAMP	Neurons	70 NF (+ MAPS 2)
		Glial	GFA + vim (20 %)
		Mesoderm	vim
	Aggregation 3 days + 10^{-7}M RA 3 days	Neurons	70 NF
		Glial	GFA + vim (20 %)
		Mesoderm	vim
PCC3/A/1	Aggregates	Neurons	70 NF

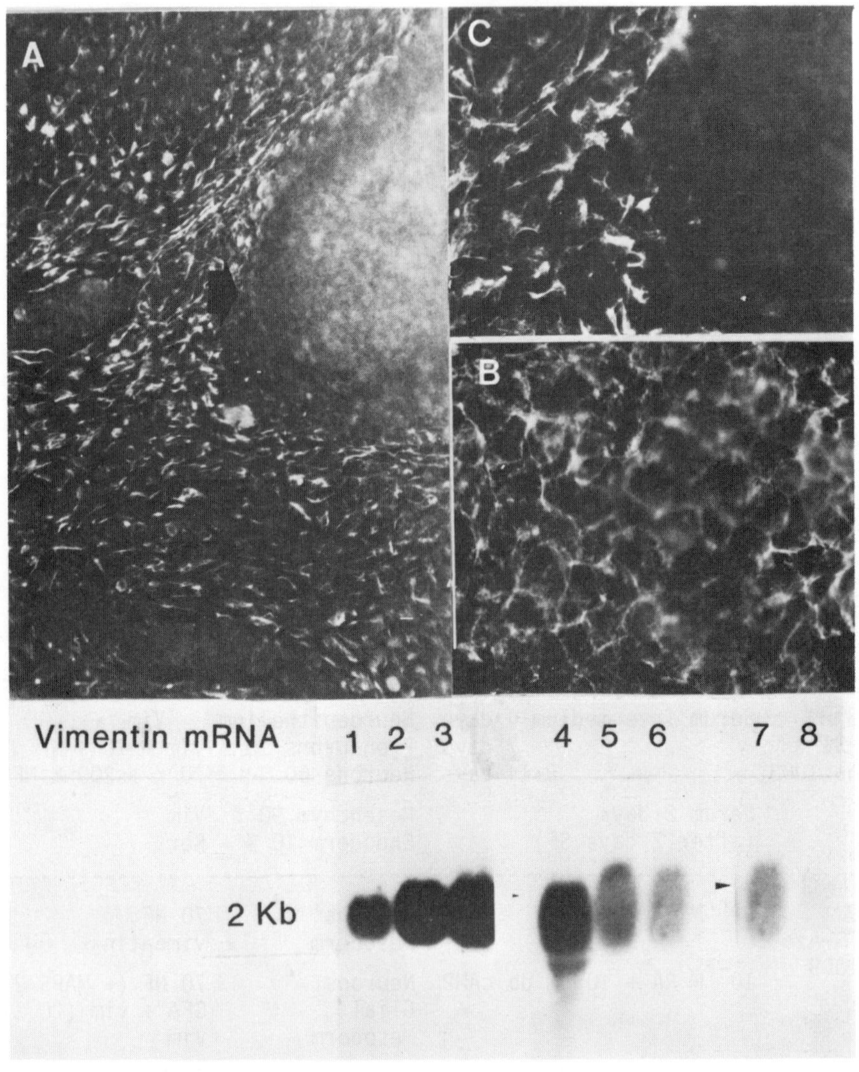

Fig.1. Vimentin synthesis in EC cells. *Immunofluorescent patterns.* (A) PCC3/A/1 cells grown 3 days as aggregate stained with vimentin antibody. (B and C) enlargement of the same portion of the centre stained with actin antibody (B), with vimentin antibody (C). Vimentin negative cells are situated in the centre. All the cells have microtubules and actin is found in surface structure. *Northern blots* : 10 µg of total RNAs extracted from (1) PYS, (2) TDM1, (3) 3T3, (4) 1009, (5) PCC4, (6-8) PCC3 were hybridized with a vimentin cDNA probe. Note the loss of vimentin mRNA for PCC3 cells (8) grown as aggregate.

CHANGES IN THE ORGANIZATION OF VIMENTIN-TYPE INTERMEDIATE FILAMENTS DURING RETINOIC ACID-INDUCED DIFFERENTIATION OF EMBRYONAL CARCINOMA CELLS

L.C. Moscinski and R.M. Evans

Department of Pathology, University of Colorado Health Sciences Center, Denver, Colorado 80262

ABSTRACT

Examination of embryonal carcinoma (EC) cells with anti-vimentin antibodies in indirect immunofluorescence reveals bright perinuclear areas of fluorescence surrounded by a diffuse cytoplasmic staining. Induction of differentiation with retinoic acid (RA) resulted in an observable alteration in specific vimentin immunofluorescence within 48 h and a progressive appearance of characteristic wavy filamentous structures throughout the cytoplasm of most cells within 6 days. Two-dimensional gel analysis of proteins from Triton-insoluble cytoskeletons of EC and RA-treated EC cells indicated that the changes in vimentin organization were accompanied by an increased amount of vimentin relative to other cellular proteins. This increase in amount of vimentin was not accompanied by any detectable alteration in phosphorylation or sensitivity of filament phosphorylation to cAMP. Vimentin filaments were also studied at the electron microscopic level using the peroxidase-labeled antibody technique. Examination of EC cytoskeletons using transmission electron microscopy demonstrated a fine network of vimentin filaments frequently with concentrations near the nucleus. Study of similar preparations from EC cells at various times following RA treatment revealed an increase in the number and size of filament bundles extending throughout the cytoplasm. These results are consistent with the hypothesis that intermediate filament proteins are developmentally regulated and may be temporally related to cytoplasmic organization during differentiation.

INTRODUCTION

Intermediate filaments are a major component of the cytoskeleton of eukaryotic cells. Although the precise function of this type of cytoplasmic filament is not well understood, both the expression and organization of intermediate filament proteins seem to be related to the state of differentiation of a cell (for review see Lazarides, 1982). Based on differences in protein subunits, it appears that there are at least five major classes of intermediate filaments. These classes include neurofilaments, glial filaments, epithelial cytokeratins, desmin filaments characteristic of muscle, and vimentin filaments typical of mesenchymally derived cells as well as most cells in culture (Lazarides, 1982). While some of the details may be in dispute, recent investigations have begun to delineate the cell type specificity of expression of intermediate filament proteins during early embryonic development (Franke et al., 1982; Franke et al., 1982; Jackson et al., 1980; Jackson et al., 1981; Lane et al., 1983; Lehtonen and Bradley, 1980; Paulin et al., 1979; Paulin et al., 1980a). However, relatively little is known about the structural organization of intermediate filaments during this process.

Embryonal carcinoma (EC) cells are stem cells derived from teratocarcinomas (Kleinsmith and Pierce, 1964) which are capable of giving rise to differentiated cell types in vitro as well as in vivo (Lehman et al., 1974; Martin, 1975). Some pluripotent EC cell lines with low rates of spontaneous differentiation can be induced to differentiate following exposure to retinoic acid (RA) (Jetten et al., 1979; Oshima, 1981; Oshima and Linney, 1980; Strickland and Mahdavi, 1978). Recent studies indicate that the expression of intermediate filament proteins in EC cells following the induction of differentiation with RA in vitro may reflect changes which occur during early embryogenesis (Paulin et al., 1980b; Lehtonen et al., 1983).

It has been suggested that intermediate filaments are important structural features in the determination of cytoplasmic organization (Lane et al., 1983). The present studies were intended to examine changes in vimentin-type intermediate filament composition and organization in EC cells during differentiation in vitro. Our results show a change in the organization of vimentin-type intermediate filaments during in vitro differentiation and suggest that this phenomenon provides a simple system to study the role of intermediate filaments in early cytoplasmic morphogenesis.

MATERIALS AND METHODS

CELL CULTURES. The EC cell line PCC4 (Jakob et al., 1973) was obtained from Dr. W. Speers and grown in minimal essential media containing 50 μg/ml penicillin and streptomycin supplemented with 10% fetal calf serum at 37°C in a humidified atmosphere containing 5% CO_2. In most experiments approximately 10^6 cells were plated into 60 mm dishes and induced to differentiate by exposure to 10^{-6} M RA (Sigma) for 24 h (Jetten et al., 1979).

Cultures were radiolabeled with 50-100 μCi/ml ^{35}S-methionine (900 Ci/mmole; New England Nuclear) for 4 h in methionine-free medium containing 5% fetal bovine serum. The cells were then gently rinsed in phosphate buffered saline (PBS) and either Triton-insoluble cytoskeletons were prepared or the cells were lysed in a solution containing 2% sodium dodecyl sulfate (SDS), 0.1% sodium deoxycholate, 1 mM EDTA, and 1 mM dithiothreitol. Triton-insoluble cytoskeletons were prepared by the procedure of Zackroff and Goldman (1979) and solubilized in the same solution as the cell lysates. Samples were immediately placed on a boiling water bath for 1 min and allowed to cool to room temperature. Solid urea was added to a final concentration of approximately 8 M prior to two-dimensional gel analysis. Samples which were not immediately analyzed were stored at -70°C in the absence of urea.

ANTISERUM. Vimentin was isolated following preparative gel electrophoresis of Chinese hamster ovary cell cytoskeletal preparations (Zackroff and Goldman, 1979). The vimentin band was cut from the gel and the protein eluted by shaking in a solution containing 0.1% SDS and 50 mM ammonium bicarbonate. The protein was precipitated in 10% trichloroacetic acid, washed with ethanol:ether 1:1 (vol/vol), and dried under a stream of nitrogen. Approximately 200 μg of protein was resuspended in complete adjuvant and injected intrapopliteally into a rabbit. The animal was again immunized 30 days and 54 days later with a similar amount of antigen in adjuvant injected subcutaneously. The rabbit was bled 95 days after initial immunization and this antiserum used for characterization of anti-vimentin antibodies.

The specificity of the anti-vimentin antibodies was established using the following criteria: 1) in indirect immunofluorescence the antibody decorated a fibrillar structure (in a variety of cultured cells) which forms perinuclear aggregates in the presence of colcemid (data not shown); 2) the antiserum does not decorate human A-431 cells which lack vimentin filaments (data not shown); 3) affinity repurified antibodies gave identical results in indirect immunofluorescence (data not shown); and 4) "Western" blotting with radiolabeled cell lysates separated by two-dimensional gel electrophoresis indicated that the antibody is specific for vimentin.

INDIRECT IMMUNOFLUORESCENCE. Cells grown on glass coverslips were rinsed with PBS and treated for 10 min with 30% methanol, 70% acetone at -20°C. Indirect immunofluorescence was carried out essentially as described by Franke et al. (1978). Anti-vimentin antiserum was diluted 1:50 and fluorescein-conjugated goat anti-rabbit serum (Miles Laboratories) was diluted 1:30 in PBS containing 1% ovalbumin. The coverslips were mounted in Aqua-mount (Lerner Laboratories). Fluorescence was viewed on either an Olympus or Zeiss photomicroscope equipped with epifluorescence optics. Photographs were taken on Kodak Tri-X pan film and developed in Kodak HC-110.

TRANSMISSION ELECTRON MICROSCOPY. Formvar-coated gold grids (Pelco) were covered with a thin film of carbon and briefly u.v. irradiated. Cells were plated on the grids at least 48 h prior to preparation for EM. The cells were extracted to prepare cytoskeletons by the method of Webster et al. (1978). The grids were removed from the culture and incubated in a solution containing 10 mM Tris-HCl pH 7.6, 0.14 M NaCl, 5 mM $MgCl_2$ and 4% polyethylene glycol 6000 for 5 min at room temperature. This solution was then replaced with the same buffer containing in addition 0.5% Triton X-100 and incubated at room temperature for 10 min. The grids were then washed in PBS and incubated with anti-vimentin IgG (protein A affinity purified) at 1.0 ng/ml diluted 1:500 in PBS containing 1% bovine serum albumin (BSA) for 30 min. The grids were then rinsed with PBS and incubated with horseradish peroxidase-conjugated goat anti-rabbit IgG (Miles Laboratory) for 30 min. After a PBS rinse, the grids were fixed for 20 min in 2% phosphate buffered glutaraldehyde, rinsed thoroughly with PBS, and then incubated for 8 min in a solution containing 0.12 mg/ml diaminobenzidine and 0.005% H_2O_2 in 0.1 M phosphate pH 6.95. The grids were rinsed three times in PBS and incubated in 1% osmium in PBS for 10 min. The grids were again rinsed, then passed through a graded ethanol dehydration series and critical point dried from CO_2. Control experiments included either no primary antiserum or pre-immune antiserum. Cytoskeletons were viewed as whole mounts without embedding or additional staining using a Philips 201 electron microscope.

POLYACRYLAMIDE GEL ELECTROPHORESIS. Two-dimensional gel electrophoresis was performed by the method of O'Farrell (1975) as previously described (Evans et al., 1979). One-dimensional SDS gel electrophoresis was performed as described by Laemmeli (1970) and modified by Studier (1972). Characterization of anti-vimentin antibodies by "Western" blotting was performed following electrophoretic transfer of ^{35}S-methionine-labeled proteins to nitrocellulose sheets by the method of Towbin et al. (1979).

RESULTS

A number of different lines of EC cells have been shown to express vimentin (Jackson et al., 1980; Lehtonen et al., 1983; Oshima, 1981; Paulin et al., 1980a) as is apparently the case for most cells in culture (Lazarides, 1982). Characteristically, cultured cells show a wavy system of cytoplasmic filaments in immunofluorescence studies with antibodies to vimentin (Franke et al., 1978; Hynes and Destree, 1978). However, when PCC4 cells are stained with anti-vimentin antibodies, the pattern of fluorescence is unlike that found in most cultured cells at this level of resolution. This may be due in part to the fact that EC cells typically have little cytoplasm and grow in colonies without prominent cell borders. However, as shown in Fig. 1, immunofluorescence with anti-vimentin antibodies produces bright perinuclear areas of fluorescence surrounded by a diffuse cytoplasmic staining and increased intensity often present at cell borders. Although there is some evidence for fibrillar arrays in a few cells, discrete filament bundles are not discernible in the vast majority of these undifferentiated PCC4 cells at this level of resolution.

The mechanism of intermediate filament assembly in living cells is unknown; however, under _in vitro_ conditions intermediate filaments can be induced to disassemble into a soluble protofilament form (Zackroff and Goldman, 1979). To pursue the possibility that the diffuse staining observed in PCC4 cells might represent vimentin in a soluble, nonfilamentous form, cells were labeled with ^{35}S-methionine and divided into Triton-soluble and Triton-insoluble fractions.

These preparations were analyzed by two-dimensional gel electrophoresis and "Western" blotting of these gels (data not shown). No detectable vimentin is found in soluble preparations from these cells. Attempts at quantitation by co-electrophoresis with unlabeled, purified vimentin, cutting out the vimentin spot and determining the ^{35}S-radioactivity in the Triton-soluble and -insoluble preparations indicate that at least 90% of the vimentin in PCC4 cells is present in a Triton-insoluble form. Moreover, immunofluorescence of Triton-extracted cells with anti-vimentin antibodies is indistinguishable from the pattern shown in Fig. 1A. Additional evidence that the organization of vimentin in these cells is filamentous is the finding that treatment of PCC4 cells with trypsin or colcemid results in perinuclear aggregation of the vimentin fluorescence, a characteristic of intermediate filaments in other cells (Blose and Chacko, 1976; Goldman and Knipe, 1973; Hynes and Destree, 1978).

While induction of differentiation of EC cells with RA is reported to result in observable morphologic changes within 24 h (Strickland and Mahdavi, 1978), little change is observed in vimentin organization by immunofluorescence as shown in Fig. 1B. However, by 48 h after the induction of differentiation with RA many of the cells have begun to exhibit large aggregates of fluorescence in the cytoplasm, which by 72 h appear as recognizable filamentous structures (Fig. 1C) in cells which have begun to increase rapidly in size. This is a progressive phenomena and by 6-7 days after induction of differentiation nearly all cells in the culture have acquired a wavy filamentous network of vimentin, characteristic of other cells in culture (Fig. 1D).

To clarify the organization of vimentin in PCC4 cells and to characterize the early changes following treatment with RA, studies were performed to identify vimentin at the electron microscopic level using the peroxidase-labeled antibody technique. It has been demonstrated that following mild extraction with Triton X-100, the three-dimensional organization of cytoskeletal elements is essentially undisturbed (Brown et al., 1976; Henderson and Weber, 1980; Lenk et al., 1977; Webster et al., 1978; Osborn and Weber, 1977).

Detergent-extracted cytoskeletons of PCC4 cells and PCC4 cells at various times following RA treatment were also labeled with anti-vimentin antibody using the indirect immunoperoxidase technique. When viewed as whole mounts in transmission electron microscopy at low magnification, undifferentiated PCC4 cells exhibit a fine network of labeled filaments as shown in Fig. 2A. These cells frequently have concentrations of filaments near the nucleus, similar to the distribution which is observed in immunofluorescence (Fig. 1A). Examination of similar whole-mount cytoskeletons following treatment with RA demonstrates a progressive increase in the amount of labeled material in the cytoplasm (Figs. 2B-2D). When sections of cytoplasm are viewed at higher magnifications this phenomena can be seen more clearly (Figs. 3A-3D). The most visible aspect is that while bundles of filaments are rare in PCC4 cells (Fig. 3A), filament bundles become larger and more abundant with time following treatment with RA (Figs. 3B, 3D). Although these large filament bundles resemble stress fibers in electron microscopy, studies with anti-actin antibodies indicate that in these cells, recognizable stress fiber formation does not occur until 7-8 days after the induction of differentiation (data not shown).

An increased amount of vimentin after RA-induced differentiation has been reported for F9 EC cells (Oshima, 1981). The immunoelectron microscopy of PCC4 cells suggests a similar increase in amount of vimentin as well as an altered organization during differentiation. This is confirmed by two-dimensional gel analysis of Triton-insoluble extracts of PCC4 cells following RA treatment. As shown in Fig. 4 there is a progressive increase in the amount of radiolabeled vimentin relative to other cytoskeletal proteins in response to RA. This increase is detectable by 72 h (Fig. 4B) and becomes prominent by 6 days (Fig. 4C) after the initiation of differentiation. Similar experiments with Triton-insoluble extracts from ^{32}P-labeled EC cells indicated that there was no detectable alteration in the level of vimentin filament phosphorylation or sensitivity to exogenous 8-bromo-cAMP during RA induced differentiation (data not shown).

DISCUSSION

Changes in cytoskeletal organization have been demonstrated to be a prominent feature of *in vitro* differentiation (Lehtonen et al., 1983; Paulin et al., 1978, 1979, 1980b). In particular, the organization of microfilaments and associated structures has been studied in teratocarcinoma cells in some detail. Experiments with the nullipotent EC cell line F9 indicate that within days of RA-induced differentiation, proteins that are characteristic of endodermal cytokeratins are expressed (Oshima, 1981) and vinculin-containing adhesion plaques form. This is followed by reorganization of some microfilaments into stress fibers (Lehtonen et al., 1983). We have found that, in addition, the pre-existing network of vimentin-type intermediate filaments also undergoes a major change in organization as a part of this same process of differentiation. This phenomenon is not a peculiarity of PCC4 cells or RA-induced differentiation alone. To support this conclusion, we have also studied another pluripotent EC cell line, 247 (Lehman et al., 1974), induced to differentiate with dimethylacetamide and have obtained identical results (data not shown).

The organization of vimentin in these EC cell lines may be considered unusual in certain respects. First is the lack of prominent or large wavy filament bundles typical of other cells that have been studied by similar techniques. Second is the presence, in many cells, of perinuclear concentrations of filaments which are apparently distinct from perinuclear aggregates produced by agents such as trypsin or colcemid.

The increase in number, size and distribution within the cytoplasm of filament bundles following RA-induced differentiation is apparently accompanied by an increased accumulation of vimentin (Fig. 4). This increase is coincidental with a dramatic increase in cell size, and while an increased amount of vimentin can be explained by cell size alone, the prominent filament bundling appears to be an independent process. It seems unlikely that these organizational changes are simply the result of increased synthesis of subunit protein. However, studies of in vitro myogenesis, where a developmentally related change in the organization of intermediate filaments is known to occur (Bennett et al., 1979; Gard and Lazarides, 1980), have shown that there are also concomitant changes in the expression of proteins believed to be filament associated (Breckler and Lazarides, 1982; Granger and Lazarides, 1980). Examination of two-dimensional gels of salt- and detergent-extracted cytoskeletons of PCC4 cells during RA-induced differentiation have thus far not revealed any similar changes in the level of additional proteins (Fig. 4).

The physiological significance of this phenomenon is at present unclear. EC cells are believed to be the neoplastically transformed counterparts of the inner cell mass cells of blastocyst stage preimplantation embryos (Brinster, 1974; Martin, 1975; Martin et al., 1978; Papaioannou et al., 1975). When some EC cell lines are allowed to differentiate in vitro, the first differentiated cells to appear are extra-embryonic endoderm (Lehman et al., 1974, Martin and Evans, 1975; McBurney, 1976). This is apparently analogous to the generation of visceral and parietal endoderm at the surface of the inner cell mass of the blastocyst. Although reports differ as to whether inner cell mass cells express vimentin at this early stage of embryogenesis (Jackson et al., 1980; Lehtonen and Bradley, 1980), there is evidence that parietal endodermal cells in early mouse embryos do express vimentin (Lane et al., 1983). Several other EC cell lines have also been shown to express vimentin (Jackson et al.,1980; Lehtonen et al., 1983; Oshima, 1981; Paulin et al., 1980a, 1980b). However, since many cells in culture appear to express vimentin filaments, it has been speculated that this may represent a general adaptation to in vitro conditions. Regardless of how closely the reorganization of intermediate filaments in cells induced to differentiate in vitro reflects an in vivo process, this phenomenon will provide a useful system to study cellular mechanisms involved in the control of vimentin filament protein synthesis, assembly and organization.

ACKNOWLEDGEMENTS

We would like to thank Dr. G. Miller for his assistance in preparing anti-vimentin antiserum, Dr. W. Speers for his advice and helpful discussions about RA-induced differentiation of PCC4 cells and A. Hartman for his expertise with electron microscopy. Special thanks are also extended to M. Tagawa for her excellent technical assistance.

This research was supported by NSF grant PCM-8120029, NIH grant CA-15823, and a departmental gift from R.J. Reynolds Industries, Inc.

REFERENCES

Bennett, G.S., Fellini, S.A., Yoyama, Y. and Holtzer, H. (1979). Redistribution of intermediate filament subunits during skeletal myogenesis and maturation _in vitro_. J. Cell Biol. 82, 577-584.

Blose, S.H. and Chacko, S. (1976). Rings of intermediate (100 Å) filament bundles in the perinuclear region of vascular endothelial cells. J. Cell Biol. 70, 459-466.

Breckler, J. and Lazarides, E. (1982). Isolation of a new high molecular weight protein associated with desmin and vimentin filaments from avian embryonic skeletal muscle. J. Cell Biol. 92, 795-806.

Brinster, R.L. (1974). The effect of cells transferred into the mouse blastocyst on subsequent development. J. Exp. Med. 140, 1049-1056.

Brown, S., Levinson, W. and Spudich, J.A. (1976). Cytoskeletal elements of chick embryo fibroblasts revealed by detergent extraction. J. Supramol. Struct. 5, 119-130.

Evans, R.M., Ward, D.C. and Fink, L.M. (1979). Asymmetric distribution of plasma membrane proteins in mouse L-929 cells. Proc. Natl. Acad. Sci. USA 76, 6235-6239.

Franke, W.W., Grund, C., Kuhn, C., Jackson, B.W. and Illmensee, K. (1982). Formation of cytoskeletal elements during mouse embryogenesis. III. Primary mesenchymal cells and the first appearance of vimentin filaments. Differentiation 23, 43-59.

Franke, W.W., Schmid, E., Osborn, M. and Weber, K. (1978). Different intermediate-sized filaments distinguished by immunofluorescence microscopy. Proc. Natl. Acad. Sci. USA 75, 5034-5038.

Franke, W.W., Schmid, E., Schiller, D.L., Winter, S., Jackson, B.W. and Illmensee, K. (1982). Differentiation-related patterns of expression of proteins of intermediate size filaments in tissues and cultured cells. Cold Spring Harbor Symp. Quant. Biol. 46, 431-453.

Gard, D.L. and Lazarides, E. (1980). The synthesis and distribution of desmin and vimentin during myogenesis _in vitro_. Cell 19, 263-275.

Goldman, R. and Knipe, D. (1973). The functions of cytoplasmic fibers in non-muscle cell motility. Cold Spring Harbor Symp. Quant. Biol. 37, 523-534.

Granger, B.L. and Lazarides, E. (1980). Synemin: A new high molecular weight protein associated with desmin and vimentin filaments in muscle. Cell 22, 727-738.

Henderson, D. and Weber, K. (1980). Immunoelectron microscopic studies of intermediate filaments in cultured cells. Exp. Cell Res. 129, 441-453.

Hynes, R.O. and Destree, A.T. (1978). 10 nm filaments in normal and transformed cells. Cell 13, 151-163.

Jackson, B.W., Grund, C., Schmid, E., Bürki, K., Franke, W.W. And Illmensee, K. (1980). Formation of cytoskeletal elements during mouse embryogenesis. I. Intermediate filaments of the cytokeratin type and desmosomes in preimplantation embryos. Differentiation 17, 161-179.

Jackson, B.W., Grund, C., Winter, S., Franke, W.W. and Illmensee, K. (1981). Formation of cytoskeletal elements during mouse embryogenesis. II. Epithelial differentiation and intermediate-sized filaments in early postimplantation embryos. Differentiation 20, 203-216.

Jakob, H., Boon, T., Gaillard, J., Nicholas, J.F. and Jacob, F. (1973). Teratocarcinome de la souris: isolement, culture et proprietes de cellules a potentialites multiples. Ann. Microbiol. (Institut Pasteur) 124, 269-282.

Jetten, A.M., Jetten, M.E.R. and Sherman, M.I. (1979). Stimulation of differentiation of several murine embryonal carcinoma cell lines by retinoic acid. Exp. Cell Res. 124, 381-391.

Kleinsmith, L.J. and Pierce, G.B. (1964). Multipotentiality of single embryonal carcinoma cells. Cancer Res. 24, 1544-1552.

Laemmelli, U.K. (1970). Cleavage of structural proteins during the assembly of the head of bacteriophage T4. Nature 227, 680-685.

Lane, E.B., Hogan, B.L.M., Kurkinen, M. and Garrets, J.I. (1983). Co-expression of vimentin and cytokeratins in parietal endoderm cells of early mouse embryo. Nature 303, 701-704.

Lazarides, E. (1982). Intermediate filaments: A chemically heterogeneous, developmentally regulated class of proteins. Ann. Rev. Biochem. 51, 219-250.

Lehman, J.M., Speers, W.C., Swartzendruber, D.E. and Pierce, G.B. (1974). Neoplastic differentiation: Characteristics of cell lines derived from a murine teratocarcinoma. J. Cell Physiol. 84, 13-28.

Lehtonen, E. and Bradley, R.A. (1980). Localization of cytoskeletal proteins in preimplantation mouse embryos. J. Embryol. Exp. Morphol. 55, 211-225.

Lehtonen, E., Lehto, V.P., Bradley, R.A. And Virtanen, I. (1983). Formation of vinculin plaques precedes other cytoskeletal changes during retioic acid induced teratocarcinoma cell differentiation. Exp. Cell Res. 144, 191-197.

Lenk, R., Ransom, L., Kaufmann, YH. and Penman, S. (1977). A cytoskeletal structure with associated polyribosomes obtained from HeLa cells. Cell 10, 67-78.

Martin, G.R. (1975). Teratocarcinomas as a model system for the study of embryogenesis and neoplasia. Cell 5, 229-243.

Martin, G.R. and Evans, M.J. (1975). Differentiation of clonal lines of teratocarcinoma cells: Formation of embryoid bodies _in vitro_. Proc. Natl. Acad. Sci. USA 72, 1441-1445.

Martin, G.R., Smith, S. and Epstein, C.J. (1978). Protein synthetic patterns in teratocarcinoma cells: Formation of embryoid bodies _in vitro_. Dev. Biol. 66, 8-16.

McBurney, M.W. (1976). Clonal lines of teratocarcinoma cells _in vitro_: differentiation and cytogenetic characteristics. J. Cell. Physiol. 89, 441-456.

O'Farrell, P.H. (1975). High resolution two-dimensional electrophoresis of proteins. J. Biol. Chem. 230, 4007-4021.

Osborn, M. and Weber, K. (1977). The detergent-resistant cytoskeleton of tissue culture cells includes the nucleus and the microfilament bundles. Exp. Cell Res. 106, 339-349.

Oshima, R.G. (1981). Identification and immunoprecipitation of cytoskeletal proteins from murine extra-embryonic endodermal cells. J. Biol. Chem. 256, 8124-8133.

Oshima, R. and Linney, E. (1980). Identification of murine extra-embryonic endodermal cells by reaction with teratocarcinoma basement membrane antiserum. Exp. Cell Res. 126, 485-490.

Papaioannou, V.E., McBurney, M.W., Gardner, R.L. and Evans, M.J. (1975). Fate of teratocarcinoma cells injected into early mouse embryos. Nature 258, 70-73.

Paulin, D., Babinet, C., Weber, K. and Osborn, M. (1980a). Antibodies as probes of cellular differentiation in the mouse blastocyst. Exp. Cell. Res. 130, 297-304.

Paulin, D., Forest, N. and Perreau, J. (1980b). Cytoskeletal proteins used as marker of differentiation in mouse teratocarcinoma cells. J. Mol. Biol. 144, 95-101.

Paulin, D., Nicolas, J.F., Yaniv, M., Jacob, F., Weber, K. and Osborn, M. (1978). Actin and tubulin in teratocarcinoma cells. Dev. Biol. 66, 488-499.

Paulin, D., Perreau, J., Jakob, H., Jacob, F. and Yaniv, M. (1979). Tropomyosin synthesis accompanies formation of actin filaments in embryonal carcinoma cells induced to differentiate by hexamethylene bisacetamide. Proc. Natl. Acad. Sci. USA 76, 1891-1895.

Schmid, E., Tapscott, S., Bennett, G.S., Croop, J., Fellini, S.A., Holtzer, H. and Franke, W.W. (1979). Differential location of different types of intermediate-sized filaments in various tissues of the chicken embryo. Differentiation. 15, 27-40.

Strickland, S. and Mahdavi, V. (1978). The induction of differentiation in teratocarcinoma stem cells by retinoic acid. Cell 15, 393-403.

Studier, F.W. (1972). Bacteriophage T7. Genetic and biochemical analysis of this simple phage gives information about basic genetic processes. Science 176, 367-376.

Towbin, H., Staehelin, T. and Gordon, J. (1979). Electrophoretic transfer of proteins from polyacrylamide gels to nitrocellulose sheets: Procedure and some applications. Proc. Natl. Acad. Sci. USA. 76, 4350-4354.

Webster, R.E., Osborn, M. and Weber, K. (1978). Visualization of the same PTK_2 cytoskeleton by both immunofluorescence and low power electron microscopy. Exp. Cell. Res. 117, 47-61.

Zackroff, R.V. and Goldman, R.D. (1979). In vitro assembly of intermediate filaments from baby hamster kidney (BHK-21) cells. Proc. Natl. Acad. Sci. USA 76, 6226-6230.

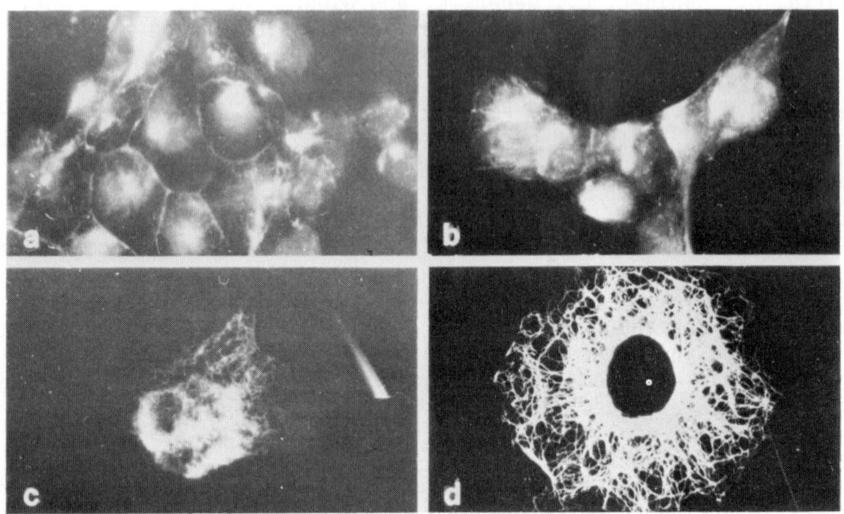

Fig. 1. Effect of retinoic-acid-induced differentiation on the indirect immunofluorescence localization of vimentin in PCC4 cells. (a) Untreated; (b) 1 d; (c) 3 d, X160; and (d) 8 d after induction of differentiation with RA, X125.

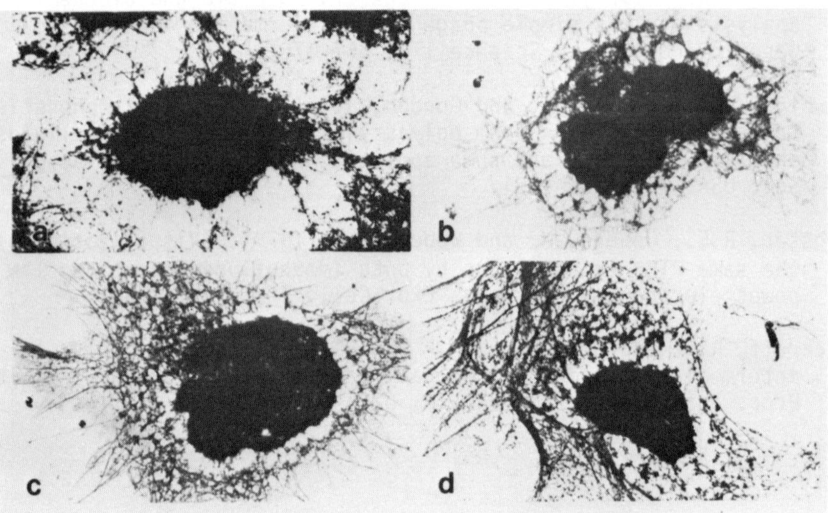

Fig. 2. Transmission electron microscopy of whole-mount cytoskeletons labeled with anti-vimentin antibody. (a) Untreated PCC4 cells; (b) 1 d,X3,000; (c) 3 d; and (d) 6 d following induction of differentiation with RA, X2,000. Nuclei are labeled nonspecifically.

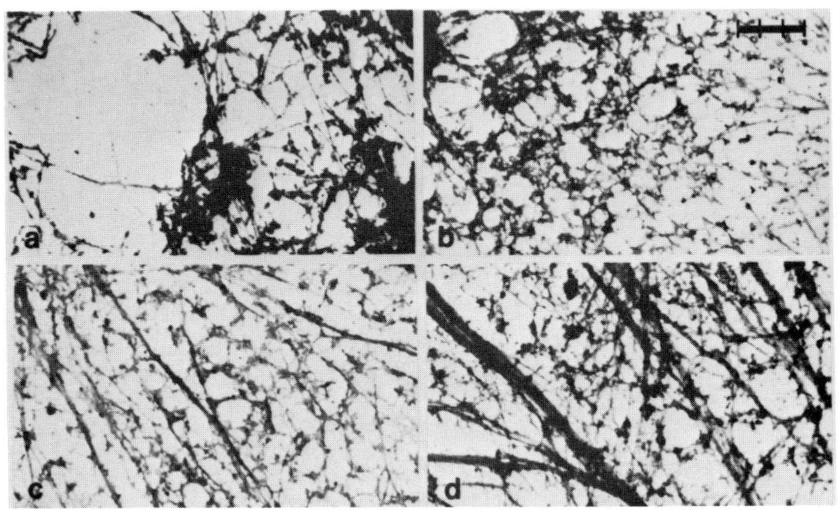

Fig. 3. Transmission electron microscopy of areas of whole-mount cytoskeletons labeled with anti-vimentin antibody. (a) Untreated PCC4 cells; (b) 1 d; (c) 3 d; and (d) 6 d following induction of differentiation with RA, X11,300. The grid in the upper right hand corner indicates actual magnification with 0.5 cm = 4,400 Å.

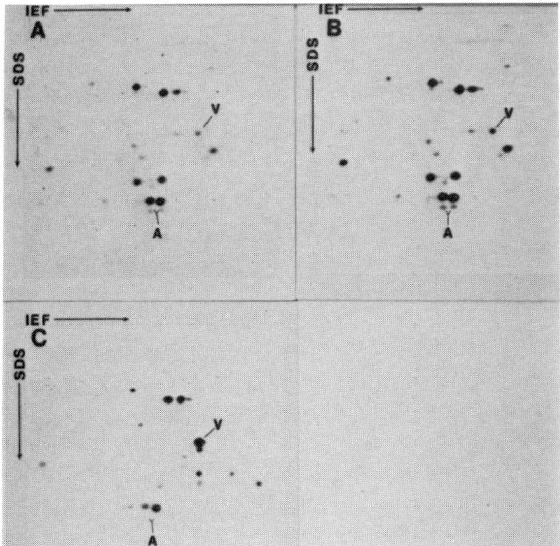

Fig. 4. Two-dimensional gel analysis of Triton-insoluble extracts from ^{35}S-methionine-labeled (a) PCC4 cells, and (b) cells 3 d and (c) 8 d after treatment with RA. The positions of (A) actin and (V) vimentin are indicated.

STRUCTURAL BIOCHEMISTRY OF MAMMALIAN SPERM OUTER DENSE FIBERS

J.C. Vera and J.H. Delgado
Department of Biochemistry, Biophysics and Genetics, University of Colorado Health Sciences Center, Denver, Colorado 80262 and Department of Pharmacology, University of Texas Health Sciences Center at Dallas, Dallas, Texas 75235, USA.

In the sperm cell of species with internal fertilization, the axonema is surrounded by nine outer dense fibers (ODF) generating a 9 + 9 + 2 cross-sectional pattern (1). These fibers are joined anteriorly to the striated columns of the connecting piece, and at their distal end are fixed to the wall of the corresponding axonemal doublet (1,2). The role of the ODF during the flagellar movement has not yet been elucidated. Immunologic and histochemical studies indicated that the ODF contained proteins similar to muscle actin and myosin (3-5). On the other hand, comparative (6), morphometric (7) and preliminary biochemical analysis (4,8) have favored a passive role for the ODF in the flexibility of the sperm tail.

As a step towards understanding the possible involvement of the ODF in the process of sperm motility, we have developed a simple procedure to isolate the ODF from mamalian sperms and have carried out a careful biochemical and immunological characterization of this important structure.

ISOLATION OF THE OUTER DENSE FIBERS. To obtain ODF, sperms were treated with cetyltrimethylammonium bromide and 2-mercaptoethanol at pH 8.0. After 30 minutes' incubation, the sperm heads and the ODF-connecting piece complexes were the only visible structures as revealed by Nomarsky optics and electron microscopy (9,l0). The ODF were purified by centrifugation on discontinuous sucrose gradients (9,l0). Electron microscopy observations confirmed that the isolated material was pure ODF without contamination from sperm heads. This procedure was successfully used to isolate ODF from rat, bull, mouse, hamster and human sperms in sufficient amount to afford further biochemical analysis.

PROTEIN CHEMICAL CHARACTERIZATION. The isolated ODF were dissolved in SDS-containing buffers and analyzed by one-dimensional SDS-polyacrylamide gel electrophoresis (PAGE-SDS). In all the species studied the ODF appear to be composed of several polypeptides, with three to six major bands of protein ranging from 87 to 11.5 kDa. The 87 kDa band is present in the ODF of all five species studied, and the same is true with respect to a polypeptide of approximately 30kDa. In addition, the 30 kDa polypeptide represents over 40% of the total protein content of the ODF as analyzed by quantitative densitometry of Coomassie blue stained gels.

The major polypeptide components of rat and bull sperm ODF were purified using gel chromatography in denaturating conditions and preparative PAGE-SDS. The purity of each polypeptide was assessed by

PAGE-SDS and NH_2-terminal analyses. Amino acid analyses of isolated polypeptides revealed a high content of serine, aspartic and glutamic acids, leucine, tyrosine,proline and cysteine. Preliminary analysis showed that the same is true with respect to the polypeptidic components of mouse and hamster sperm ODF. Hence, the amino acid composition of the proteins present in the ODF is clearly different from those of muscle contractile proteins. Furthermore, the purified polypeptides showed a large charge heterogeneity as analyzed by isoelectric focusing and two-dimensional polyacrylamide gel electrophoresis. Phosphoamino acid analysis showed that the six major components of rat sperm ODF and the 87 and 33 kDa polypeptides of bull sperm ODF were phosphorylated at serine residues.

IMMUNOLOGICAL STUDIES. Antibodies against the major polypeptidic components of rat sperm ODF were produced in rabbits. Immunoblot analyses carried out using affinity purified antibodies revealed a total absence of cross-reactivity with rat myosin, actin, tubulin, keratin and collagen. Also, polyclonal antibodies raised against tubulin, keratin and collagen showed absence of reactivity with rat ODF polypeptides.

CONCLUSIONS. These results indicate that the major components of mammalian sperm ODF are a unique family of phosphoproteins, clearly different from muscle actin and myosin by biochemical and immunological criteria. On the other hand, their biochemical characteristics raise several questions in regard to their possible involvement in the process of sperm motility, especially if we consider the close relationship between protein phosphorylation and sperm motility (11).

ACKNOWLEDGEMENTS. We thank L. Burzio for providing us with facilities in his laboratory at U.Austral of Chile to develop the initial part of this investigation (partially supported by a Grant from the International Foundation for Science).

REFERENCES

(1) Fawcett, D.W. (1975) Dev. Biol. 44, 394-436.
(2) Fawcett, D.W. and Phillips, D.M. (1969) Anat.Rec. l65, 153-184.
(3) Nelson, L. (l958) Biochim.Biophys.Acta 27, 634-641.
(4) Baccetti, B., Palliini, V. and Burrini, A.G. (1973) J.Submicrosc.Cytol. 5,237-256
(5) Baccetti, B., Bigliardi, E., Burrini, A.G., Gabbiani,G., Jackosch,B.M. and Leoncini,P. (1984) J.Submicrosc.Cytol. l6,79-84.
(6) Phillips,D.M. (1972) J.Cell Biol. 53,56l-573.
(7) Serres,C., Escalier,D. and David,G. (1983) Biol.Cell 49,l53-162.
(8) Olson,G.E. and Sammons, D.W. (1980) Biol.Reprod. 22,3l9-332.
(9) Vera,J.C., Brito,M., Zuvic,T. and Burzio, L. (1984) J.Biol.Chem. 259,5970-5977.
(l0) Brito,M., Figueroa,J., Vera,J.C., Cortes,P., Hott,R. and Burzio, L. (1986) Gamete Res. 15 327-336.
(11) Tash, J.S. and Means,A.R. (1983) Biol.Reprod. 28, 75-104

A WALDESTRöM MACROGLOBULIN WITH ANTI-INTERMEDIATE FILAMENTS ACTIVITY

C. Osuna, A, Sánchez. Centro de Biología Molecular. CSIC-UAM. Madrid. Canto Blanco. 28049 Madrid. Spain
J. Segui, C. Montalbán, Y. Revilla. Servicio de Inmunologí y de Medicina Interna del Hospital Ramón y Cajal. Madrid.

Waldeström macroglobulinemia (WM) is characterized by over production of monoclonal antibodies of the IgM class (1). It has been discussed whether those IgM proteins have a functional activity. Several reports indicate that IgM from WM patients have natural antibody-like activity, fundamentally against cytoeskeletal proteins (2) including intermediate filaments (IF) proteins (3). In this report we have studied whether IgM obtained from Waldeström patient rects with any known protein. Our results suggest that this WM serum binds to IF proteins.

(I) The WM antibodies react with smooth muscle. A serum from a Waldeström patient was analized by immnunofluorescence in rat smooth muscle showing the pattern indicated in Fig. 1. This pattern could be due to the reaction of the serum with muscle action or actin binding proteins such as α-actinin as previously reported (4). To test whether the serum binds to microfilaments, we have studied the immunofluorescence in fibroblast since those cells contain microfilaments in which actin or actin binding proteins are present.

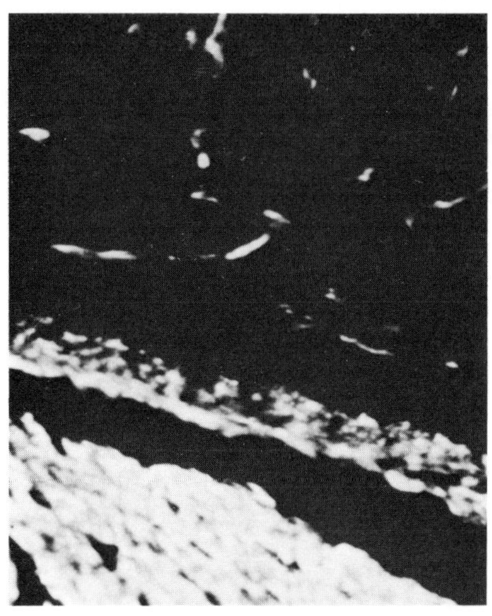

Fig. 1 Serum from WM analized at 1:50 dilution by indirect immunofluorescence stains the muscularis (smooth muscle) in rat stomach thin section

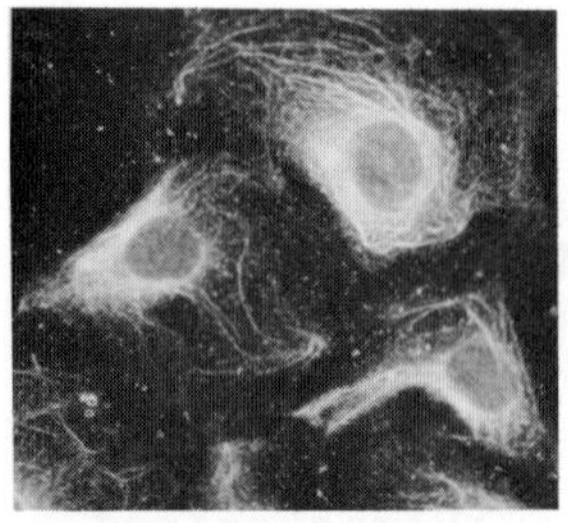

Fig. 2 Indirect immunofluorescence on freshly grown and fixed Ptk2 cells, labeled with WM serum (1:50) and goat antihuman FITC, showing staining of a network of cytoplasmic filaments.

(II) The WM activity reacts with IF network in cultured cells. Fig. 2 shows that the WM serum reacts on cultured cells with a filamentous structure wich is not related to microfilaments but to intermediate filament, newtwork. Reaction with microtubules was discared by studying the immunofluorescence pattern on cells at low temperature (not shown). Since IF from Ptk2 are composed of vimentin (see i.e. 5), the WM serum may react against vimentin.

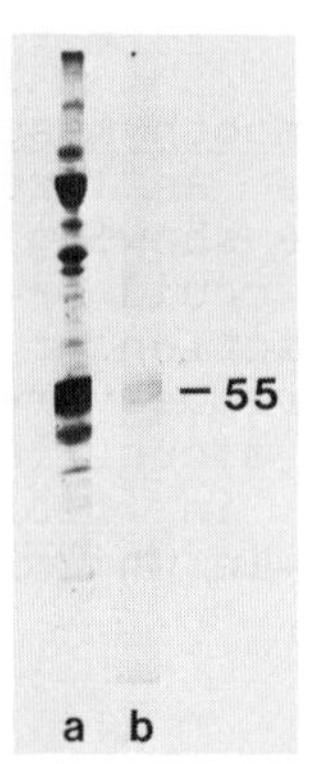

Fig. 3. Interaction of the WM patient serum with desmin-enriched fractions analyzed by SDS-polyacrylamide gel electrophoresis and transfered to NC paper (Wester blots).

(III) The WM serum reacts with desmin. Since the main IF protein in muscle is desmin we have isolated desmin-enriched fractions from rabbit stomach. We have found by immunoblotting (Fig. 3), that the WM serum recognizes a 55 kd protein that has been identified as desmin by peptide mapping (not shown). These results suggest that the WM serum reacts with both vimentin and desmin. Therefore, the natural antigen that is recognized by this WM serum could be located in the region which is related in all the IF proteins.

REFERENCES

(1) Waldeström, J. (1968) Monoclonal and Polyclonal Hypergammaglobulinemia (University, Press, London).

(2) Dighiero, G., Guilbert, B. and Avrames (1982) J. Immunol. 128, 2788-2792.

(3) Dellagi, K., Brouet, J.C., Perrau, J. and Paulin, D. (1982). Proc. Natl. Acad. Sic. USA. 79, 446-430.

(4) Nicouin, C. et. al. (1984) Clin. exp. Immunol. 58, 677-684.

(5) Ulrike, E., Troub, W., Nelson, J. and Troub, P. (1983) J. Cell. Sci. 62, 129-147.

EVALUATION OF THE LOCOMOTION OF 3T3/A31 CELLS AND ITS DERIVED MALIGNANT SUBLINES WITH CYTOSKELETAL DISARRAYS

A. Baroja, F. Vidal-Vanaclocha, E. Barbera-Guillem
Dpt. of Histology and Cell Biology. Faculty of Medicine. University of the Basque Country. Spain.

INTRODUCTION. Important steps of the metastatic behavior of malignant tumor cells are related to cytoskeleton-dependent properties such as cell shaping and deformation, cell adhesion to the extracellular matrix and cell locomotion (1,2). Recently (3) it has been established a new angiosarcoma tumor system composed of the untransformed cloned 3T3/A31 cells and three selected sublines derived from this parental line which expressed increasing malignancy: transformed (A31/Tr), tumorigenic (A31/Tu) and metastatic (A31/Me) cell lines. Parental and transformed cell lines showed actin filaments organized as stress fibers distributed longitudinally near the ventral cell membrane and vinculin deposited mostly in patches throughout the underlying membrane. Contrary, tumorigenic and metastatic cell lines showed actin structures organized as tightly packed short filament bundles distributed near the inner cell circumference and vinculin in small spots and patches. In this work, we take advantage of this cell line series with different cytoskeleton patterns to evaluate their locomotive behaviour on different culture substrata.

MATERIALS AND METHODS. The BALB-3T3/A31 untransformed cloned cell line and its derived sublines were obtained from Dr. A. Raz (Weizmann Institute of Science, Israel). These cells were maintained at 37°C in a humidified atmosphere of 7% CO2 in a DMEM medium to which glutamine, non-essential aminoacids, vitamins, antibiotics and 10% fetal calf serum (Flow Labs, Irvine, Scotland) were added. For each locomotive assay, 1 x 10^5 cells were plated on plastic, type I collagen-coated (30 µg/ml, Biochrom KG, Berlin, WG), or fibronectin-coated (15 µg/ml, Prof. Pierre Martin, Fac. Medicine, Marseille) 35 mm. tissue culture dishes (Nunc, Denmark). Then, they were maintained in the standard culture conditions for 12 hours to allow the cells to attach. Subsequently, dishes were filmed by means of a computer-driven system of time-lapse microcinematography, which controls temperature, CO2 intake and taking of frames. A frame was taken every 60 secs. in Agfa Moviechrome 40 film. Quantitative analysis of time lapse films was carried out by a dynamic recording of cell tracks on a computer-aided digitizer table.

RESULTS AND DISCUSSION. The recording of spontaneous locomotive patterns of 3T3/A31 cell variants cultured on plastic in similar conditions (Table I; Fig. 1a y 1c) reveals a decreasing locomotive speed which goes parallel to both the progressive cytoskeleton disarray and the increasing in vivo malignancy in immunodeppressed

Table I. Speed (µm/hour; mean ± SD) of 3T3/A31 cell variants

Cell lines	Plastic	Fibronectin
3T3/A31-Parent	43.2 ± 10.2	37.4 ± 9.2
3T3/A31-Transformed	40.8 ± 10.8	---
3T3/A31-Tumorigenic	25.2 ± 7.2	---
3T3/A31-Metastatic	14.4 ± 10.2	35.7 ± 9.4

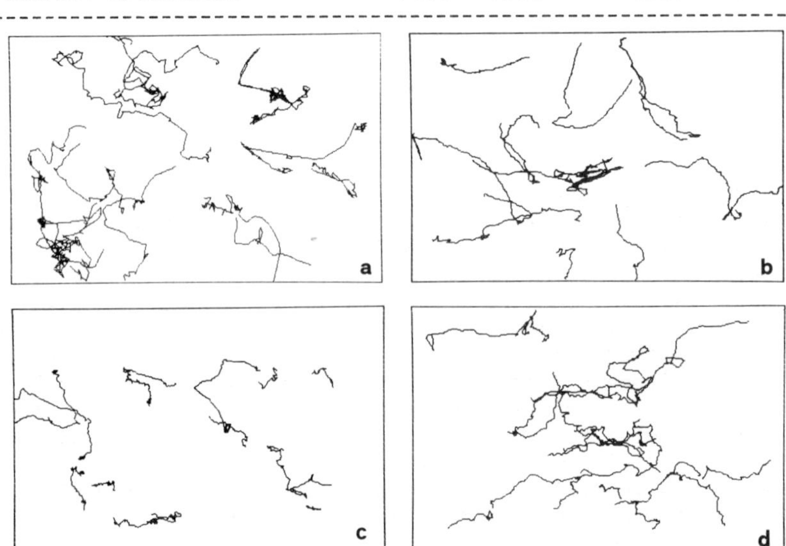

Fig. 1. Plots of cell motility patterns from 3T3/A31 cell variants. Top: Parental line (a: plastic; b: fibronectin). Bottom: Metastatic line (c: plastic; d: fibronectin)

mice as shown previously (3). When the two extreme cell lines (parental and metastatic) were examined on type I collagen, no differences were observed between them since all the cells slowly attached, a small percentage divided and spread forming discrete colonies and no locomotion of cells were observed. However, when cultured on fibronectin matrix, cell movement was significantly stimulated in the metastatic variant while parental cells slightly reduced their motility with respect to that on plastic (Table I; Fig. 1**b** and 1d). Thus, motile properties of 3T3/A31 cell lines seem not to be dependent on the cytoskeleton disarray. Further studies would be directed to elucidate whether reorganizing cytoskeleton responses take place at specific sites of tumor cell invasion which allow increased expression of the metastatic phenotype.

REFERENCES

(1) Liotta, L.A. (1986) Cancer Res. 46, 1-7.
(2) Strauli, P. and Haemmerli, G. (1984) Cancer Met. Rev. 3, 127-141.
(3) Zvibel, I. and Raz, A. (1985) Int. J. Cancer, 36, 261-272.

THE COMPARISON OF CYTOSKELETAL PROTEINS FROM FOUR DIFFERENT FIBROBLAST CELL LINES OF DIFFERING DEGREES OF TRANSFORMATION

L.A. Girao and B.C. Davidson

Department of Medical Biochemistry, University of the Witwatersrand Medical School, York Road, Parktown, Johannesburg, 2193, South Africa

INTRODUCTION

The cell cytoskeleton is involved in a wide variety of functions ranging from chromosome movement during mitosis to endocytosis, exocytosis and especially plasma membrane stabilisation (1). It has been previously shown that there are differences in plasma membrane fluidity between cells (2). It is possible, therefore, that there may be differences in both type and amount of protein found in the cytoskeletons from a range of cells of differing degrees of transformation. A common difficulty is the very slow growth of most normal cells in culture. Skin fibroblasts, however, grow well in culture (3). We thus have examined the four fibroblast cell lines of differing degree of transformation available in this laboratory. At the same time an attempt was made to exacerbate any membrane related changes by incubating the cell lines with polyenoic fatty acids, since such fatty acids are known to be associated with membrane fluidity processes (2).

METHODS

Four different adherent fibroblast cell lines were used in this study. One was a normal human cell line, whilst the other three were a 3T3 benign, 3T6 benign, and a malignant line derived from 3T6. Cells were seeded at 100,000 cells/ml in medium consisting of 90% DMEM and 10% FBS. The cell cultures were incubated for 72 hours at 37°C in an atmosphere of 5% carbon dioxide:95% air, and the cell viability determined by the Trypan Blue exclusion method (4). Cells were detached from the tissue culture flasks by trypsinisation. The suspended cells were then centrifuged at 1000 x g for 10 minutes, washed in 10 ml phosphate-buffered isotonic saline (PBS) and recentrifuged at 1000 x g for 10 minutes. Washed cell pellets were resuspended in 5 ml PBS pH 7.4 containing glycerol, Triton X-100, phenylmethylsulphonyl fluoride and pepstatin. Following a 10 minute incubation at room temperature, samples were centrifuged for 15 minutes at 700 x g, and the supernatants discarded (5). Protein content was assayed by the Lowry-Folin method (6). Pellets were diluted 1:1 with a solution of 0.0625M Tris-HCl pH 6.8 containing SDS, glycerol, mercaptoethanol, bromophenol blue and phenylmethylsulphonyl fluoride (7). Samples were boiled for

2 minutes to solubilize the pellets, and then centrifuged for 30 seconds. Aliquots containing 60 μg protein were loaded onto 10% resolving polyacrylamide gels (PAG) using a 5% stacking gel, and the cytoskeletal proteins resolved into their different components by electrophoresis.

RESULTS AND DISCUSSION

Table 1 shows the results of the electrophoretic analyses as relative percentages and amounts of the cytoskeletal components from 60 μg protein loaded onto the gels. The number of cytoskeletal components was the greatest in the human skin fibroblast (21), whilst the number in the 3T3 and the 3T6 benign were shown to be 17 and 16 respectively; the 3T6 derived only had 10 components. The 87,77,72,67,26 and 18 kD moeities were found to be common to all four of the fibroblast cell lines. The amount of cytoskeletal protein per cell was found to be the same irrespective of the cell line, and thus the μg cytoskeletal protein amounts shown in Table 2 are directly comparable with respect to specific molecular weight components. The components absent from the benign and malignant cell lines appeared mainly to be lost in groups closely related on a molecular weight basis. This effect was most pronounced in the 3T6 derived cells, but was still noticeable in the 2 benign lines. It is possible that the changes observed relate to different plasma membrane stabilisation requirements of transformed cells, and also to the altered plasma membrane fluidity demonstrated in such cells (2). This is only a part reason, however, since cells grown with polyenoic fatty acids exhibit decreased cell viability but no further differences in cytoskeletal proteins. A further factor may be differences in the nuclear production of RNA coding for cytoskeletal proteins.

CONCLUSION

1. The degree of transformation is inversely related to the number of different molecular weight protein components.

2. Exogenous unsaturated fatty acids do not influence the range of cytoskeletal proteins expressed.

REFERENCES

1. Lux S A, Nature, Lond., 1979, 281: 426-429.
2. Spector A A and Yorek M A, J.Lipid Res.,1985,26: 1015-1035.
3. Spector A A, Denning G M, Stoll L L, In Vitro,1980,16:932-940.
4. Paul J, Cell & Tissue Culture, 5th Edition, London, Churchill-Livingstone Press, 1975, p.368.
5. Giometti C S, Willard K E and Anderson N L, Clin.Chem., 1982, 28: 955-961.
6. Lowry D H, Rosenbrough N J, Farr A L and Randall R J, J.Biol, 1951, 193: 265-275.
7. Laemmli U K and Fauvre M, J.Mol.Biol., 1973, 80: 575-579.

Table 1. Quantities of cytoskeletal proteins detected in each of the four fibroblast cell lines used

Mr	Cell Type: HSF		3T3B		3T6B		3T6D	
	RP	Pn(µg)	RP	Pn(µg)	RP	Pn(µg)	RP	Pn(µg)
87	2.8[0.7]	1.7	3.9[0.5]	2.3	3.9[0.5]	2.3	3.4[0.7]	3.4
82	2.4[0.3]	1.4	4.6[0.7]	2.8	-	-	-	-
80	2.5[0.5]	1.5	-	-	4.7[0.3]	2.8	-	-
77	3.5[0.5]	2.1	4.2[0.9]	2.5	3.4[0.9]	2.0	5.4[0.5]	3.2
72	3.7[0.6]	2.2	5.0[0.5]	3.0	4.5[1.0]	2.7	9.7[0.2]	5.8
67	4.3[0.9]	2.6	4.7[0.5]	2.8	6.0[0.6]	3.6	4.0[0.5]	2.4
61	12.9[1.6]	7.7	-	-	13.9[1.4]	8.3	8.8[1.8]	5.3
57	4.9[1.0]	2.9	4.6[0.2]	2.8	-	-	-	-
53	13.4[0.9]	8.0	14.0[0.7]	8.4	6.2[0.3]	3.7	-	-
48	2.3[0.2]	1.4	-	-	12.2[1.0]	7.3	-	-
46	2.0[0.4]	1.2	6.8[0.8]	4.1	-	-	-	-
41	3.0[0.9]	1.8	-	-	2.7[0.3]	1.6	-	-
38	2.1[0.4]	1.3	5.1[1.0]	3.1	-	-	-	-
36	3.1[0.2]	1.9	-	-	-	-	12.7[0.9]	7.6
34	1.6[0.2]	1.0	5.2[1.4]	3.1	-	-	-	-
31	1.7[0.3]	1.0	2.8[0.3]	1.7	-	-	-	-
28	3.1[0.3]	1.9	4.0[0.4]	2.4	10.6[0.6]	6.4	-	-
26	2.6[0.8]	1.6	6.1[0.7]	3.7	3.6[0.6]	2.2	10.0[0.5]	6.0
24	-	-	5.2[0.6]	3.1	3.2[0.9]	1.9	-	-
21	-	-	3.4[0.5]	2.0	4.0[0.9]	2.4	-	-
18	14.6[1.9]	8.8	3.1[1.0]	1.9	2.1[0.4]	1.3	22.4[0.6]	13.4
15	3.0[0.2]	1.8	-	-	14.4[1.2]	8.6	7.8[0.6]	4.7
13	10.8[1.1]	6.5	17.6[1.7]	10.6	4.9[0.8]	2.9	15.0[1.7]	9.0

Mr values expressed as kiloDaltons. Relative Percentages (RP) with sem in square brackets []. n = 8 in all cases. Protein (Pn) = wt. of protein from 60µg applied to the gels. Each cell type contained 1.45µg protein per 10^5 cells except HSF (1.48 µg).
HSF = human skin fibroblast. 3T3B = 3T3 benign. 3T6B = 3T6 benign. 3T6D = 3T6 derived.

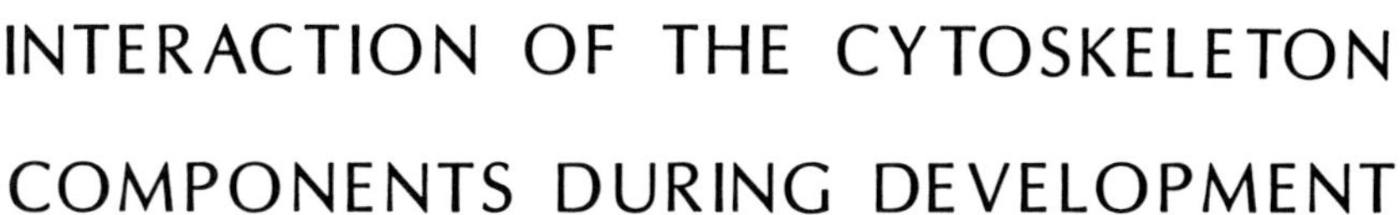

INTERACTION OF THE CYTOSKELETON COMPONENTS DURING DEVELOPMENT

MATRIX CYTOSKELETON INTERACTIONS DURING ODONTOBLAST AND AMELOBLAST DIFFERENTIATION

J.V. RUCH, H. LESOT and D. KUBLER

Institut de Biologie Médicale, UER Médecine 11, rue Humann 67085 Strasbourg Cedex, France.

ABSTRACT

Odontoblast and ameloblast terminal differentiation requires specific plasma membrane-mediated matrix-cytoskeleton interactions. Redistribution of actin, α-actinin and vinculin occured in a similar manner during polarization of these cells. Vinculin interacted with talin and probably with 3 high molecular weight membrane associated proteins. A monoclonal antibody (MC16A16) directed against a high molecular weight membrane associated protein produced a decrease in vinculin containing focal contacts and disorganization of stress fibers of dental epithelial cells in primary culture. This antibody also inhibited the polarization of odontoblasts.

INTRODUCTION

Histomorphogenesis and cytodifferentiation of most embryonic tissues and organs are under the control of epithelial-mesenchymal matrix mediated interactions (Kemp and Hinchliffe, 1984), however, the nature and mechanisms of these interactions are still under investigation.

The embryonic tooth constitutes an accurate model for the analysis of such interactions (for reviews see Slavkin, 1974 ; Thesleff and Hurmerinta, 1981 ; Ruch, 1984, 1985). The tooth consists of two interacting tissues : the epithelial enamel organ and the neural crest derived dental papilla. The overt differentiation of odontoblasts (secreting the predentin-dentin components) and ameloblasts (secreting enamel proteins) takes place according to a tooth specific timing and spatial pattern. Cell differentiation starts from the area of the cuspal tips and proceeds in a cervical direction. The terminal differentiation of odontoblasts preceeds that of the ameloblasts. The preodontoblasts underlining the inner dental epithelium become post-mitotic, polarize and secrete predentin. The preameloblasts, (cells of the inner dental epithelium) become post-mitotic shortly after the odontoblasts, but they polarize only after a layer of predentin has been secreted. The enamel secretion by ameloblasts occurs at the onset of mineralization of dentin.

The terminal differentiation of odontoblasts is triggered by specific interactions with the mesenchymal face of a stage and space specific basement membrane. Temporal and spatial modifications of the dental basement membrane exist during terminal differentiation of odontoblasts : collagen type III disappears from the epithelial mesenchymal junction and fibronectin which surrounds preodontoblasts is restricted to the apical pole of odontoblasts. ^{3}H-glucosamine and ^{35}sulfate radioautography allowed also for the detection of differential accumulation and turnover of extracellular matrix components at the epithelial mesenchymal junction during terminal differentiation of odontoblasts. Secretion of predentin by odontoblasts is a prerequisite to the differentiation of ameloblasts.

Polarization of odontoblasts and ameloblasts requires the integrity of the cytoskeleton. Lesot et al., (1982) demonstrated a redistribution of vimentin during polarization of odontoblasts and an apical accumulation of prekeratin and actin during ameloblast differentiation. Plasma membrane modifications of polarizing odontoblasts have also been suggested.

We hypothesyzed that stage and space specific informations are encoded in the dental extracellular matrix and that the plasma membrane of preodontoblasts and preameloblasts recognizes these signals. Recognition probably requires the existence of different cell membrane domains able to interact specifically with matrix components. Cell membrane organization will permanently reflect the architecture of the matrix and plasma membrane conformation probably controls directly or indirectly the functional state of the cytoskeleton, which may affect morphological and functional aspects of cytodifferentiation (Ruch et al., 1984). According to this working hypothesis we have analyzed the dental cell interactions with type I collagen and fibronectin (Lesot et al., 1985). Isolated dental papillae and enamel organs were cultured for increasing periods of time in the presence of iodinated fibronectin or type I collagen and both matrix molecules were observed to preferentially bind to dental papillae. Membrane proteins were prepared from isolated enamel organs and dental papillae. After separation by electrophoresis on SDS-polyacrylamide gels these proteins were transferred onto nitrocellulose and then incubated in the presence of either iodinated fibronectin or iodinated type I collagen. Autoradiography confirmed the preferential interaction of fibronectin with the membrane of dental papilla cells : Fibronectin interacted strongly with three high molecular weight proteins (145Kd, 165Kd and 185Kd) and weakly with a protein of 56Kd . These proteins were not detected when membranes were prepared from enamel organs. These membrane associated proteins probably correspond to the fibronectin receptor identified by Hughes et al. (1981), Akiyama et al. (1986).

In view of the increasing evidence for transmembranous matrix-cytoskeleton interactions (Horwitz et al., 1986 ; Singer and Paradiso, 1981 ; Izzard et al., 1986), we have studied 1) the distribution patterns of actin, α-actinin and vinculin in dental cells ; 2) the interaction of membrane associated proteins with α-actinin and vinculin and 3) the action of antibodies directed against membrane associated proteins on microfilament organization.

RESULTS

1) <u>Localization of actin, α-actinin and vinculin in dental cells.</u>

The localization of actin was studied using rhodamine and fluorescein labelled phalloidin. The localization of α-actinin and vinculin was performed by indirect immunofluorescent staining using rabbit anti-α-actinin and mouse anti-vinculin (BioYeda) and FITC-conjugated anti-rabbit and anti-mouse immunoglobulins (Cappel, Nordic Immunology).

The localization of these components in epithelial and mesenchymal cells of primary cultures of trypsin isolated enamel organs and dental papillae are in agreement with the well known distribution patterns in cultured cells (Fig. 1, 2, 3). The distribution patterns of actin, α-actinin and vinculin were also studied on frozen sections of mouse first lower molars.

Actin : Intense fluorescence was observed in preameloblasts and in the stratum intermedium as well as in preodontoblasts and dental papilla cells (Fig. 4). A redistribution of actin occured during terminal differentiation of odontoblasts and ameloblasts (Fig. 5). The apical and distal extremities of ameloblasts exhibited strong fluorescence corresponding to the terminal webs and intracellular fibers were observed (Fig. 6). Differentiating odontoblasts demonstrated an apical staining corresponding to the terminal web. The cell bodies were not stained but fluorescent filaments were observed in odontoblastic cell processes of predentin-dentin (Fig. 7). This data is in agreement with observations by Nishikawa and Kitamura (1986).

α -actinin : Preameloblasts were strongly stained while preodontoblasts and dental papilla cells exhibited faint fluorescence (Fig. 8). Codistribution with actin was observed during terminal differentiation of odontoblasts and ameloblasts. In ameloblasts, α-actinin accumulated at both extremities and fibrillar structures were seen (Fig. 9). During polarization of odontoblasts a weak but significant apical accumulation of α-actinin was observed (Fig. 10) and cell processes were stained (Fig. 11). A control using preimmune globulins is shown in Fig. 12.

Vinculin : The preameloblasts and the cells of the stratum intermedium exhibited more fluorescence than the preodontoblasts and dental papilla cells (Fig. 13). During polarization of ameloblasts we observed an accumulation in the region of the terminal webs (Fig. 14). The apical pole of polarized odontoblasts demonstrated faint fluorescence (Fig. 15) and the odontoblastic cell processes exhibited a weak staining (Fig. 16).

2) Interaction of membrane associated proteins with α-actinin and vinculin. Membrane were prepared according to Cates and Holland (1978) from 15-17 days old mouse embryos and proteins analysed by SDS-polyacrylamide gel electrophoresis (SDS-PAGE). For blotting experiments, samples were transferred onto nitrocellulose (Fig. 17) which was saturated with 4% BSA and incubated with purified, iodinated α-actinin and vinculin (Fig. 18, b,a). After washing the strips were exposed for autoraiography.

α-actinin bound strongly to a 110Kd protein and weakly to a 220Kd protein (Fig. 19b). Iodinated vinculin interacted strongly with 220Kd

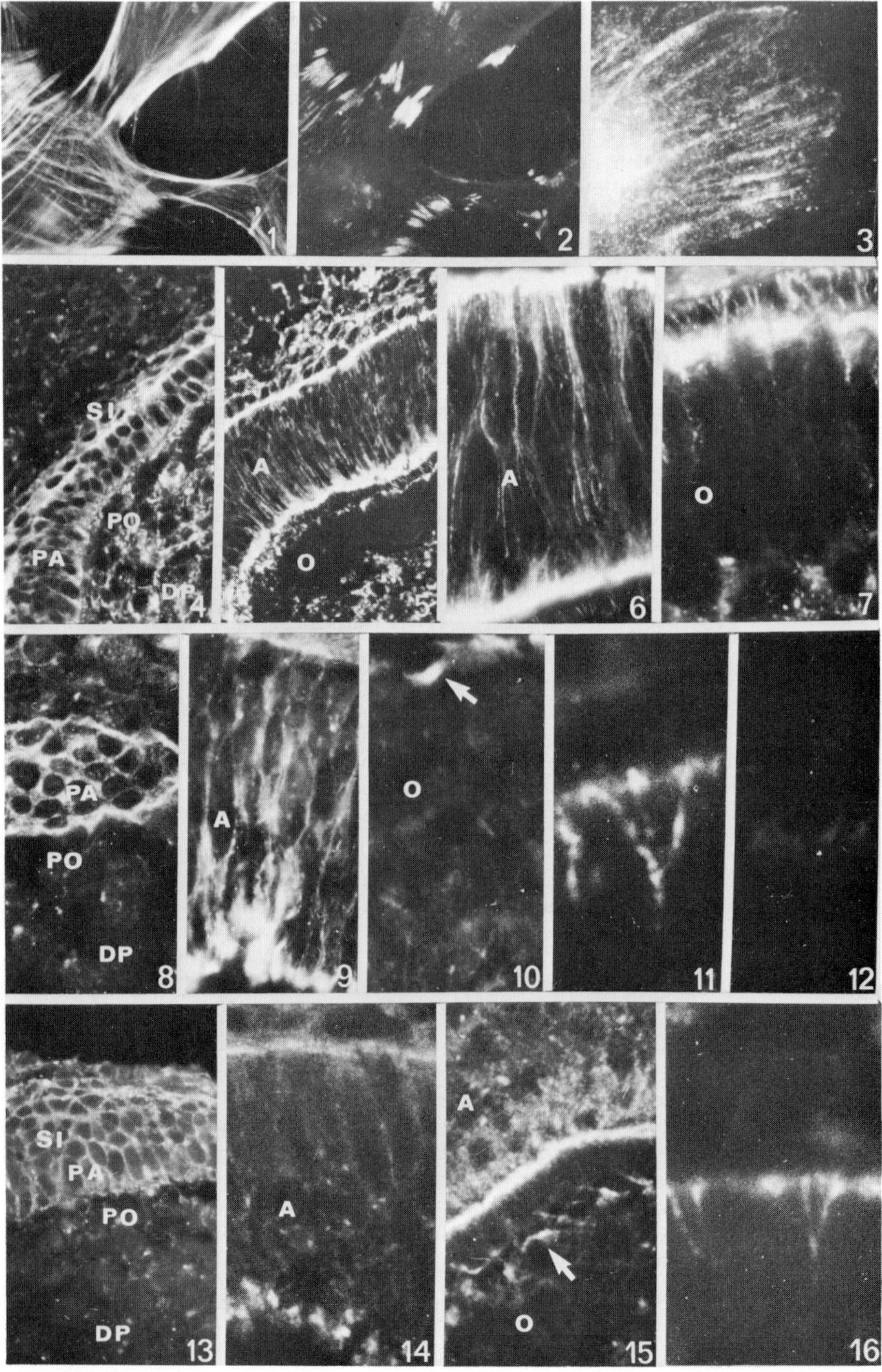
1
2
3
SI
PO
PA
DP
4
A
O
5
A
6
O
7
PA
PO
DP
8
A
9
O
10
11
12
SI
PA
PO
DP
13
A
14
A
O
15
16

Fig. 1-3. Primary cultures of dental cells : double staining with rhodamine-phalloidin (Fig. 1) and anti-vinculin (Fig. 2) of paraformaldehyde (4%) fixed and Triton X-100 (0.1%) permeabilized cells. Fig. 3 : Triton X-100 permeabilized cells stained with anti α-actinin.

Fig. 4-16. Triton X-100 permeabilized, mouse molar frozen sections. Fig. 4-7 : Sections stained with phalloidin. Fig. 8-11 : Sections stained with rabbit anti-α-actinin. Fig. 12 : Control using rabbit preimmune globulins. Fig. 13-16 : Sections stained with anti-vinculin. SI : stratum intermedium ; PA : preameloblast ; PO : preodontoblast; DP : dental papilla ; A : ameloblast ; O : odontoblast. Arrows indicates the apical fluorescence of odontoblasts.

and 190Kd proteins and weakly with three other proteins (170, 160 and 150Kd) (Fig. 19a).

Preliminary control experiment using purified fibronectin (Fig. 20b), and talin (Fig. 20a) showed that iodinated vinculin interacted with talin (Fig. 20c). The presence of 190Kd band is in agreement with O'Halloran and Burridge (1986) and corresponds probably to a degradation product of talin. No interaction was observed with fibronectin (Fig. 20d).

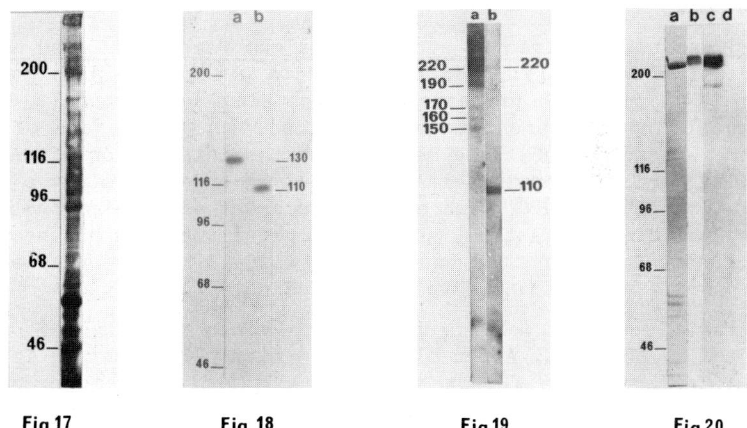

Fig. 17. Membrane proteins extracted from mouse embryo transferred onto nitrocellulose, and stained with amidoblack.

Fig. 18. SDS polyacrylamide gel electrophoresis of purified iodinated vinculin (Fig. 18a) and α-actinin (Fig. 18b) detected by radioautography.

Fig. 19. Binding of iodinated vinculin (Fig. 19a) and α-actinin (Fig. 19b) to plasma membrane proteins transferred onto nitrocellulose.

Fig. 20. SDS-polyacrylamide gel electrophoresis of purified fibronectin (Fig. 20b) and talin (Fig. 20a) stained with Coomassie blue. Vinculin did not bind to transferred fibronectin (Fig. 20d) but interacted with talin (Fig. 20c).

Fig. 21–24. Frozen sections of mouse molars (Fig. 21, 22, 23) and permeabilized dental epithelial cells in primary culture stained with MC16A16 antibodies (Fig. 24) and detected with FITC-rabbit anti-rat IgG. A : ameloblast ; BV : blood vessel ; DM : dental matrix ; Od : odontoblast ; PA : preameloblast ; PO : preodontoblast ; SI : stratum intermedium.

Fig. 25–30. Dental epithelial cells grown in primary culture. Cells were permeabilized with 0.1% Triton X-100, fixed with 4% paraformaldehyde and stained either with rhodamine-phalloidin (Fig. 25, 26, 27) or with anti-vinculin antibodies (Fig. 28, 29). To test the effects of MC16A16 on the cytoskeleton, cultures were incubated for 30 min. in the presence of this antibodies before staining (Fig. 26, 29) and compared to control cultures (Fig. 25, 28). In treated cultures, intracellular MC16A16 was visualized by staining permeabilized and fixed cultures with FITC-rabbit anti-rat antibodies (Fig. 30). The effect of MC16A16 on the stress fiber organization could be reversed by washing away the excess of MC16A16 and postculture of the cells for 30 min. in the absence of antibody (Fig. 27).

Fig. 31–32. Effect of MC16A16 supernatant on dental epithelial cells in primary culture : phase contrast microscopy. Well spread cells at the periphery of the explant (Fig. 31) rapidly detached and had a rounded shape (Fig. 32).

3) Antibodies directed against cell membrane-associated antigens: biological effects. Fibronectin interacts with three high molecular weight (HMW) membrane proteins (140Kd-complex) and recently, the attachment of microfilament termini to the cytoplasmic face of plasma membrane was suggested to be mediated by the interaction of talin with the 140Kd complex (Horwitz et al., 1986). In order to study a possible implication of such HMW membrane and membrane associated proteins in odontoblast differentiation, an immunological approach has been developed. Antibodies directed against HMW (110-190Kd) membrane and membrane-associated proteins have been prepared.

Plasma membrane were prepared from mouse embryos and proteins separated by SDS-PAGE. After blotting on nitrocellulose and incubation with ^{125}I-fibronectin (Lesot et al., 1985) this protein preparation was found to contain the 140Kd complex. Fisher rats were immunized by three injections of polyacrylamide gel containing proteins with MW ranging from 110 to 190Kd. The splenocytes of immunized rats were fused with NS1 myeloma cells. Hybridoma were selected and cultures screened for antibody production by immunodot assay and by indirect immunofluorescence either on frozen sections of mouse molars or on dental cells grown in primary cultures; In this report we summarize preliminary data obtained with a supernatant produced by a hybrid clone (MC16A16).

Western immunoblotting allowed the detection of a 165Kd antigen.

By indirect immunofluorescence, a strong staining of the stratum intermedium and of the preameloblasts (Fig. 21) was observed. Filaments were found within dental papillae cells (Fig. 22, 23) and blood

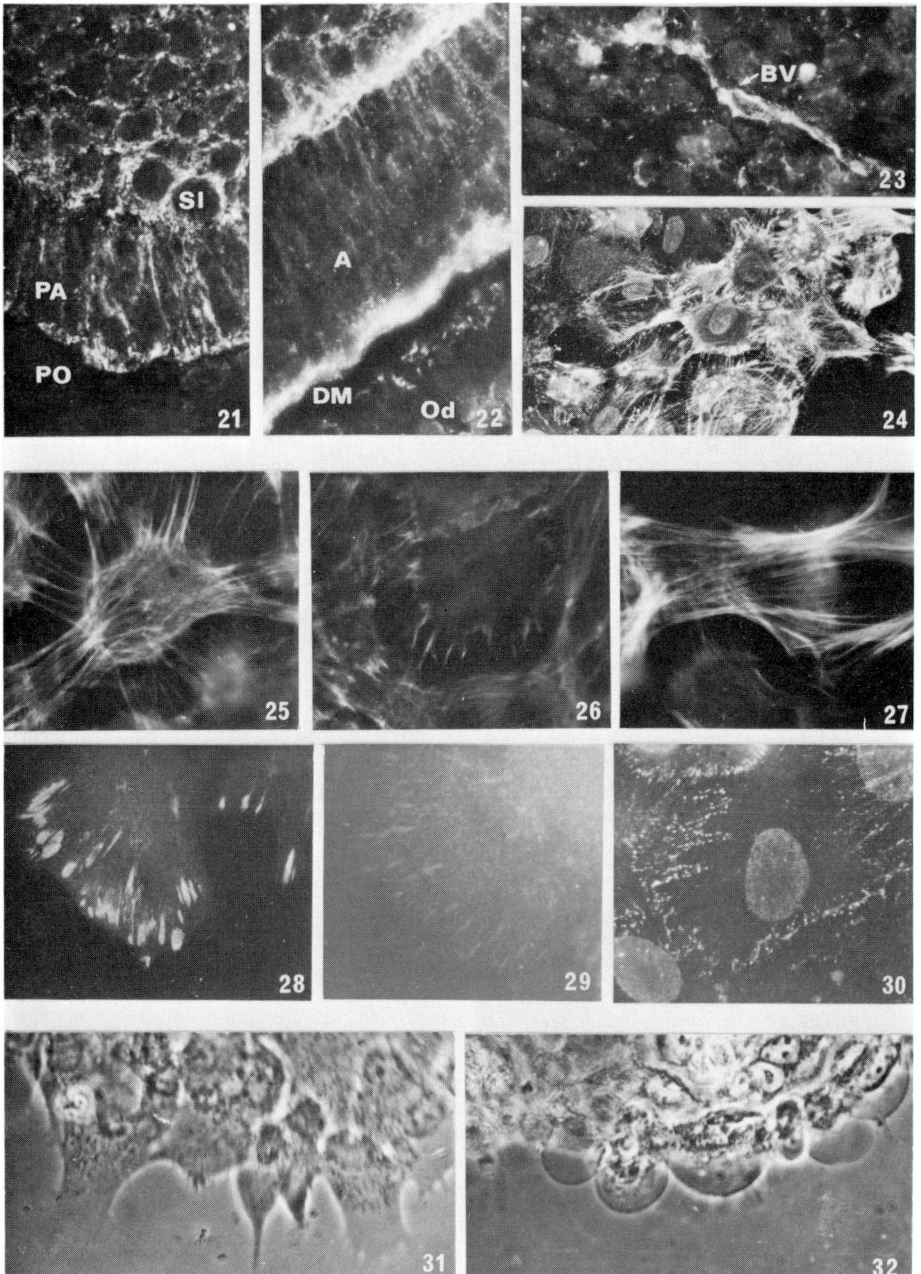
SI
PA
PO
21
A
DM
Od
22
BV
23
24
25
26
27
28
29
30
31
32

vessels were heavily stained (Fig. 23). During the polarization of ameloblasts and odontoblasts, MC16A16 accumulated in the region of the terminal webs (Fig. 22). MC16A16 stained fibrillar structures of permeabilized cells grown in primary culture (Fig. 24).

When added to the culture medium of dental epithelial cells grown in primary culture, the supernatant of MC16A16 hybridoma produced a rapid dissociation and retraction of the cells localized within the explant. Cells at the periphery of the explant (Fig. 31) had retracted already 2 min. after the addition of antibodies and then started to round up (Fig. 32).

MC16A16 produced a disorganization of actin containing stress fibers (Fig. 25, 26). Actin was observed in a diffuse pattern and accumulated at the periphery of the cells (Fig. 26). This effect required the internalization of MC16A16 which after permeabilization was localized preferentially at the periphery of the cells (Fig. 30) and could be reversed (Fig. 27). Incubation of dental epithelial cells in the presence of MC16A16 also resulted in a decrease of vinculin-containing focal contacts (Fig. 28, 29). However other cytoskeletal constituents such as microtubules and keratin intermediate filaments were not affected. Control supernatants had no effect on the behaviour of epithelial cells.

Day-16 molars containing only preodontoblasts were cultured for 4 days in control medium (RPMI supplemented with 15% fetal calf serum) or in the presence of the same medium containing MC16A16 hybridomas (10^5 cells/ml).

The control teeth demonstrated normal polarization of odontoblasts (Fig. 33). In the presence of MC16A16 the differentiation of odontoblasts did not occur (Fig. 34), although mitotic activity was not decreased. Hybridoma producing non-related antibodies had no effect.

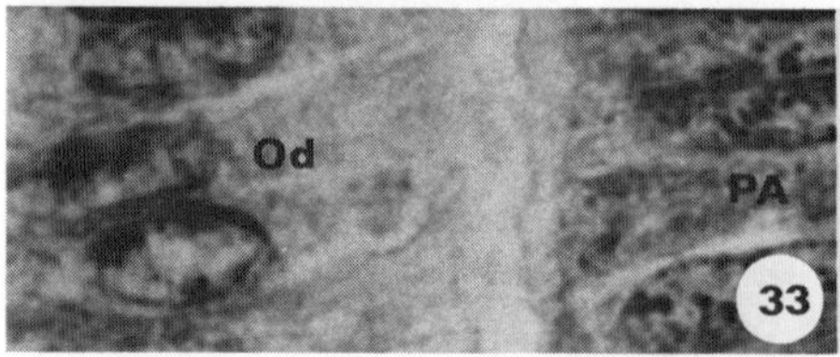

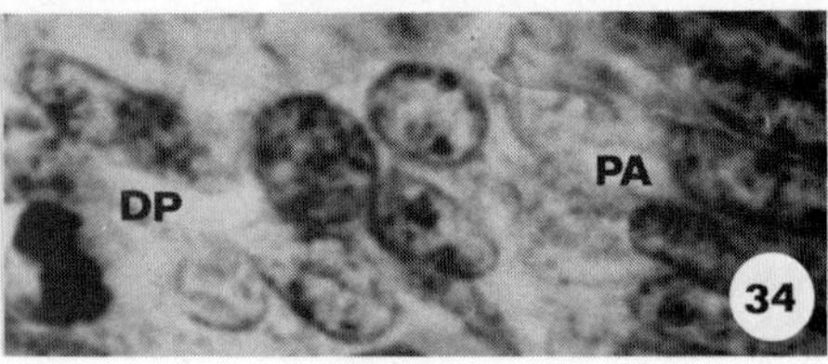

Fig. 33-34. Day 16 mouse molars cultured for 4 days in RPMI synthetic medium supplemented with 15% FCS. Fig. 33 : control culture. Fig. 34: tooth germ cultured in the presence of MC16A16 producing hybridoma. PA : preameloblast ; DP : dental papilla ; Od : odontoblast.

The distribution pattern of fibronectin in teeth cultured in control medium was the same as it would have been in vivo. Fibronectin surrounded the preodontoblasts and dental papilla cells, but was restricted to the apical pole of polarized odontoblasts. In teeth grown in the presence of the hybridoma producing MC16A16 the fibronectin fibers appeared to be disorganized.

DISCUSSION

During development the extracellular matrix demonstrates compositional and structural diversity, which is reflected in the morphology of different cells and tissues (Watt, 1986). Matrix influences on the organization of the cytoskeleton will affect both morphogenesis and cytodifferentiation (Ben Ze'ev, 1986). However the understanding of the molecular mechanisms which allow the membrane relationship between the matrix and the cytoskeleton needs further study. The best documented data concern the adherens junctions, where a transmembranous linkage between matrix molecules (fibronectin) and actin-associated components exists. Talin or vinculin might interact with the fibronectin membrane receptor (Singer and Paradiso, 1981 ; Horwitz et al., 1986).

As far as the differentiation of odontoblasts or ameloblasts is concerned the importance of specific cell-matrix interactions is well documented. Fibronectin might be involved in the polarization of odontoblasts (Lesot et al., 1985). During polarization of both odontoblasts and ameloblasts a redistribution of actin and associated α-actinin and vinculin occurs.

The preliminary data on interactions between plasma membrane associated proteins and vinculin indicates an interaction between vinculin and talin, which is in agreement with the data of Otto (1983) and Wilkins et al. (1983). The interaction of vinculin with three other high molecular weight proteins (150, 160, 170Kd) raises the question of their nature. It will be important to determine whether fibronectin also interacts with these proteins.

Another approach to the study of membrane-cytoskeleton interaction involved the production of antibodies directed against high molecular weight membrane associated proteins. One of these antibodies (MC16A16) produced a dissociation and retraction of cultured dental epithelial cells resulting from a decrease in vinculin containing focal contacts and disorganization of the stress fibers. MC16A16 also inhibited directly or indirectly the polarization of odontoblasts. This monoclonal antibody constitutes an interesting tool for further analysis of the matrix-cytoskeleton interactions involved in dental cytodifferentiation.

ACKNOWLEDGMENTS

These studies were supported by CNRS, INSERM and the Fondation pour la Recherche Médicale.

REFERENCES

Akiyama, S.K., Yamada, S.S. and Yamada K.M. (1986). Characterization of a 140Kd avian cell surface antigen as a fibronectin binding molecule. J. Cell Biol. 102, 442-448.

Ben Ze'ev, A. (1986). The relationship between cytoplasmic organization gene expression and morphogenesis. Trends Biochem. Sci. 11, 478-481.

Cates, G.A. and Holland, P.C. (1978). Biosynthesis of plasma-membrane proteins during myogenesis of skeletal muscle in vitro. Biochem. J. 174, 873-881.

Horwitz, A., Duggan, K., Buck, C., Beckerle, M.C. and Burridge, K. (1986). Interaction of plasma membrane fibronectin receptor with talin: a transmembrane linkage. Nature 320, 531-533.

Hughes, R.C., Butters, T.D. and Aplin, J.D. (1981). Cell surface molecules involved in fibronectin-mediated adhesion. A study using specific antisera. Eur. J. Cell Biol. 26, 198-207.

Izzard, C.S., Radinsky R. and Culp, L.A. (1986). Substratum contacts and cytoskeletal reorganization of BALB/c 3T3 cells on a cell-binding fragment and heparin-binding frangments of plasma fibronectin. Exp. Cell Res. 165, 320-336.

Kemp, R.B. and Hinchliffe, J. (eds) (1984). Matrices and Cell Differentiation. A.R. Liss, Inc., New York.

Lesot, H., Meyer, J.M., Ruch, J.V., Weber, K., Osborn, M. (1982). Immunofluorescent localization of vimentin, prekeratin and actin in odontoblast and ameloblast. Differentiation 21, 133-137.

Lesot, H., Karcher-Djuricic, V., Mark, M., Meyer, J.M. and Ruch, J.V. (1985). Dental cell interaction with extracellular-matrix constituents : type I collagen and fibronectin. Differentiation 29, 176-181.

Nishikawa, S. and Kitamura, H. (1986). Localization of actin during differentiation of the ameloblast, its related epithelial cells and odontoblasts in the rat incisor using NBD-phallacidin. Differentiation 30, 237-243.

O'Halloran, T. and Burridge, K (1986). Purification of a 190Kda protein from smooth muscle : relationship to talin . Biochemica et Biophysica Acta 869, 337-349.

Otto, J.J. (1983). Detection of vinculin-binding proteins with an ^{125}I-vinculin gel overlay technique. J. Cell Biol. 97, 1283-1287.

Ruch, J.V. (1984). Tooth morphogenesis and differentiation. In : Dentin and Dentinogenesis (ed. A. Linde), pp. 47-49. CRC Press, Boca Raton.

Ruch, J.V. (1985). Odontoblast differentiation and the formation of the odontoblast layer. J. Dent. Res. 64, 489-498.

Ruch, J.V., Lesot, H., Karcher-Djuricic, V., Meyer, J.M. (1984). Extracellular matrix-mediated interaction during odontogenesis. In: Matrices and Cell differentiation (ed. R.B. Kemp & J.R. Hinchliffe), pp. 103-114. A.R. Liss, Inc., New York.

Slavkin, H.C. (1974). Embryonic tooth formations. A tool for developmental biology. In : Oral Sciences Reviews (ed. A.H. Melcher). Munksgaard, Copenhagen.

Singer, I.I. and Paradiso, P.R. (1981). A transmembrane relationship between fibronectin and vinculin (130Kd protein) : serum modulation in normal and transformed hamster fibroblasts. Cell 24, 481-492.

Thesleff, I. and Hurmerinta, K. (1981). Tissue interactions in tooth development. Differentiation 18, 75-88.

Watt, F.M. (1986). The extracellular matrix and cell shape. Trends Biochem. Sci. 11, 482-485.

Wilkins, J.A., Chen, K.Y. and Lin, S. (1983). Detection of high molecular weight vinculin binding proteins in muscle and non muscle tissues with an electroblot-overlay technique. Biochem. and Biophys. Res. com. 116, 1026-1032.

STUDIES OF CENTROMERE REGION WITH HUMAN ANTIKINETOCHORE SERUM

M.M. Valdivia[1], A. Tousson[2], R.D. Balczon[2], J.M. Hall[3], S.L. Brenner[3] and B.R. Brinkley[2]

[1]Department of Biochemistry, Facultad de Ciencias, Universidad de Cadiz, Puerto Real, Cadiz, Spain.

[2]Department of Cell Biology and Anatomy, University of Alabama, Birmingham AL 35294, USA.

[3]Department of Cell Biology, Baylor College of Medicine, Houston TX 77030, USA.

ABSTRACT

The term Microtubule Organizing Centre (MTOCs) implies a general site for gathering microtubules into arrays. They represent intracellular locations where microtubule assembly is initiated but they may also be sites where free ends of microtubules are attracted or captured. Two major MTOCs have been described in eukaryotic cells : centrosomes and centromeres. In this report human sera from CREST patients were used in two different studies to investigate the distribution and association of centromeres in mammals. In the first study we used a kinetochore-specific human serum as an immunofluorescent probe to examine centromeres regions in two muntjac species. In Indian muntjac (2N=6 ♀ , 7 ♂) centromeres in mitosis are composed of a linear beadlike arrays of subunits. In Chinese muntjac (2N=46) centromeres consisted of minute fluorescent dots in each chromosome. However in interphase cells the centromeres of the Chinese muntjac clustered into aggregates reminiscent of the beadlike arrays organization seen in the Indian muntjac. It can be concluded from this study that centromeres of the Indian muntjac evolved by linear fusion of unit centromere of the Chinese muntjac. In the second study we investigate the organization of centromeres in mouse meiotic cells. The immunofluorescent pattern in different stages during meiosis was analyzed with a CREST serum. In pachytene cells each bivalent contains one fluorescent spot and these are distributed around the periphery of the nucleus. In posterior stages of prophase the number of fluorescent spots increased from 21 to 40 corresponding to the diploid number of chromosomes. In metaphase II 20 pairs of kinetochores were observed. During spermiogenesis the number of kinetochores correlated with the haploid chromosome number. An association of centromeres into a chromocenter reduces the number of fluorescent spots in mid-spermatid cells. In mature spermatozoa a total absence of staining could be observed. Both studies are illustrative examples of the usefulness of human kinetochore autoantibodies as tools for uncovering the molecular organization of centromeres in mammals.

INTRODUCCION

During cell division a major mitotic event is the separation of sister chromatids of metaphase chromosomes at their centromeres. This movement is made possible by the attachment of spindle microtubules to condensed chromatin. One specific centromere locus, the kinetochore, is the binding site of tubulin or tubulin associated proteins to chromosomes in mitosis and meiosis. At the electron microscopy level, the organization of centromere/ kinetochore region has been the subject of numerous studies. It is known that the morphology of the kinetochore, changes throughout the cell cycle. The kinetochore in prophase presents a globular structure of condensed fibrils and it shifts from a trilamellar plate in metaphase to a deformed plate in anaphase and further it changes into a fibrous material similar to that of chromatin arms in telophase. The association of microtubules with kinetochores occurs in prometaphase. It has not been proved yet whether kinetochores nucleate their own microtubules or attach to microtubules radiating from the centrosome (Mitchinson et al. 1986). The role of kinetochores as a MTOC needs to be elucidated. At the biochemical level some specific components of kinetochores have been indicated (Rieder, 1982). Most of the sera from patients suffering from the CREST syndrome, a variant of systemic sclerosis, stain the centromere/kinetochore region of metaphase chromosomes by immunofluorescent (Moroi et al. 1980) and Fig.1. By using immunoperoxidase and immunogold at the electron microscopy level these sera bind specifically to the kinetochore plates (Brenner et al. 1981),(Valdivia et al. 1986). Using different sera from patients with the CREST disease, several groups have indicated the putative antigens located on the kinetochore by the Western transfer technique (Earnshaw et al. 1984),(Guldner et al. 1984), (Spowart et al. 1985),(Valdivia et al. 1985). In our studies, we have used kinetochore-specific serum derived from human patients with the CREST autoimmune disease as an immunofluorescent probe to examine the distribution and arrangements of centromere regions in two related species, the Indian and Chinese muntjac. We further utilized kinetochore sera to investigate the association and distribution of centromeres in mouse cells during meiosis and spermiogenesis.

EVOLUTION OF COMPOUND KINETOCHORES

Indian muntjacs may be a unique specie among mammals due to their low diploid chromosome number. Cytogenetic studies have suggested that the unusual karyotype has evolved by Robertsonian fusion of chromosomes from the ancestral related deer Chinese muntjac. We have used kinetochore serum to examine centromere/ kinetochore of both muntjac species (Brinkley et al. 1984).

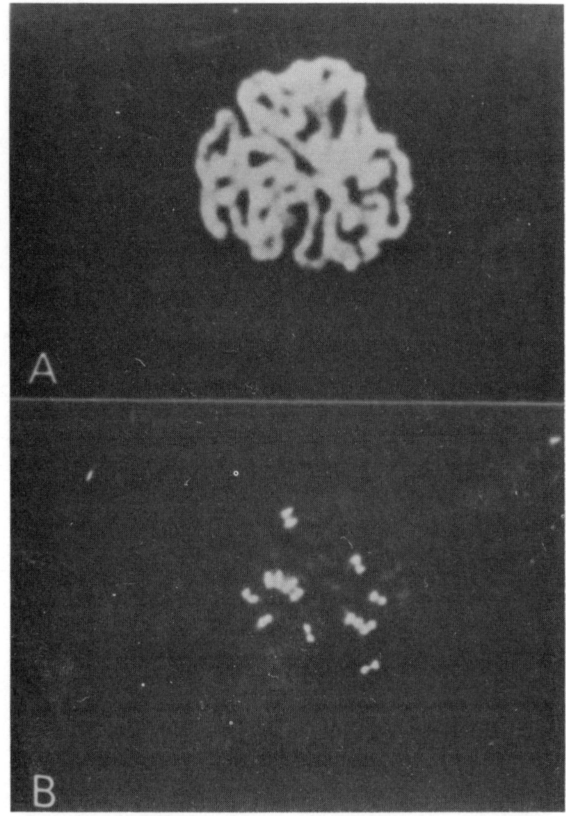

Fig.1. Indirect immunofluorescence of a mammalian cell stained with human kinetochore antiserum. A) DNA of the cell as shown by Hoechst stain. B) Kinetochores of the same cell stained with CREST serum.

It was possible to examine the kinetochores throughout the cell cycle because those autoantibodies bind to mitotic kinetochores as well as to interphase prekinetochores (Brenner et al. 1981). During metaphase the kinetochores of Indian muntjac appear intensely fluorescent when stained by the CREST serum (Fig. 2a). Staining was confirmed to each side of the primary constriction (presumed to be the kinetochore) but not to the entire centromere. In S phase an unexpected pattern of fluorescent was observed. Although some nuclei displayed seven discrete spots (Fig. 2b), most exhibited a much more complex staining pattern. The stained structures were in the form of small dots arranged into large clusters (Fig. 2c) or long threads, beadlike arrays (Fig. 2d) and finally in G2 phase of the cell cycle as double fluorescent spots. We suggested that the variation in staining was due to changes in chromatin organization in the centromere region during the cell cycle.

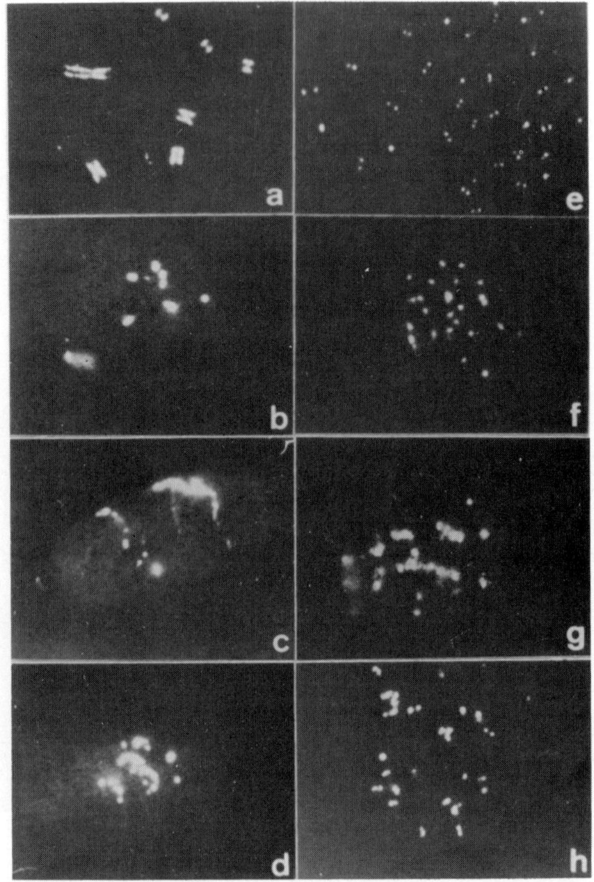

Fig.2. Kinetochore cell cycle in Indian (a,b,c,d,) and Chinese muntjac (e,f,g,h,). A mitotic cell is in a ; b represents a G1-phase pattern; c is a long,thin beadlike strands form ; d shows short strands of beadlike structures ; e represents a late G2-phase patter ; f shows a G1-phase nuclei with single dots ; g beadlike strands form ; h shows paired dots indicating duplicated kinetochores but still some remains as linear arrays of dots. (Reproduced by permission of Chromosoma, Springer Verlag, from Brinkley et al. 1984).

That was indicated by measuring the amount of DNA content and correlates it with the prekinetochore staining pattern. During metaphase the kinetochores of Chinese muntjac appear as minute pairs of fluorescent dots located at the tips of the small telocentric chromosomes (a cell in G2 phase is shown in Fig. 2e). Obviously these kinetochores were much smaller that those of the Indian muntjac and also were all of uniform size. Although metaphase kinetochores of Indian and Chinese muntjac were different in size and staining, their interphase staining were amazingly similar. As shown in Fig.2, the interphase nuclei of the Chinese muntjac displayed a familiar range of staining pattern. Nuclei in Gl consisted of 46 identical stained dots representing prekinetochores in a stage similar to that of Indian muntjac in the Gl phase (compare Fig.2b and 2f). In some other stages prekinetochores appeared to aggregate into fewer but larger arrays (Fig.2g), and short linear arrays (Fig.2h) identical to some configurations adopted by prekinetochores in Indian muntjac (Fig.2c, 2d). The ultrastructure of both Indian and Chinese muntjac kinetochores was also investigated at the electron microscopy level. Serial sections indicated similar structures in both species than in others mammals (Rieder, 1982). Kinetochore stained material was equivalent in both species and supports the idea of the conservation of kinetochores . On previous studies it was shown that the DNA content of Indian muntjac is 20% less than in Chinese muntjac (Wurster and Atkin, 1982). However Schmidtke et al. (1981) found a 2% nucleotide different in single copy sequence of the two species. On the other hand, retention of chromosome arms is supported by the G-banding pattern in the chromosomes of the two muntjac species. Further, highly repetitive DNA sequences are present in all centromere regions of the Indian muntjac chromosomes and also in Chinese muntjac DNA (Yu et al.1986). These data together support the idea of nonrandom association of centromeres during interphase. Our interpretation of Indian muntjac supports the notion that the large compound kinetochores observed in this species evolved by linear fusion of smaller units of an ancestral species like the Chinese muntjac. Such fusion could have been facilitated by the arrangements of prekinetochores during the interphase stage of the cell cycle.

ORGANIZATION OF MEIOTIC KINETOCHORES

The centromeres of mouse chromosomes are characterized by the presence of heterochromatin and satellite DNA (Pardue and Gall, 1970). Very little is known about the replication and distribution of those centromeres during meiosis and gametogenesis. In the present study we have used human autoantibodies from CREST patients as an immunofluorescent probes to investigate the organization of meiotic centromeres during meiosis and spermiogenesis in the mouse (Brinkley et al.1986). Cell preparation and immunofluorescent techniques have been described in recent papers (Valdivia and Brinkley, 1985) (Brinkley et al.1986). In meiosis I, cells in pachytene were the most common stage seen in our preparations. When antikinetochore serum was used to stain pachytene cells, these showed a clustered arrangements and peripheral distribution of fluorescent spots within each nucleus. Usually 21 individual fluorescent spots could de seen in pachytene cells. These fluorescent kinetochores were distributed in 4-6 distint clusters with each containing 3 or 4 individual foci. These 21 centromeres regions correspond with 19 autosomal bivalents and the sex pair. The presence of only 21 fluorescent spots suggested

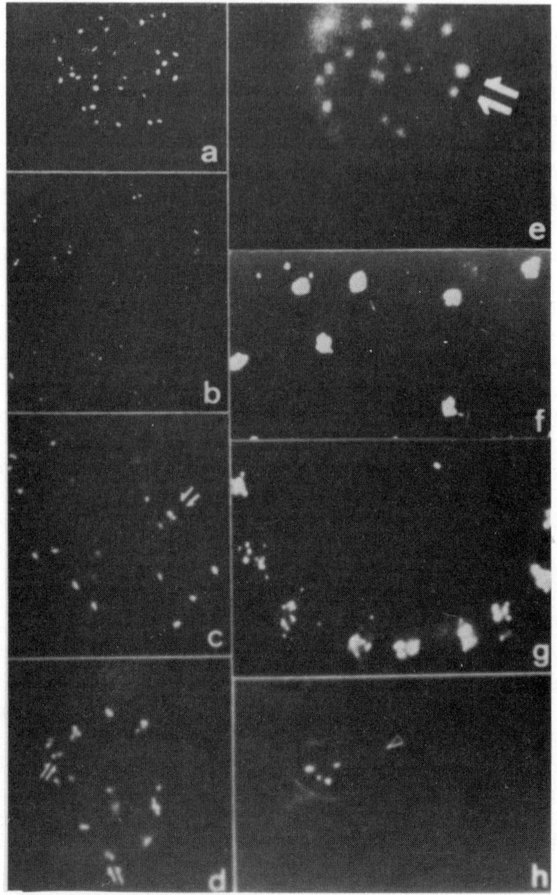

Fig.3. Immunofluorescence of kinetochore in mouse meiotic cells. a: nuclei in diplotene stained with antikinetochore serum. b: diakinesis showing paired centromeres widely separated. c: metaphase I cell. d: metaphase II. e: early spermatid. f: mid-spermatid. g: late-spermatid. h: a late-spermatid cell stained and some mature sperm not stained. Reproduced partially from Brinkley et al. 1986, by permission of Chromosoma, Springer Verlag.

that either each autosomal bivalent contains one unduplicated centromere or the centromeres already duplicated are to close each other so that both kinetochores appear as a single fluorescent spot. In diplotene the number of fluorescent spots doubled in each nuclei. The bivalent at this stage shows a visible chiasmata. The fluorescent spots are paired and presumably each of the 40 spots represents the kinetochore of an individual univalent. The distribution of kinetochores pairs are random throughout the nucleus (Fig.3a). Same number of fluorescent spots are seen in diakinesis. The 40 spots are no longer clustered but exist in pairs randomly distributed with the chromosomes in the nuclei (Fig.3b). When the cell reached metaphaseI a total of 40 stained centromeres could be seen and a fluorescent spot was localized at the centromeric ends of each chromosome of the bivalent (Fig.3c). Really each spot was clearly double representing a pair of sister kinetochores. During meiosis II, each metaphase II cell has a total of 20 chromosomes. The staining of these cells with antikinetochore serum shows that each chromosome contained a pair of fluorescent spots at the centromere (Fig.3d), (a total of 40 kinetochores are present). After chromatid separation in very early spermatid, 20 discrete fluorescent spots representing the complete haploid chromosome number could be seen (Fig.3e). However after this stage a dramatic change in chromatin organization occurs when spermiogenesis starts. The spermatid chromatin condense and the fluorescent centromeres migrated to the center of the nucleus. A fluorescent chromocenter develop at this stage and it contains fewer but larger fluorescent spots (Fig.3f). A total of 10 discrete spots was distinguished at this stage. The number and localization of centromeres has changed during the process of spermatid differentiation. With a final condensation of chromatin the stained chromocenter of late spermatid nuclei started to disappear and the mature sperm head was totally unstained with antikinetochore serum (Fig.3h). Centromeric association clearly accompanies the differentiation of spermatids into mature spermatozoa. The lack of immunofluorescent staining in the mature sperm suggest that kinetochore antigens were either lost during differentiation or blocked by condensing chromatin. Using immunoblotting techniques and antikinetochore sera, we identified a kinetochore antigen in extract of mature sperm cells. This result indicated that kinetochores in those cells are indeed present even though they are not reactive by immunofluorescence (Fig.4M). Therefore the lack of staining by immunofluorescence was probably due to the blocking of kinetochore antigen epitope during the final condensation of chromatin during spermiogenesis. In HeLa cells extract two polypeptides of 18 kd and 80 kd were identified as putative kinetochore antigens. However in extracts from mouse mature sperm only the 80 kd was present in our blots. The absence or decreased of the 18 kd antigen in mature sperm suggest that this polypeptide behaves as a histone-like protein becoming altered in chromatin reorganization during spermatogenesis. However, the 80 kd polypeptide is probably a structural component which like the lamins is present throughout spermatogenesis (Maul et al. 1986).

Specific biological probes are needed to elucidate the mechanism of chromosome movements in mitosis and meiosis. Understanding how microtubules and centromeres interact will be necessary in order to identify the molecular components of both cellular structures involved in that interaction. Human autoantibodies have proved to be an excellent tools in several areas of research specifically, CREST antikinetochore sera are a powerful probe for studies on centromere/kinetochore organization, as was presented in this report. The role of cytoskeletal proteins in some aspects of development and differentiation, specially microtubules, clearly depends on the requirement of the genome (centromere function during mitosis).

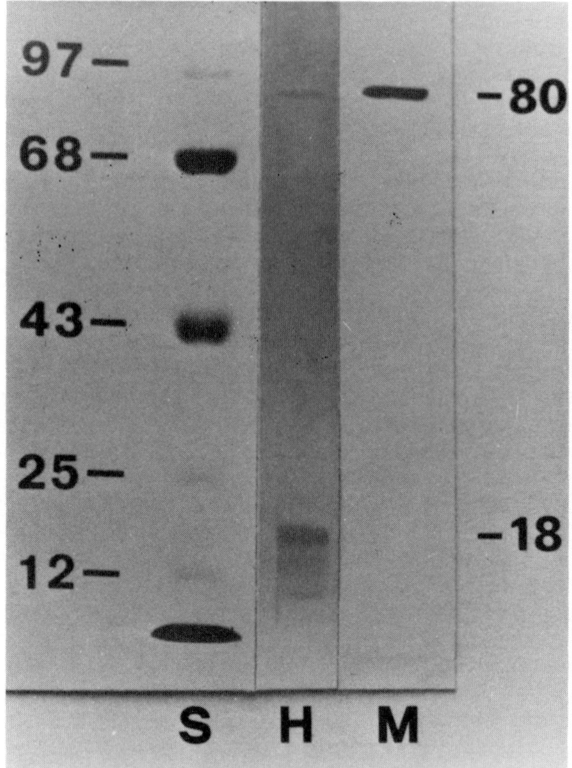

FIG.4. Immunoblotting of human cells (HeLa) and mouse meiotic cells with human antikinetochore serum. Two polypeptides of 18 kd and 80 kd are reactive in HeLa cells with the CREST serum (lane H). In mouse mature sperm cells a single 80 kd polypeptide is recognized by the CREST serum (lane M). Molecular weight markers are in lane S.

ACKNOWLEDGEMENTS

These studies were supported in part by a grant from the HHS, National Cancer Institute CA 23022 to B.R.B..

REFERENCES

Brenner, S.L., Pepper, D., Berns, M.W., Tan, E., and Brinkley, B.R. (1981). Kinetochore structure, duplication and distribution in mammalian cells: analysis by human autoantibodies from scleroderma patients. J. Cell Biology 91, 95-102.

Brinkley, B.R., Valdivia, M.M., Tousson, A. and Brenner, S.L. (1984). Compound kinetochores of Indian muntjac: evolution by linear fusion of unit kinetochores. Chromosoma 91, 1-11.

Brinkley, B.R., Brenner, S.L., Hall, J.M., Tousson, A., Balczon, R.D. and Valdivia, M.M. (1986). Arrangements of kinetochores in mouse cells during meiosis and spermiogenesis. Chromosoma 94, 309-317.

Earnshaw, W.C., Halligan, N., Cooke, C. and Rothfield, N. (1984). The kinetochore is part of the metaphase chromosome scaffold. J. Cell Biology 98, 352-357.

Guldner, H.H., Lakomek, H-J. and Bautz, F.A. (1984). Human anti-centromere sera recognize a 19.5 kd nonhistone chromosomal protein from HeLa cells. Clin. Exp. Immunol. 58, 13-20.

Maul, G.G., French, B.T. and Bechtol, K.B. (1986). Identification and redistribution of lamins during nuclear differentiation in mouse spermatogenesis. Develop. Biol. 115, 68-77.

Mitchinson, T., Evans, L., Schulze, E. and Kirschner, M. (1986). Sites of microtubule assembly and disassembly in the mitotic spindle. Cell 45, 515-527.

Moroi, Y., Peebles, C., Fritzler, M.J., Steigerwald, J. and Tan, E.M. (1980). Autoantibody to centromere (kinetochore) in scleroderma sera. Proc. Natl. Acad. Sci. USA. 77, 1627-1631.

Pardue, M.L. and Gall, J.G. (1970). Chromosomal localization of mouse satellite DNA. Science 168, 1356-1358.

Rieder, C.L. (1982). The formation, structure and composition of the mammalian kinetochore and kinetochore fiber. Inter. Rev. Cytol. 79, 1-58.

Schmidtke, J., Brennecke, H., Schmid, M., Neitzel, H. and Sperling, K. (1981). Evolution of muntjac DNA. Chromosoma 84, 187-193.

Spowart, G., Forster, P., Dunn, N. and Cohen, B.B. (1985). Clinical and biochemical studies on anti-kinetochore antibody in patients with rheumatic diseases: a diagnostic marker for CREST. Disease Markers 3, 103-112.

Valdivia, M.M. and Brinkley, B.R. (1985). Fractionation and initial characterization of the kinetochore from mammalian metaphase chromosomes. J.Cell Biology 101, 1124-1134.

Valdivia, M.M., Tousson, A. and Brinkley, B.R. (1986). Human antibodies and their use for the study of chromosome organization. In: Meth. Achiev. Exp. Pathol. (ed. G. Jasmin and R. Simard) pp 200-223. S.Karger, Basel.

Wurster, D.H. and Atkin, N.B. (1972). Muntjac chromosomes. A new karyotype for Muntiacus muntjac. Experientia 28, 972-973.

Yu, L-C., Lowensteiner, D., Wong, E.F-K., Sawada, I., Mazrimas, J. and Schmid, D. (1986). Localization and characterization of recombinant DNA clones derived from the highly repetitive DNA sequences in the Indian muntjac cells: their presence in the Chinese muntjac. Chromosoma 93, 521-528.

STRUCTURE AND CELL-TYPE SPECIFIC EXPRESSION OF THE CYTOKERATIN MULTIGENE FAMILY

M. Blessing[1,3], L. Domenjoud[2], A. Knapp[1], A. Alonso[2] and J.L. Jorcano[1,3]

[1]Division of Membrane Biology and Biochemistry, Institute of Cell and Tumor Biology, German Cancer Research Center,

[2]Institute of Experimental Pathology, German Cancer Research Center,

[3]Center of Molecular Biology, University of Heidelberg, D-6900 Heidelberg, FRG.

ABSTRACT

The cytokeratins are a family of differentially expressed cytoskeletal proteins that form the intermediate filaments (IF) of epithelial cells. Cloning of different cytokeratin genes from several species has allowed us to analyze the structure and evolution of the genes as well as the chromosomal organization of the family and the identification of sequences relevant for the control of the tissue-specific expression of this multigene family. Transfection of these genes into adequate recipient cells has been used to study *in vivo* assembly of the cytokeratin IF cytoskeleton.

INTRODUCTION

Cytokeratins are a family of highly insoluble proteins that constitute the cytoskeleton of intermediate-sized filaments (IF; 8-10 nm in diameter) characteristic of the epithelial cells of vertebrates and tumors derived therefrom (Franke et al., 1982; Moll et al., 1982; Osborn and Weber, 1983). Recently, IF have also been detected in epithelia of invertebrates (Bartnik et al., 1986) but final proof of their keratin nature awaits amino acid sequence data. In humans, 19 different cytokeratins have been identified (Moll et al., 1982) which, on the basis of biochemical, immunological and protein sequence data, have been subdivided into two subfamilies: the basic (type II) cytokeratins, with a relatively basic isoelectric pH and usually larger size, and the acidic (type I) cytokeratins, more acidic and usually smaller (Moll et al., 1982; Kim et al., 1983; Sun et al., 1984). In spite of their diversity, all cytokeratins have a common structure that is also shared by the other non-epithelial IF polypeptides: a central α-helical rod region of conserved length and sequence and that can assume a coiled-coil conformation is flanked by non-α-helical amino (head)- and carboxy(tail)-terminal regions of variable length and sequence (Weber and Geisler, 1984; Steinert et al., 1985). Unlike other IF proteins, such as vimentin, desmin or GFAP, that can form IF structures composed of a single polypeptide (homopolymers), assembly of cytokeratin IF requires equimolar amounts of type I and type II polypeptides because the basic structural filament subunit is a heterotetramer containing two molecules of each subfamily (Woods, 1983; Quinlan et al., 1984).

Cytokeratin expression is an exquisitely regulated process such that the different epithelia can be characterized by the pattern of cytokeratins they synthesize (Moll et al., 1982). Although reconstitution (Hatzfeld and Franke, 1985) and mRNA microinjection experiments (Franke et al., 1984) indicate that either in vitro and in vivo almost any combination of basic and acidic cytokeratins can polymerize into IF, normal epithelia have a strong preference for particular pairs of basic and acidic polypeptides (Moll et al., 1982; Sun et al., 1984). This observation has led to the concept of the "expression pairs" as markers for the different routes of epithelial differentiation (Sun et al., 1984). We are interested in two major problems: a) What are the mechanisms controlling the cell-type specific expression of this multigene family. b) What is the cellular function of this cytoskeleton and does it depend on the cytokeratin composition of the filaments. We have approached these questions by cloning and analyzing different cytokeratin genes from several species.

STRUCTURE AND EVOLUTION OF CYTOKERATIN GENES

In the last few years, the genes coding for several cytokeratins have been isolated and characterized (for references, see Blessing et al., 1987). Most of these genes are expressed in differentiated epithelia and are represented once per haploid genome. Intronless pseudogenes have been only reported for cytokeratin 8 which is also expressed in embryonal epithelia from which germ cells develop (Vasseur et al., 1985; Franke et al., 1982). As Fig. 1 shows, genes coding for basic cytokeratins (represented by H6) are interrupted by 8 introns which disrupt the coding sequences practically at the same positions.

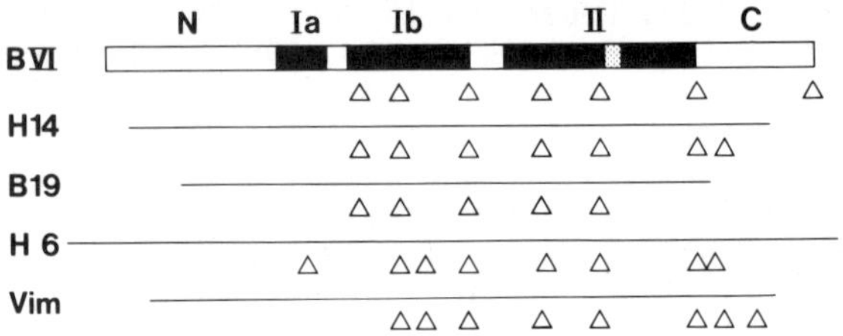

N Ia Ib II C
BVI
H14
B19
H 6
Vim

Fig. 1. Conservation of intron positions in IF genes. The polypeptide domains of bovine cytokeratin VI (BVI) are shown schematically. Black boxes (Ia, Ib, II) represent α-helical regions. Open boxes indicate non-α-helical regions (N, C: amino(head)- and carboxy(tail)-terminal domains). Triangles mark intron positions. H14: human cytokeratin 14. H6: human cytokeratin 6. B19: bovine cytokeratin 19. Vim: hamster vimentin.

Genes for acidic cytokeratins (BVI, H14 in Fig. 1) are interrupted by 7 introns, except cytokeratin 19 (B19 in Fig. 1) which has only 5 introns. The 5 introns common to all acidic cytokeratin genes are also placed at identical positions. Comparison of the genes coding for cytokeratins with those coding for other IF proteins (represented by vimentin in Fig. 1; Quax et al., 1983) reveals that the positions at which the introns interrupt the gene region for the α-helical central domain are considerably conserved during evolution from frogs to humans. These data suggest that the cytokeratin genes and probably the genes for all types of IF arose from a common ancestor (Lehnert et al., 1984; Marchuk et al., 1984; Steinert et al., 1985), although the neurofilament genes could be an exception (Lewis and Cowan, 1986).

The conservation of the rod region is most likely explained by its importance for the formation of cytokeratin tetramer subunits (Gruen and Woods, 1983; Weber and Geisler, 1984; Magin et al., 1987). On the contrary, the hypervariable tail regions seem to be dispensable for tetramer and filament assembly and it has been hypothesized that they could specify or modulate the functions of cytokeratin IF (Weber and Geisler, 1984; Steinert et al., 1985; Bader et al., 1986). Interestingly, the introns interrupting this domain are placed at different positions in each type of IF gene (Fig. 1). This heterogeneity is particularly notable when one considers the acidic cytokeratin subfamily: the genes coding for the glycine-rich subtype (Jorcano et al., 1984; BVI in Fig. 1) have the last intron immediately before the stop codon; the genes encoding the serine- and valine-rich subtype (Jorcano et al., 1984; H14 in Fig. 1) have it shortly after the end of the α-helical domain; and the gene for cytokeratin 19 (B19 in Fig. 1) lacks the C-terminal domain and has lost the last two introns (Bader et al., 1986). Therefore, the α-helical rod region differs not only from the hypervariable C-terminal domain in the degree of DNA and protein sequence conservation, but the structure of the genes in these two domains also seems subject to very different evolutionary pressures.

CHROMOSOMAL ORGANIZATION OF CYTOKERATIN GENES

Some multigene families are dispersed throughout the genome, but frequently their members are clustered in one or a few chromosomal loci. The latter case seems to be the true for the cytokeratin family. Two tandems of type II genes have been analyzed in the bovine genome (Fig. 2; Blessing et al., 1987). The genes in each tandem have the same polarity and are separated by ∿ 11 Kbp. On the other hand, linkage of type I genes has been reported in X. laevis (Miyatani et al., 1986) and in the human genome (Chaudhury et al., 1986), and genes coding for polypeptides of each type have found linked in the sheep genome (Powell et al., 1986). Together, these data suggest that the genes coding for basic or acidic cytokeratins are clustered but that the two subfamilies are not. This clustering of different subfamilies into separate loci is also found in other multigene families (e.g. the α- and β-globins) which, like the cytokeratins, code for two types of polypeptides whose stoichiometric interaction is required to produce the functional protein. This chromosomal organization is probably the

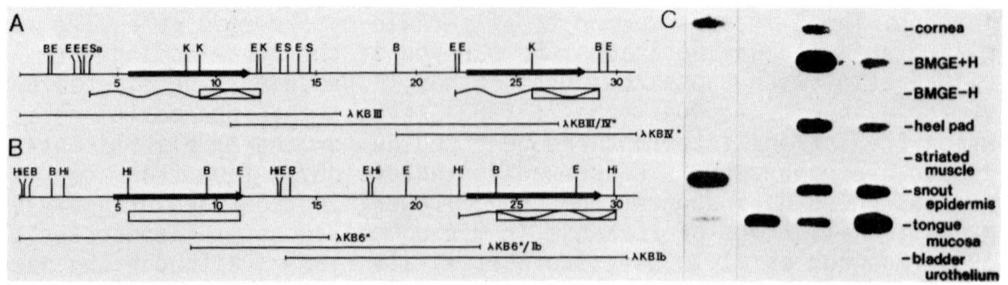

Fig. 2. Linkage (A,B) and expression (C) of bovine type II cytokeratin genes. Maps of the loci containing the genes for cytokeratins III and IV (A) and 6* and Ib (B). The position and polarity of the genes is indicated by the horizontal arrows. Distances are given in Kbp. For a more detailed description of these loci see Blessing et al. (1987). (C) 10 μg of total cellular RNA extracted from the tissues and cell lines indicated were bound to nitrocellulose filters and hybridized with specific probes from linked genes. From left to right: Ib, 6*, III and IV*.

result of the history of these gene families which may have been generated by a duplication of a single ancestor gene able to form by itself the functional multimeric protein, followed by the insertion of one of the copies at a new chromosomal locus. Subsequent rounds of amplification and divergence gave rise to the two clustered gene subfamilies. However, it is not clear whether such relatively simple mechanisms are sufficient to provide each cytokeratin gene with a tissue-specific promoter which probably has to interact with the tissue- and developmental stage-specific factors that activate or repress it.

Unlike other clustered multigene families, the cytokeratin genes do not seem to be arranged in an order related to their expression during development and differentiation. As shown in Fig. 2C, genes linked in tandem are co-expressed in some tissues but differentially expressed in others. On the other hand, genes that are co-expressed are not necessarily linked. Moreover, because the basic and acidic cytokeratin genes are situated at different loci it is clear that genes coding for polypeptide partners in the "expression pairs" (Sun et al., 1984) are also not linked. Thus, the linkage of certain cytokeratin genes seems to be an evolutionary consequence rather than a means to control the coordinate expression of this multigene family (Blessing et al., 1987).

SEQUENCES AT THE 5'-FLANKING REGION REGULATE THE TISSUE-SPECIFIC EXPRESSION OF CYTOKERATIN GENES

The 5'-flanking regions have been shown to play an important role in the transcriptional regulation of many eukaryotic genes. Chaudhury et al. (1986) have reported that the human genes encoding two acidic (nos. 14 and 17) and one basic (no. 6) cytokeratins that are co-expressed in cultured epidermal keratinocytes exhibit a significant degree of sequence homology in the 5'-upstream region. Blessing et al. (1987) have shown that this region is also highly conserved in the genes coding for a given cytokeratin in different species (bovine, murine, human). These authors have also detected the presence of an octanucleotide AAPuCCAAA upstream of the TATA box in the genes coding for human, bovine and murine epidermal cytokeratins and in the involucrin gene, that is also expressed in epidermal keratinocytes. This motif is found neither in the genes coding for cytokeratins expressed in simple epithelia nor in the genes coding for other IF proteins (Blessing et al., 1987). All these results suggest that the 5'-flanking regions could be important for the control of the activity of the cytokeratin genes. Indeed, when tested in chloroamphenicol acetyltransferase (CAT) assays (Gorman et al., 1982), the 5'-upstream region of bovine cytokeratin gene IV* directs CAT synthesis in a cell-type specific manner (Fig. 3). Using BMGE+H cells (a bovine cell line expressing gene IV* (Blessing et al., 1987)) for the transfections, a recombinant in which the 2.2 Kbp upstream of the IV*-mRNA cap site have been placed preceeding the CAT gene stimulates the synthesis of CAT enzyme approximately 200 fold (Fig. 3; BglII lane) as compared to the reference CAT plasmid alone (CAT lane). The critical sequences for this stimulation seem to lie in a region between 600 bp (HindIII fragment, still fully active) and 180 bp (XmnI fragment, inactive) upstream the cap site. This HindIII-XmnI region stimulates the thymidin kinase (TK) promoter, a basal heterologous promoter (compare lanes Hind-Xmn/TK and TK) as efficiently as the SV40 early region enhancer (lane SV40/TK). In addition, a XhoII-XmnI smaller fragment still conserves the capacity to stimulate the TK promoter when placed either upstream (Xho-Xmn/TK lane) or downstream (TK/Xho-Xmn lane) of it. These results suggest that the transcription of cytokeratin gene IV* is regulated by an enhancer element similar to those found in other viral and eukaryotic genes (Voss et al., 1986).

The above described fragments are inactive in MDBK cells (Fig. 3), an epithelial line synthesizing the cytokeratins characteristic of simple epithelia but not cytokeratin IV, the product of gene IV*. This indicates that cytokeratin gene IV* expression is controlled by an enhancer that is activated in a tissue-specific manner and we have evidence that other cytokeratin genes are regulated in a similar way.

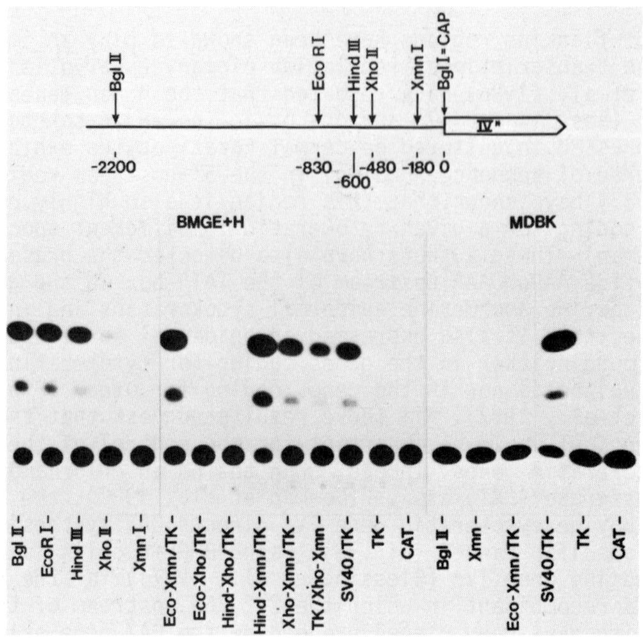

Fig. 3. Identification of sequences controlling the tissue-specific expression of cytokeratin gene IV*. Upper part: map of the region 5'-upstream of IV* indicating the fragments used in the CAT-assays. Lower part: Levels of CAT expression directed by the corresponding restriction fragment alone or in combination with the thymidine kinase (TK) promoter. BMGE+H cells express gene IV* whereas MDBK do not.

TRANSFECTION OF CYTOKERATIN GENES INTO NON-EPITHELIAL CELLS

To study the assembly of IF _in vivo_ as well as the fate of cytokeratin polypeptides expressed in unusual combinations or cells, we have transfected 3T3 fibroblasts with plasmids carrying the cDNAs coding for the mouse type I cytokeratin 18 (Alonso et al., 1987) and the frog type II keratin 8 (Franz and Franke, 1986) under the control of the SV40 early promoter. The expression of the transfected cytokeratin sequences was followed by indirect immunofluorescence using specific antibodies. As shown in Fig. 4 cytokeratin 18 alone did not form filaments but granular aggregates of variable size were observed throughout the cytoplasm. Similar aggregates of acidic cytokeratins have also been detected by Giudice and Fuchs (1987) in NIH 3T3 fibroblasts transfected with the gene for human cytokeratin 14. These results agree with the observation that assembly of IF _in vitro_

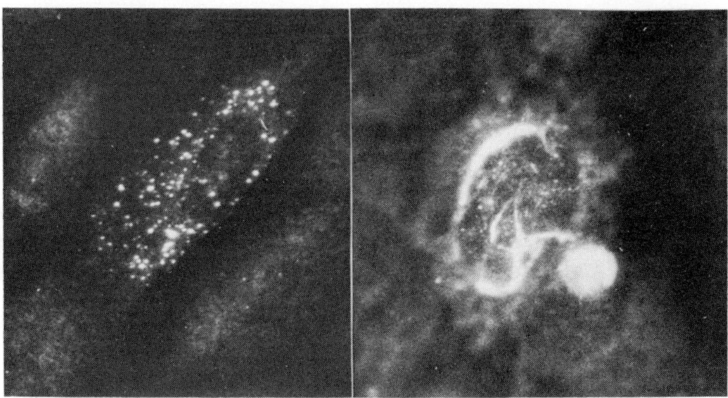

Fig. 4. Examples of transfected 3T3 fibroblasts synthesizing cytokeratin 18 alone (left) or in combination with cytokeratin 8 (right).

requires the cooperation of basic and acidic cytokeratins (Hatzfeld and Franke, 1985; Eichner et al., 1986).

Cotransfection with the cDNAs for the two cytokeratins produced filamentous structures (Fig. 4) similar to those found in non-epithelial cells microinjected with cytokeratin mRNAs (Krieg et al., 1983). Since these structures have been shown to consist of typical IF, these results demonstrate that a basic and an acidic cytokeratin are able to form IF in the absence of other special factors specific for epithelial cells, even if the two cytokeratins come from evolutionarily very distant species. However, the fibrillar meshwork assembled in 3T3 fibroblasts does not resemble the cytoskeleton characteristic of epithelial cells. Therefore, the correct organization of this tridimensional structure might require the presence of other elements such as desmosomes to which the cytokeratin bundles could anchor as they do in most types of epithelial cells (Cowin et al., 1985).

ACKNOWLEDGEMENTS

We are indebted to M. Freudenmann for skillful technical assistance, to Dr. M. Little for critically reading the manuscript and to Dr. W.W. Franke for encouragement and support. The work has been supported by the German Ministry of Research and Technology (ZMBH Project) and by the Deutsche Forschungsgemeinschaft (DFG).

REFERENCES

Alonso, A., Weber, T. and Jorcano, J.L. (1987). Cloning and characterization of keratin D, a murine endodermal cytoskeletal protein induced during in vitro differentiation of F9 teratocarcinoma cells. Roux's Arch. Dev. Biol. 196, 16-21.

Bader, B.L., Magin, T.M., Hatzfeld, M., and Franke, W.W. (1986). Amino acid sequence and gene organization of cytokeratin no. 19, an exceptional tail-less intermediate filament protein. EMBO J. 5, 1865-1875.

Bartnik, E., Osborn, M. and Weber, K. (1986). Intermediate filaments in muscle and epithelial cells of nematodes. J. Cell Biol. 102, 2033-2041.

Blessing, M., Zentraf, H. and Jorcano, J.L. (1987). Differentially expressed bovine cytokeratin genes. Analysis of gene linkage and evolutionary conservation of 5'-upstream sequences. EMBO J. 6, 567-575.

Chaudhury, A.R., Marchuk, D., Lindhurst, M. and Fuchs, E. (1986). Three tightly linked genes encoding human type I keratins: Conservation of sequence in the 5'-untranslated leader and 5'-upstream regions of coexpressed keratin genes. Mol. Cell. Biol. 6, 539-548.

Cowin, P., Franke, W.W., Grund, C., Kapprell, H.P. and Kartenbeck, J. (1985). The desmosome-intermediate filament complex. In The Cell in Contact (ed. G.E. Edelman and J.-P. Thiery), pp. 427-460. J. Wiley & Sons. London.

Eichner, R., Sun, T.-T. and Aebi, U. (1986). The role of keratin subfamilies and keratin pairs in the formation of human epidermal intermediate filaments. J. Cell Biol. 102, 1767-1777.

Franke, W.W., Schmid, E., Schiller, D.L., Winter, S., Jarasch, E.-D., Moll, R., Denk, H., Jackson, B.W. and Illmensee, K. (1982). Differentiation-related patterns of expression of proteins of intermediate-size filaments in tissues and cultured cells. Cold Spring Harbor Symp. Quant. Biol. 46, 431-453.

Franke, W.W., Schmid, E., Mittnacht, S., Grund, C. and Jorcano, J.L. (1984). Integration of different keratins into the same filament system after microinjection of mRNA for epidermal keratins into kidney epithelial cells. Cell 36, 813-825.

Franz, J.K. and Franke, W.W. (1986). Cloning of cDNA and amino acid sequence of a cytokeratin expressed in oocytes of Xenopus laevis. Proc. Natl. Acad. Sci. USA 83, 6475-6479.

Giudice, G.J. and Fuchs, E. (1987). The transfection of epidermal keratin genes into fibroblasts and simple epithelial cells: evidence for inducing a type I keratin by a type II gene. Cell 48, 453-463.

Gorman, C., Moffat, L. and Howard, B. (1982). Recombinant genomes which express chloramphenicol acetyltransferase in mammalian cells. Mol. Cell. Biol. 2, 1044-1051.

Gruen, L.C. and Woods, E.F. (1983) Structural studies on the microfibrillar proteins of wool. Biochem. J. 209, 587-595.

Hatzfeld, M. and Franke, W.W. (1985) Pair formation and promiscuity of cytokeratins: Formation in vitro of heterotypic complexes and intermediate-sized filaments by homologous and heterologous recombinations of purified polypeptides. J. Cell Biol. 101, 1826-1841.

Jorcano, J.L., Rieger, M., Franz, J.K., Schiller, D.L., Moll, R. and Franke, W.W. (1984). Identification of two types of keratin polypeptides within the acidic cytokeratin subfamily I. J. Mol. Biol. 179, 257-281.

Kim, K.H., Rheinwald, J.G. and Fuchs, E.V. (1983). Tissue specificity of epithelial keratins: differential expression of mRNAs from two multigene families. Mol. Cell. Biol. 3, 495-502.

Kreis, T.E., Geiger, B., Schmid, E., Jorcano, J.L. and Franke, W.W. (1983). De novo synthesis and specific assembly of keratin filaments in nonepithelial cells after microinjection of mRNA for epidermal keratin. Cell 32, 1125-1137.

Lehnert, M.E., Jorcano, J.L., Zentgraf, H., Blessing, M., Franz, J.K. and Franke, W.W. (1984). Characterization of bovine keratin genes: similarities of exon patterns in genes coding for different keratins. EMBO J. 3, 3279-3287.

Lewis, S.A. and Cowan, N.J. (1986). Anomalous placement of introns in a member of the intermediate filament multigene family: an evolutionary conundrum. Mol. Cell. Biol. 6, 1529-1534.

Magin, T.M., Hatzfeld, M. and Franke, W.W. (1987). Analysis of cytokeratin domains by cloning and expression of intact and deleted polypeptides in Escherichia coli. EMBO J., in press.

Marchuk, D., McCrohon, S. and Fuchs, E. (1984). Remarkable conservation of structure among intermediate filament genes. Cell 39, 491-498.

Miyatani, S., Winkless, J.A., Sargent, T.D. and Dawid, I.B. (1986). Stage-specific keratins in Xenopus laevis embryos and tadpoles: The XK81 gene family. J. Cell Biol. 103, 1957-1965.

Moll, R., Franke, W.W., Schiller, D.L., Geiger, B. and Krepler, R. (1982). The catalog of human cytokeratin polypeptides: patterns of expression of specific cytokeratins in normal epithelia, tumors and cultured cells. Cell 31, 11-24.

Osborn, M. and Weber, K. (1983). Tumor diagnosis by intermediate filament typing: a novel tool for surgical pathology. Lab. Invest. 48, 372-394.

Powell, B.C., Cam, G.R., Fietz, M.J. and Rogers, G.E. (1986). Clustered arrangement of keratin intermediate filament genes. Proc. Natl. Acad. Sci. USA 86, 5048-5052.

Quax, W., Egberts, V.W., Hendriks, W., Quax-Jeuken, Y. and Bloemendal, H. (1983). The structure of the vimentin gene. Cell 36, 215-223.

Quinlan, R.A., Cohlberg, J.A., Schiller, D.L., Hatzfeld, M. and Franke, W.W. (1984) Heterotypic tetramer (A_2D_2) complexes of non-epidermal keratins isolated from cytoskeletons of rat hepatocytes and hepatoma cells. J. Mol. Biol. 178, 365-388.

Steinert, P.M., Steven, A.C. and Roop, D.R. (1985). The molecular biology of intermediate filaments. Cell 42, 411-419.

Sun, T.-T., Eichner, R., Schermer, A., Cooper, D., Nelson, W.G. and Weiss, R.A. (1984). Classification, expression and possible mechanisms of evolution of mammalian epithelial keratins: a unifying model. In Cancer Cells 1, The Transformed Phenotype, Vol. 1 (ed. A. Levine, G.F. Vande Woude, W.C. Topp and J.D. Watson), pp. 169-176. Cold Spring Harbor Laboratory, Cold Spring Harbor.

Vasseur, M., Duprey, P., Brulet, P. and Jacob, P. (1985). One gene and one pseudogene for the cytokeratin Endo A. Proc. Natl. Acad. Sci. USA 82, 1155-1159.

Voss, S.D., Schlokat, U. and Gruss, P. (1986). The role of enhancers in the regulation of cell-type specific transcriptional control. TIBS 11, 287-289.

Weber, K. and Geisler, N. (1984). Intermediate filaments from wool α-keratins to neurofilaments: a structural overview. In Cancer Cells 1, The Transformed Phenotype, Vol. 1 (ed. A. Levine, G.F. Vande Woude, W.C. Topp, J.D. Watson), pp. 153-159. Cold Spring Harbor Laboratory, Cold Spring Harbor.

Woods, E.F. (1983). The number of polypeptide chains in the rod domain of bovine epidermal keratin. Biochem. Int. 7, 769-774.

CHARACTERIZATION OF A NUCLEAR-MITOTIC SPINDLE PROTEIN IMMUNOLOGICALLY RELATED TO THE MICROTUBULE-ASSOCIATED PROTEIN MAP-1B

J. Díaz-Nido[1], J.S. Bonifacino[2], I.V. Sandoval[2] and J. Avila[1]. [1]Centro de Biología Molecular. (CSIC-UAM) Canto blanco. 28049 Madrid. Spain. [2]Cell Biology and Metabolism Branch. NICHHD. Bethesda. Md 20205. USA.

It has previously been reported that the monoclonal antibody 8D12 recognized a nuclear-mitotic spindle protein referred to as p280 which was widely distributed in mammalian tissues and cultured cells (1).

(I) Subcellular distribution of p280 in Ptk2 cells. By using indirect immunofluorescence, we have observed that this protein is mainly associated with the residual nuclear matrix which is obtained after treatment with detergents, nucleases and high salt extraction (see fig. 1). On the other hand, digestion of RNA in the presence of EDTA followed by high salt extraction results in almost complete solubilization of p280. This points to an association of p280 with the internal ribonucleoprotein network of interphase nuclei. In mitotic cells, p280 is localized to the mitotic spindle. High salt extraction of mitotic cells results in the solubilization of all the antigen except for that at the mitotic poles (see fig. 1). Furthermore, when cells are treated with cold Ca^{2+} solutions to extract microtubules, the P280 staining also disappears. This suggest that P280 is actually bound to mitotic spindle microtubules.

(II) Relationship between p280 and MAP-1B. Because of the binding of p280 to microtubules in mitotic Ptk2 cells, we have examined the relationship between this protein and brain MAPs that have also

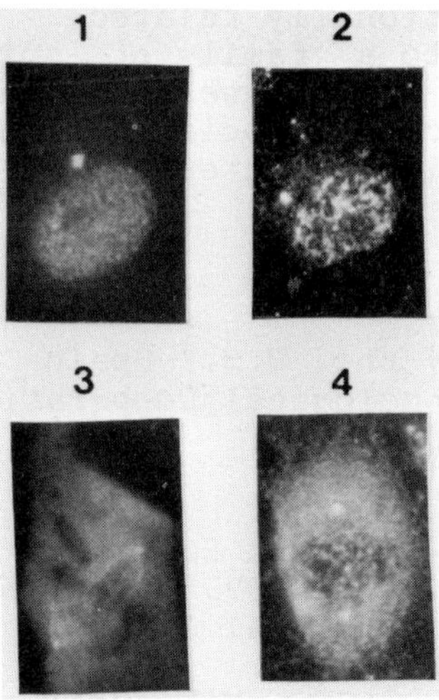

Fig. 1. Indirect Immunofluorescent staining patterns of Ptk2 cells incubated with monoclonal antibody 8D12. It is shown the staining of intact cells (1 and 3) and residual skeletons (2 and 4) in both interphase (1 and 2) and mitosis (3 and 4). Residual skeletons were obtained "in situ" as described.

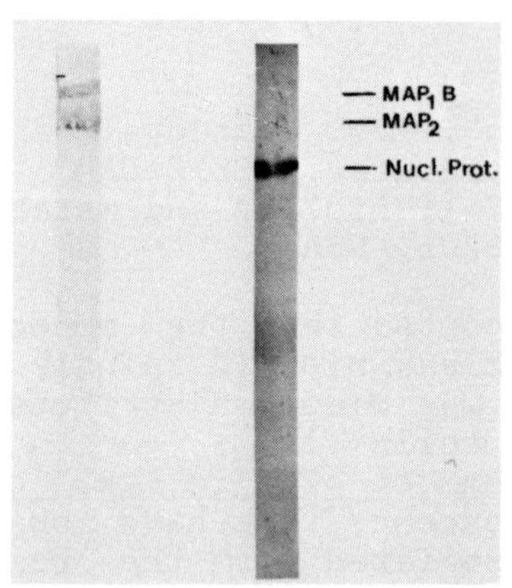

Fig. 2. Immunoblotting analysis showing the reaction of 8D12 with MAP-1B and p280. Rat brain microtubule protein (left) or rat liver nuclei (right) were subjected to SDS-PAGE, transferred to nitrocellulose and probed with the monoclonal antibody 8D12. It is observed the reaction of 8D12 with MAP-1B and its degradation product in microtubule protein whereas in liver nuclei is reacting with p280.

been localized to mitotic spindles (see 3 for a review). It had already been shown that p280 and the microtubule-associated protein MAP-1 were immunologically related (1). Since MAP-1 really includes several polypeptides (3), we have tested which of these proteins react with the monoclonal antibody 8D12 (see fig. 2). In this way we have found that MAP-1B is the protein related to p280. Also we have performed the comparison of their amino acid analyses and their S. aureus V8 proteolytic profiles. The amino acid composition of both proteins, p280 and MAP-1B, are quite similar but their digestion patterns are somewhat different. On the other hand, preliminary data indicate that both p280 and MAP-1B have a similar ability to bind to taxol-polymerized tubulin suggesting that these proteins are at least functionally related.

Thus, p280 seem to belong to a family of spindle-associated proteins (SAPs) that are sequestered in the nucleus during interphase and become associated with the mitotic spindle through a direct interaction with tubulin.

ACKNOWLEDGEMENTS

This work was supported by the U.S. Spain Joint Committee for Scientific and Technological Cooperation.

REFERENCES

(1) Bonifacino, J.S., Klausner, R.D. and Sandoval, I.V. (1985) Proc. Natl. Acad. Sci. USA. 82, 1146-1150.
(2) Staufenbiel, M. and Deppert, W. (1984) J. Cell. Biol. 98, 1886-1894.
(3) Vallee, R.B. and Bloom,. G.S. (1984) Mol. Cell. Biol. 3, 21-75.

THE CYTOSKELETON IN SOME CILIATED PROTOZOA

R. Gil
Centro de Investigaciones Biológicas, Velázquez 144, 28006-Madrid, Spain

INTRODUCTION. Great number of cytoskeletal structures (1, 2) and cytoskeletal proteins (3) have been studied in ciliated protozoa by several investigators. As it is known, when detergent extracting methods are used, the less stable cytoplasmic components disappear and all the filaments become more visible. Schliwa (4) suggests that it would be interesting to study protozoa extracted with Triton X-100 to determine the existing relation among the different structures forming the cytoskeleton. In the present work done with extracted cells of Frontonia depressa, Frontonia leucas and Paramecium putrinum, I observed that these ciliated have some similar cytoskeletal structures.

MATERIAL AND METHODS. Permeabilization was obtained by transferring small pools of cell into 5% Triton X-100 in PHEM buffer (5), pH 6,9. Cells were washed several times. These cells and control cells were fixed for 3 min in a freshly prepared mixture of 2% glutaraldehyde, 1% OsO_4 and 0,5% tannic acid, all in PHEM, then embedded in Epon. A Philips E.M. 300 (60-80 KV) was used.

RESULTS. The cytoskeletal components remaining within the cells after Triton X-100 treatment in F. depressa, F. leucas and P. putrinum are the following: In the somatic cortex the inner and the outer alveolar membranes and the epiplasm are structures that remain. The cellular membrane is removed. The epiplasm is constituted by a loose filaments net. The infraciliary lattice also remains and is in contact with the epiplasm and with the Y-shaped microfibrillar net that is situated all over the cytoplasm. The trichocysts are trapped in the net with Y-shaped filaments. Kinetodesmal fibers, postciliary fibers, microtubules that are associated with kinetosomes, and subpellicular microtubules remain after Triton X-100 treatment. In the macronucleus I observed three different microfibrillar nets: a microfibrillar net formed by Y-shaped filaments with a diameter of 10 nm; a net with diameter filaments of 20 nm, and the third net formed by more tenuous filaments than the two previous ones. In the oral region the deep microfibrillar net constituted by the postciliary fibers of the peniculi and the net with condensation knots situated in the ribs along the ribbed wall also remains after Triton X-100 treatment.

DISCUSSION. The net of filaments in Y-shape, well developed in the macronucleus and that follows the same net through the cytoplasm is similar in Y-shape and size to the net of intermediate filaments in vertebrate cells (5). It would be possible that by trichocysts exocytosis there were specific receptors in the Y-shaped filaments net becouse the trichocysts are trapped in this net. There are continuity in the oral skeleton of the three studied ciliated protozoa between the left and the right side by several microfibrillar nets and the skeleton of the oral zone is connected by means of microfibrillar nets with the skeletal components of the macronucleus, cytoplasm and cortex. I can conclude that there are continuity among all cytoskeletal structures.

REFERENCES

(1) Gil, R. (1984) Trans. Am. Microsc. Soc., 103, 353-364.
(2) Gil, R. (1986) Microbiología, 2, 47-54.
(3) Williams, N.E. (1986) Progress in Protistology, Vol. I, 309-324.
(4) Schliwa, M. (1986) Cell Biology Monographs, Vol. 13.
(5) Schliwa, M. and van Blerkom, J. (1981) J. Cell Biol., 90, 222-235.

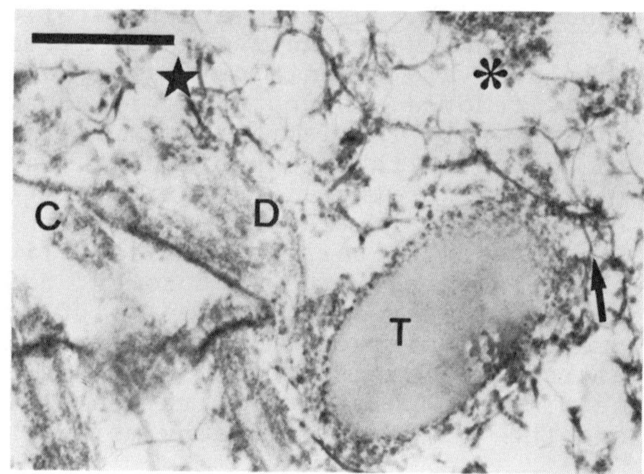

A section in the oral zone of *Paramecium putrinum* treated with Triton X-100 showing the trichocysts (T) trapped in the Y-shaped microfibrillar net (↑). C, oral cilium; D, deep net; ★, microfibrillar net with dense filaments; ✱, microfibrillar net with tenuous filaments. Scale bar represents 0.5 μm.

CYTOPLASMIC ORGANIZATION DURING DEVELOPMENT

Experimental Analysis of Cytoskeletal Function in Early *Xenopus laevis* Embryos.

Daniel Chu & Michael W. Klymkowsky*
Molecular, Cellular & Developmental Biology, University of Colorado at Boulder
Boulder, Colorado 80309-0347 U.S.A.

Except for an apparent role in the assembly of frog virus 3 cytoplasmic assembly sites (1), antibody-injection studies of intermediate filament function have failed to reveal **any** apparent active role for this major cytoskeletal system in cultured cells (2-6). These observations suggest that in normal cultured cells, the functions of intermediate filaments are simply too subtle to be apparent. We have therefore turned to a developing system, the clawed frog *Xenopus* in order to study the function of intermediate filaments in a more rigorous developmental and organismic context.

One of the major problems in studying intermediate filament function in *Xenopus* is the large size of the *Xenopus* egg & embryo, which makes visualizing the effects of injected antibodies on intermediate filaments and cell behavior less than straightforward. We have overcome this problem by developing whole-mount immunolabeling methods that enable us to visualize the cortical cytokeratin system (7); by using Andrew Murray's embryo clearing method, we are able to visualize internal antigens as well (J.A. Dent & M.W. Klymkowsky, work in progress).

Our first studies on intermediate filaments in *Xenopus* have focused on the organization and function of the cortical cytokeratin system in the *Xenopus* oocyte, egg and early embryo. There appear to be three cytokeratin proteins in the *Xenopus* oocyte and early (pre-midblastula transition) embryos (8) and they form a predominantly cortical system (7-13). Using whole-mount immunolabeling methods, we have discovered i) that the mature oocyte contains a unique and polar "geodesic"-type cortical cytokeratin system; ii) that a dramatic disruption of this cytokeratin system accompanies egg maturation; iii) that fertilization begins a process, apparently triggered by internal Ca^{2+}, of cytokeratin reorganization; iv) that cytokeratin reorganization within the egg is insensitive to antimicrotubule and antimicrofilament drugs, or the inhibition of protein synthesis; and iv) that this reorganization of the cytokeratin filaments leads to the formation of a polarly asymmetric embryonic cytokeratin system (7; Klymkowsky & Maynell, ms. in prep.).

Our goal is to use monoclonal antibodies to disrupt specific intermediate filament systems within the embryo and to then look for reproducible defects in development, which should reflect the functional roles of intermediate filaments. A prerequisite to such a study is the characterization of antibodies that can be used in control experiments. We have previously demonstrated that monoclonal antibodies *per se* are benign and without apparent non-specific effects when injected into cultured cells (14). However, there is an important difference between injection studies on cultured cells and those on developing systems. In cultured cells, the injected cell either survives injection or dies; those cells that survive rapidly (within 15 to 30 minutes depending on the cell type (6)) become indistinguishable from their uninjected neighbors. In contrast, in the *Xenopus* embryo, an injected blastomere can appear to recover completely from injection only to produce a stage-specific abnormalities later in development (15; unpubl. obs.).

An ideal control antibody would be a monoclonal antibody derived from the same species and of the same immunoglobulin class as the experimental antibody; in addition, it would be helpful if the antibody could be used independently of whether or not its specific antigen was present within the injected cell or its progeny. In the course of studying the effects of various monoclonal antibodies within cells, we have identified just such an antibody - a monoclonal antiβ-tubulin antibody generated by Mike McCutcheon and Sean Carroll (UC Boulder) from a mouse immunized with an affinity-purified, *Escherichia coli*-derived, fusion protein composed of the *E. coli* -galactosidase protein and the *Drosophilia* fushi `tarazu (*ftz*) gene product. The monoclonal antibody, E7 (IgG_1), is specific for the *ftz* portion of the fusion protein, as determined by ELISA assay. However, in immunolabeling and western immunoblot analyses, the antibody reacts specifically with -tubulin from a wide range of species ranging from *Chlamydomonas* to human. The exact nature of the apparent cross-reactivity between the nuclear *ftz* protein and the cytoplasmic -tubulin protein is unclear, but may be due to limited regions of shared amino acid homology in the C-terminal regions of the two proteins (A. Laughon & S. Carroll, per. comm.).

The work of Wehland & Willingham (16) and Blose et al., (17) had previously shown that the intracellular injection of monoclonal and polyclonal antitubulin antibodies caused the disruption of microtubules and in many ways mimicked the effects of the antimicrotubule drug colchicine, in that the antibody-induced disruption of microtubule organization caused the collapse of the intermediate filament network and the fragmentation of the Golgi apparatus. We were therefore surprised to find that the intracellular injection of the E7 anti-tubulin antibody had no apparent effect on microtubule function

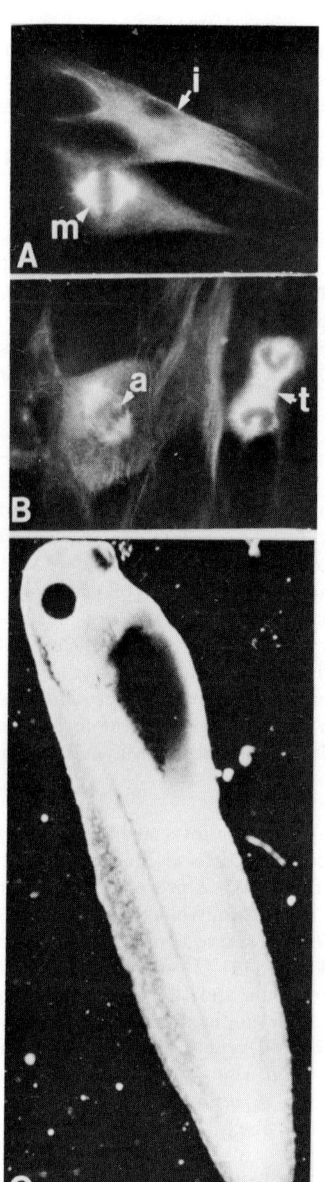

Fig.1
A & B. A_6 cells injected with 20mg/ml antiβ-tubulin antibody were fixed 20 hours later; cells were found in all stages of mitosis ("i" - interphase, "m" - metaphase, "a"- anaphase, "t" - teleophase). C. Fertilized eggs injected with antiβ-tubulin developed into apparently normal tadpoles.

when injected into either baby hamster kidney 21 (BHK) cells or *Xenopus* kidney epithelial A_6 cells. Within minutes the antibody decorated the entire microtubule system of the injected cell. Even when the antibody was used at concentrations of 20mg/ml and higher, the microtubules within the injected cell remained intact and continued to function normally as judged by the continuation of saltatory motion. Intermediate filament and Golgi apparatus organization were unaffected. Most surprisingly, cells went through mitosis normally (Fig. 1A,B) without any obvious slowing of the mitotic rate.

For a more rigorous test of the antibody's apparent lack of effect on microtubule function, we injected it (20-25nL at 20mg/ml) into fertilized *Xenopus* eggs. When injected prior to first cleavage (25-45 minutes post-fertilization), the antibody was found to diffuse throughout the embryo and all blastomeres contained antibody at the 8 to 16 cell stage. We obtained normal embryos from these anti-tubulin injected eggs (Fig. 1C) at the same percentage obtained from buffer (10mM Tris, pH 7.6)-injected eggs, indicating that the antibody had no deleterious effect on cell behavior or subsequent development.

In summary, this monoclonal antiβ-tubulin antibody appears to be a completely "non-invasive" reagent; we have submitted it to the American Type Cell Culture collection and it should be available through them. We are currently using it in conjugation with a number of anticytokeratin and antivimentin antibodies to study intermediate filaments *in situ* and hopefully resolve some of the questions concerning intermediate filament function.

Acknowledgements.

We thank Sean Carroll (UC Boulder) for the gift of the antitubulin hybridoma. This work was supported by grants from the Pew Biomedical Scholars Program, the March of Dimes and the National Institutes of Health.

References

(1) Murti, K.G., R. Goorha & M.W. Klymkowsky. 1986. J. Cell Biol. 103:416a.
(2) Klymkowsky, M.W. 1981. Nature 291:249-251.
(3) Lin, J.J.-C. & J.R. Feramisco. 1981. Cell 24:185-193.
(4) Gawlitta, W., M. Osborn & K. Weber. 1981. Eur. J. Cell Biol. 25:83-90.
(5) Lane, E.B. & M.W. Klymkowsky, M.W. 1982. Cold Spring Harb. Symp. Quant. Biol. 46:387-402.
(6) Klymkowsky, M.W., R.H. Miller & E.B. Lane. 1983. J. Cell Biol. 96:494-509.
(7) Klymkowsky, M.W., L.A. Maynell & A.G. Polson. 1987. Development. in press.
(8) Franz, J.K., L. Gall, M.A. Williams, B. Picheral & W.W. Franke. 1983. Proc. Natl. Acad. Sci. USA 6254-6258.
(9) Franke, W.W., P.C. Rathke, E. Seib, M.F. Trendelenburg, M. Osborn & K. Weber. 1976. Cytobiologie 14:111-130.
(10) Gall, L., B. Picheral & P. Gounon. 1983. Biol. Cell. 47:331-342.
(11) Godsave, S.F., C.C. Wylie, E.B. Lane & B.H. Anderton. 1984. J. Embryol. exp. Morph. 83:157-167.
(13) Wylie, C.C., D. Brown, S.F. Godsave, J. Quarmby & J. Heasman. 1985. J. Embryol. exp. Morph. 89 suppl. 1-15.
(14) Klymkowsky, M.W. 1982. EMBO J. 1:161-165.
(15) Rebagliati, M.R. & D.A. Melton. 1987. Cell 48:599-605.
(16) Wehland, J. & M.C. Willingham. 1983. J. Cell Biol. 97:1476-1490.
(17) Blose, S.H., D.I. Meltzer & J.R. Feramisco. 1984. J. Cell Biol. 98:847-858.

THE ROLE OF ACTIN FILAMENTS IN MIGRATION OF BASAL BODIES DURING MACROCILIOGENESIS IN *BEROË*

Sidney L. Tamm and Signhild Tamm

Boston University Marine Program, MBL, Woods Hole, MA, USA

ABSTRACT

The development of macrocilia and actin bundles was studied by electron microscopy of regenerating macrociliary cells on the lips of the ctenophore *Beroë*. Basal bodies arise *de novo* in association with dense fibrogranular bodies near the nucleus and Golgi. Long striated rootlets develop from the proximal ends of newly-formed basal bodies. Strands of parallel microfilaments assemble and attach to the sides of the basal bodies and striated rootlets, forming an anastomosing skein of actin filaments running toward the apical cell surface. Basal body-rootlet units migrate to the surface at the heads of trails of microfilaments. The number and length of the microfilaments increase as more and more basal bodies are transported to the apical surface. Basal body migration thus seems to be caused by elongation of assembling actin filaments.

INTRODUCTION

In many ciliated epithelia, centrioles (basal bodies) arise deep in the cell and migrate to the apical surface to initiate ciliogenesis. The mechanisms responsible for the directional movements of centrioles is not known. Use of cytoskeletal-disrupting drugs has suggested that actin filaments are involved in the positioning and mobility of centrosomes in neutrophils (Euteneuer and Schliwa, 1985), and in centriole migration during ciliogenesis in the quail oviduct (Boisvieux-Ulrich et al., 1984). However, structural evidence for participation of actin microfilaments in centriole motility has been difficult to obtain.

We have used a new system for studying the mechanism of centriole migration. unlike previous studies, we have been able to directly visualize the association of actin filaments with migrating centrioles. The morphological relationship strongly suggests that microfilaments play an important role in centriole movements.

SYSTEM

Beroë is a ctenophore, or comb jelly, and is a very voracious member of the marine zooplankton. Around the inside of the lips of

Beroë is a band of unusual ciliary organelles called macrocilia, which are used in feeding (Horridge, 1965; Swanberg, 1974). Macrocilia are finger-shaped compound cilia, 40 µm long and 5 µm thick, consisting of 250 cross-linked ciliary axonemes surrounded by a common membrane (Tamm and Tamm, 1984, 1985). Macrocilia arise from the broad end of slipper-shaped epithelial cells. A massive bundle of actin filaments extends from the base of the macrocilium to the opposite end of the cell, where the bundle terminates at a junction with underlying smooth muscle cells (Fig. 1) (Tamm and Tamm, 1987). The actin filaments run parallel to the long axis of the cell. The filament bundle is thought to act as an intracellular tendon to couple the motor activities of macrocilia and muscle during prey ingestion (Tamm and Tamm, 1987).

Cutting off the lips of *Beroë* induces regeneration of macrociliary cells (Franc, 1970). Differentiation proceeds in a spatial gradient along the oral-aboral axis. We fixed regenerating lips for electron microscopy and examined the sequence of events during one pattern of development of macrociliary cells (Fig. 2). A second pattern of differentiation will be described elsewhere.

BASAL BODY FORMATION

Basal bodies arise *de novo* near the nucleus and Golgi apparatus. Procentrioles are arranged in layered groups in close association with dense granules of fibrous and particulate material (Fig. 3). The dense granules are 60-80 nm in diameter, and are often surrounded by ribosomes. The dense granules are arranged in plaques, and lie close to the proximal, or cartwheel ends of the developing basal bodies. The procentrioles often are positioned side by side in two layers, with their cartwheel ends facing toward the outside (Fig. 4). No pre-existing centrioles are seen near the procentriole groups. Microtubule development proceeds in a conventional fashion: hub-and-spoke cartwheel and A microtubules appear first, then B microtubules, and finally C microtubules to complete the triplet wall.

STRIATED ROOTLET AND ACTIN FILAMENT FORMATION

Newly-formed basal bodies disband from the groups and develop long, blade-shaped striated rootlets from their proximal ends. The rootlet consists of laterally-aligned fine fibrils with periodic transverse densities.

At the same time, strands of parallel actin microfilaments appear in the surrounding cytoplasm near the newly-formed basal bodies. The microfilaments run obliquely toward the apical cell surface.

BASAL BODY MIGRATION

Basal body-rootlet units are transported to the cell surface in close association with the assembling actin bundle. Microfilaments appear to attach laterally to the sides of the basal bodies and striated rootlets, and trail behind the migrating basal bodies like the tail of a comet (Figs. 5,7). The filaments form a three-dimensional anastomosing skein directed toward the surface. Basal body-rootlet units are oriented parallel to the paths of the microfilaments during their upward movement, which occurs independently of one another (Figs., 5,7). The number and length of the actin microfilaments increase during basal body migration.

INITIATION OF CILIOGENESIS

The growing actin bundle curves and runs parallel to and under the cell surface. Basal body-rootlet units which have reached the cortical region tilt upward out of the actin bundle to contact the overlying plasma membrane (Fig. 6). The basal bodies emerge in a uniform orientation with the flat sides of their striated rootlets parallel to the paths of the microfilaments. Basal feet appear on the lower wall of upwardly tilting basal bodies, and serve as focal points of cortical microtubules (Fig. 6). A small number of single cilia initially grow out. The tufts of cilia elongate and later fuse to form long slender macrocilia. The developing macrocilia gradually increase in diameter as more and more basal bodies reach the cell surface.

During ciliogenesis, the actin bundle continues to increase in length and thickness, until it attains its long, wedge-shaped form. Concomitant with enlargement of the actin bundle, the macrociliary cells change from cuboidal to an highly elongated shape.

CONCLUSIONS

The present study offers the most dramatic structural evidence to date for the participation of actin in centriole migration. The close spatial association of migrating basal body-rootlet units with growing actin filaments, and the parallel direction of basal body transport and microfilament orientation, suggest that actin plays an important role in basal body migration in macrociliary cells.

These results confirm previous studies, using cytoskeletal drugs, that actin is involved in motility of centrioles and basal bodies (Euteneuer and Schliwa, 1985; Boisvieux-Ulrich et al., 1984).

How might actin drive the directional migration of basal bodies? Two possible mechanisms are (1) an actomyosin sliding mechanism like muscle, (2) elongation of assembling actin filaments.

An active sliding model does not seem likely in macrociliary cells because we do not see organized tracks of microfilaments <u>ahead</u> of the migrating basal bodies. Nor have we detected thick filaments associated with basal bodies or actin filaments.

Instead, elongation of polymerizing actin filaments seems a more likely mechanism to drive basal body movements in our system. Assembly and growth of the actin bundle coincides temporally and spatially with the migration of basal bodies toward the apical surface. Moreover, the directions of migration and filament orientation coincide.

Further experiments using actin-depolymerizing drugs and immunocytochemistry should help determine the role of actin in basal body migration in macrociliary cells.

ACKNOWLEDGEMENTS

This work was supported by NSF Grant 83-14317 and NIH Grant GM 27903.

REFERENCES

Boisvieux-Ulrich, E., Laine, M.-C., and Sandoz, D. (1984). Effets de la cytochalasine D sur la ciliogenese dans l'oviducte de caille. Biol. Cell 52, 66a.

Euteneuer, U. and Schliwa, M. (1985). Evidence for an involvement of actin in the positioning and motility of centrosomes. J. Cell Biol. 101, 96-103.

Franc, J.-M. (1970). Evolutions et interactions tissulaires au cours de la regeneration des levres de Beroë ovata. Cah. Biol. Mar. 11, 57-76.

Horridge, G.A. (1965). Macrocilia with numerous shafts from the lips of the ctenophore Beroë. Proc. R. Soc. Lond. B162, 351-364.

Swanberg, N. (1974). The feeding behavior of Beroë ovata. Mar. Biol. 24, 69-76.

Tamm, S.L. and Tamm, S. (1984). Alternate patterns of doublet microtubule sliding in ATP-disintegrated macrocilia of the ctenophore Beroë. J. Cell Biol. 99, 1364-1371.

Tamm, S.L. and Tamm, S. (1985). Visualization of changes in ciliary tip configuration caused by sliding displacement of microtubules in macrocilia of the ctenophore Beroë. J. Cell Sci. 79, 161-179.

Tamm, S.L. and Tamm, S. (1987). Massive actin bundle couples macrocilia to muscles in the ctenophore Beroë. Cell Motility and Cytoskel. 7, 116-128.

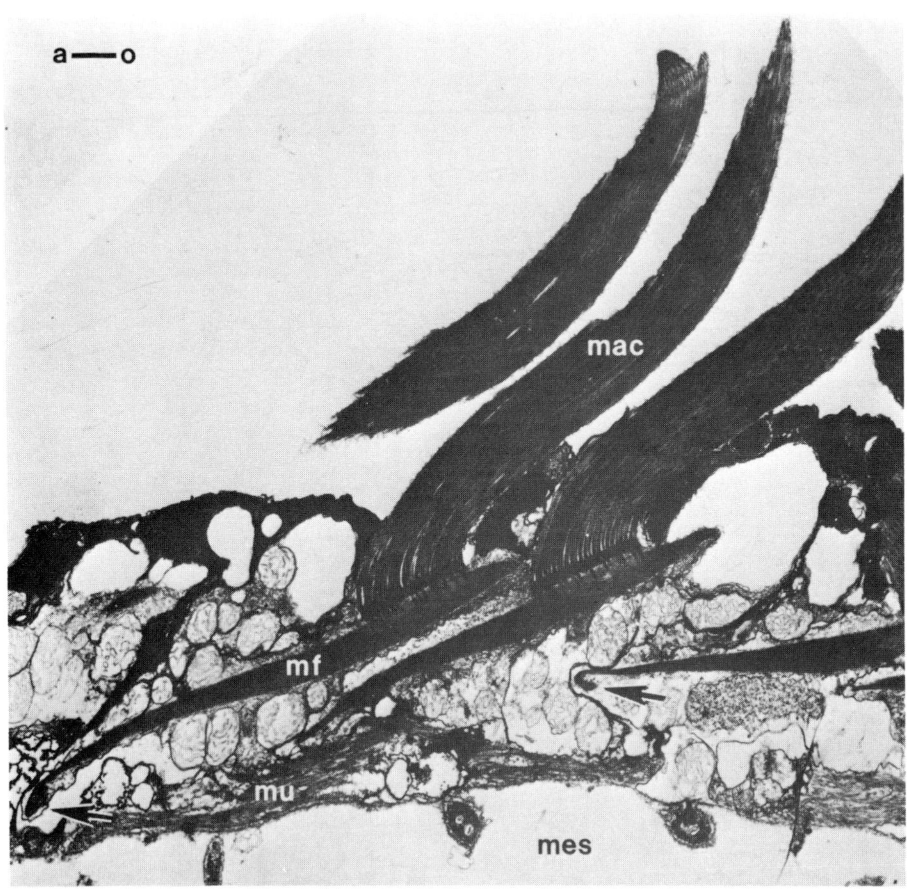

Fig. 1. Longitudinal survey view of macrociliary cells in the lip epithelium of non-regenerating Beroë. A massive wedge-shaped bundle of actin filaments (mf) extends from the base of the macrocilium (mac) to the lower end of the cell facing the mesoglea (mes) and underlying muscles (mu). a-o, aboral-oral axis. x3,700 (from Tamm and Tamm, 1987).

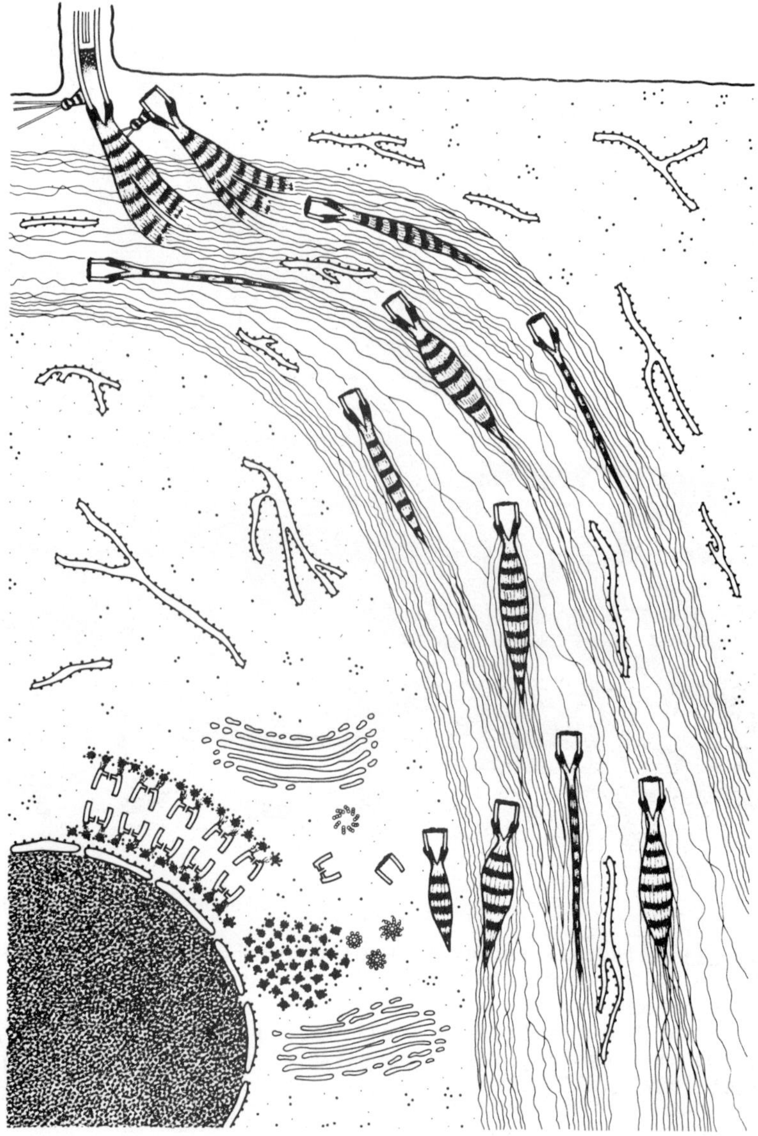

Fig. 2. Summary diagram of basal body origin and migration in differentiating macrociliary cells. Procentrioles arise in ordered groups adjacent to plaques of dense granules (lower left). Striated rootlets develop and basal body-rootlet units become associated with assembling strands of microfilaments which run toward the cell surface. Basal bodies migrate upward at the heads of trails of growing microfilaments. Basal body-rootlets tilt out of the actin bundle to contact the cell surface and initiate ciliary growth.

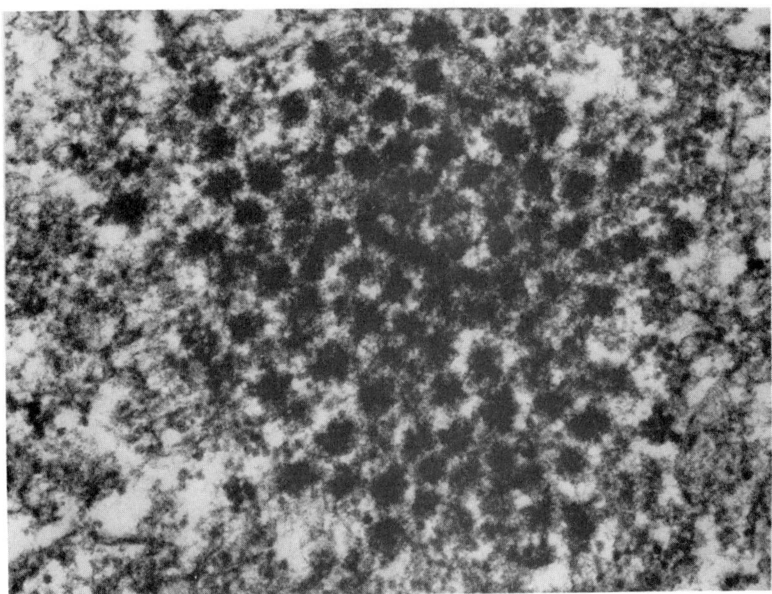

Fig. 3. A plaque of dense granules is the earliest sign of basal body formation in regenerating macrociliary cells. x72,700.

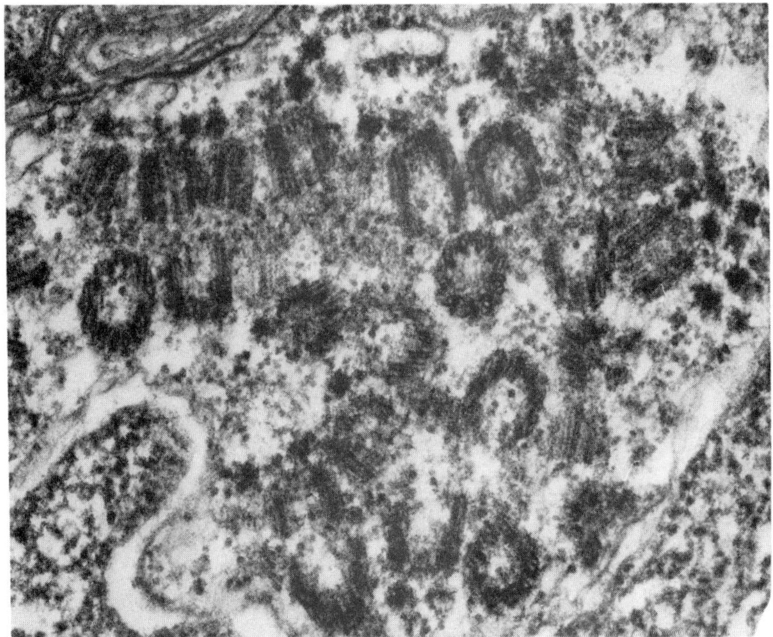

Fig. 4. Procentrioles usually lie side-by-side in layers with their cartwheel ends adjacent to the dense granules. x65,200.

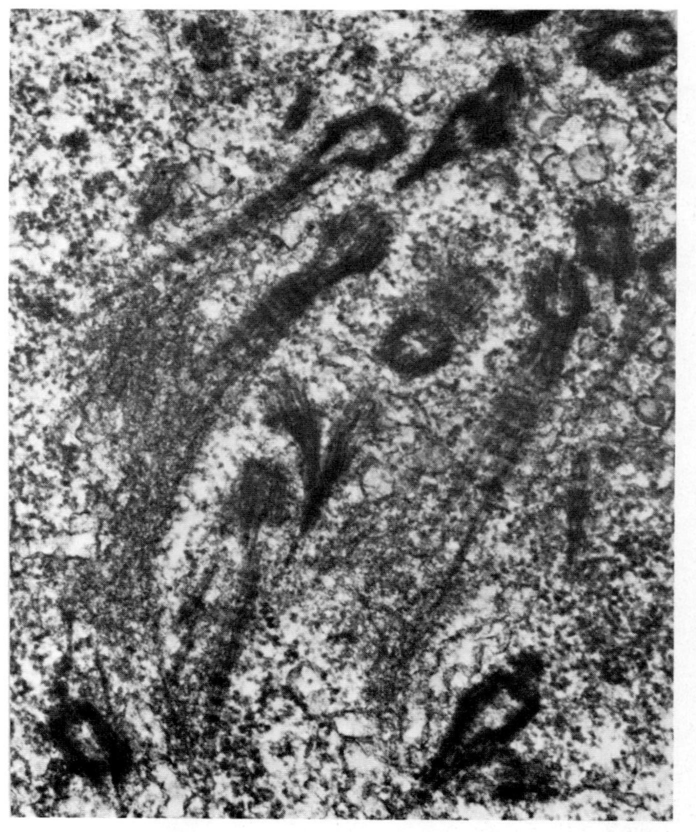

Fig. 5. Migrating basal bodies with microfilaments trailing from their striated rootlets. x55,300.

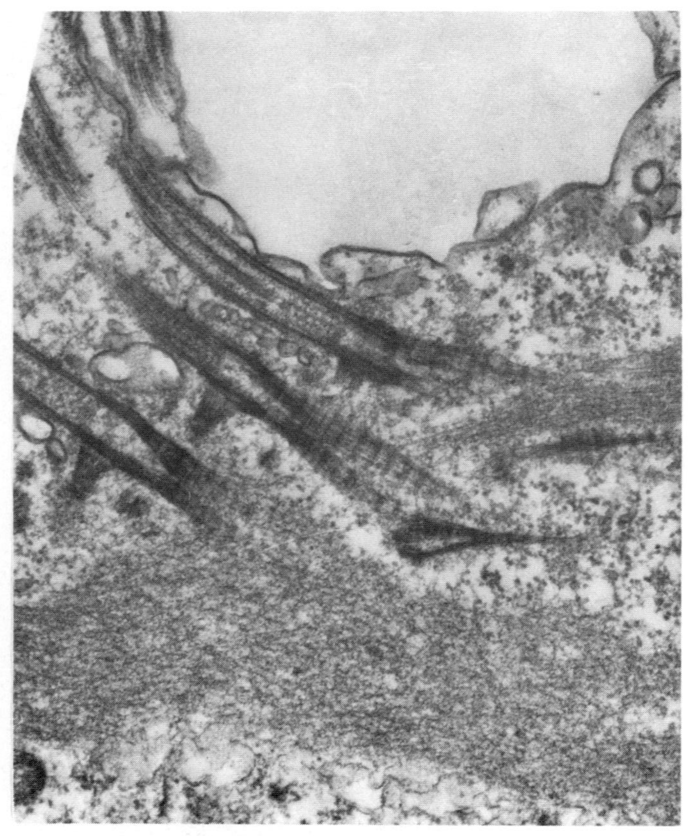

Fig. 6. Basal body-rootlet units tilt from the actin bundle to initiate ciliogenesis. x48,300.

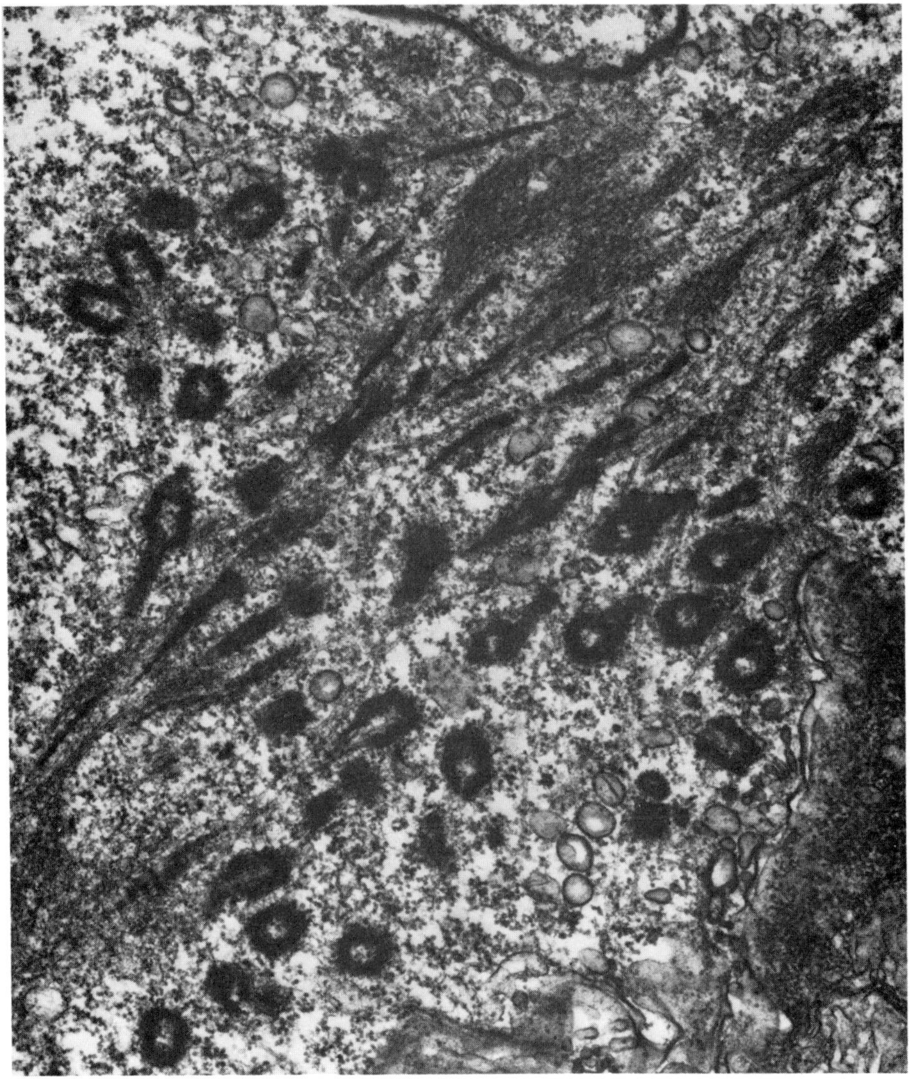

Fig. 7. Oblique section near the apical surface showing migrating basal bodies with trails of microfilaments. x38,700.

PRELIMINARY IDENTIFICATION, SUBCELLULAR LOCALIZATION AND DEVELOPMENTAL PROFILE OF A 94 K POLYPEPTIDE, A PUTATIVE COMPONENT OF THE DROSOPHILA MELANOGASTER EXTRACELLULAR ENVIRONMENT.

Cervera,M., Domingo,A., González-Jurado,J., Vinós,J. & Marco,R.

Instituto de Investigaciones Biomédicas del CSIC and Departamento de Bioquímica UAM . Facultad de Medicina . Universidad Autónoma de Madrid, 28029 MADRID. Spain.

Extracellular components are difficult to isolate, separate and identify, a fact that has limited the studies on their physiological roles and their fates during development. However, in the last ten years, several extracellular matrix components have been isolated and intensely studied (1).

In the case of high invertebrates, very little is known about the extracellular components, for example in *Drosophila melanogaster*, in spite of the great interest of their characterization in this organism, in view of the potentiality of the genetical and the increasing knowledge of its developmental properties. This is even more a pity, since the role of the extracellular matrix in cell-cell interaction in development is the current focus of the work of many research groups (2).

During an investigation of the presence of intermediate-like filaments from *Drosophila* (3) we have partially purified from extracts of whole adults flies a 94K polypeptide, which is associated with the cytoskeleton Triton X-100 insoluble matrix. It is solubilized by treatment with relatively low salt (0.2 M NaCl) and separated by DEAE chromatography under denaturing conditions and ammonium sulphate salting out. By extraction from an SDS-polycrylamide gel, and injection into a rabbit, a polyclonal antibody has been prepared which recognizes this polypeptide in whole extracts. As it is well known, antibodies supply an excellent tool for detecting specific proteins and particularly unknown ones associated with the extracellular matrix. Using inmunofluorescence and inmunotransfers of $NaDodSO_4$ gels we have studied the properties of solubilization, subcellular localization and distribution during development of the 94K polypeptide.

The major reactivity appears associated with the exterior of the cells in different tissue, surrounding the fat body cells (A, phase contrast; B, immunofluorescence), the muscle fibers (C, phase contrast; D, immunofluorescence) and the intestinal organs cardias (E, immunofluorescence) and gut (F, immunofluorescence). These results are consistent with the antigen being extracellular closely associated with the cell surface and/or basal lamina.

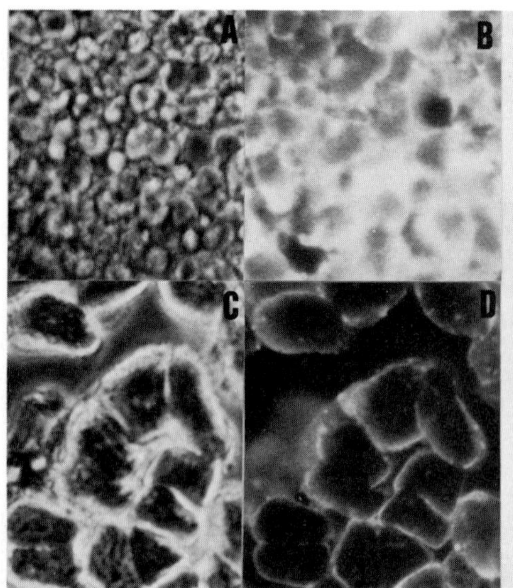

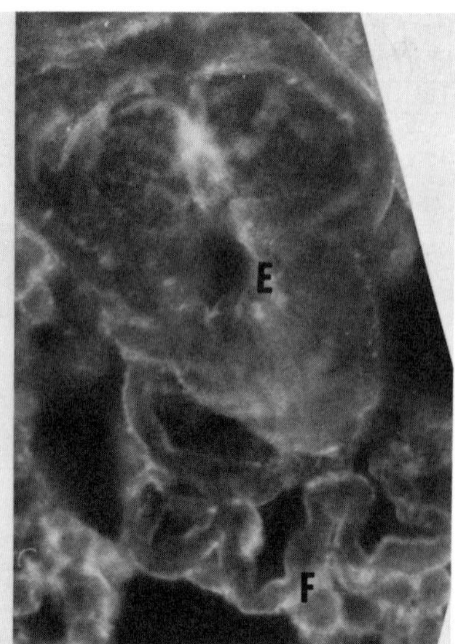

In conclusion, among the polypeptides present in a 0,5% Triton X 100 insoluble pellet of Drosophila melanogaster homogenates from adult flies we have identified a 94K insoluble polypeptide which can be extracted with 0.2 M salt. It is solubilized and further fractionated by DEAE Sephacel chromatography in the presence of 8 M Urea eluting at 0.075 M NaCl. The polypeptide can be purified by ammonium sulphate salting-out (0-45% ammonium sulphate) and salt extraction. Using a polyclonal antibody which recognizes this polypeptide its developmental profile has been studied (it increases from embryos to adults). Moreover, the subcellular distribution of the 94 K protein has been preliminary studied (the antibody labels structures which surround muscles, gut and fat body). In accordance with this finding, PAS staining indicates that it is a glycoprotein. We are currently studying more of its properties to clarify to which of the known extracellular components, if any, this polypeptide may correspond. In this conection it is intriguing that recently (4) it has been found that a 92k component is often coprecipitated with the common 110kd Drosophila position-specific glucopeptidic antigen similar to the vertebrate fibronectin receptor.

1) Ekblom, P., Vestweber, D. and Kemler, R. Ann.Rev.Cell Biol. 2, 27-47 (1986).
2) Thiery, J.P., Duband, J.L. and Tucker, G.C. Ann.Rev.Cell Biol. 1,91-113.
3) Cervera, M., Domingo, A., Vinós, J. and Marco, R. Biochem Biophys.Res.Comm. (in the press).
4) Leptin, M., Aebersold, R. & Wilcox, M. EMBO J. 6, 1037-1043 (1987).

ORGANIZATION OF THE CYTOSKELETON DURING ESTABLISHMENT OF CYTOPLASMIC DOMAINS IN EGGS OF THE LEECH *Theromyzon rude*.

J. Fernández, N. Olea and C. Matte.
Department of Biology, Faculty of Sciences, University of Chile.
Casilla 653, Santiago. Chile.

It is well established that several different kinds of eggs are subjected to profound spatial reorganization prior to the initiation of cleavage. Furthermore, there is growing evidence that such reorganization affects many parts of the egg (1-5). In annelids, spatial reorganization of the uncleaved egg leads to the accumulation of conspicuous masses of organelle-rich cytoplasm at its poles (poleplasms). So far, such studies are mostly based on the oligochaete *Tubifex* (6). We have found that the large eggs of some glossiphoniid leeches present many advantages for the study of similar processes (7, 8). These eggs are easily cultured and manipulated and form 3 large ooplasmic domains along the animal/vegetal axis: the perinuclear plasm at the egg center and the teloplasms at the poles. During cleavage the animal and vegetal teloplasms are first confined to the D blastomere. Upon division of this cell, the teloplasms are separately funneled into the first and second somatoblasts. With the proliferation of the somatoblasts, the teloplasms are finally sequestered into 10 large stemlike cells called teloblasts. Highly unequal division of the latter cells allows their teloplasm to be gradually passed into small blast cells, that become organized into 5-paired row of cells called germinal bandlets. The bandlets constitute cell lines involved in the formation of ectodermal and mesodermal structures (9). In this manner, the teloplasms represent egg territories that may not only be accurately traced throughout early development, but also may enclose morphogenetic determinants responsible for the marked mosaicism of the leech embryo. The teloplasms form as result of poleward translocation of organelles, vesicles and granules across the egg surface. This process is accompanied by stereotyped deformation movements of the egg manifested in the form of polar rings and meridional bands of contraction. Constriction of the rings and shortening of the meridians appear to provide the force required for bipolar displacement of large amounts of ectoplasm. At 20°C, the whole process is completed in about 1 h (10).

MATERIAL AND METHODS. Fertilized eggs of the duck leech *T. rude* were removed from the ovisacs and cultured in a simple saline medium. To explore the manner in which the cytoskeleton participates in teloplasm formation we followed the migration of live mitochondria, loaded with the fluorochrome rhodamine 123, and determined the organization of the ectoplasmic F-actin, as visualized with the fluorescent probe rhodamine-labelled phalloidin. Furthermore, the arrangement of microtubules and microfilaments was determined in sectioned or in detergent-extracted eggs examined under the electron microscope. The role of microtubules and actin filaments was also assessed by studying the motility and structure of eggs incubated from 0 h of development in colchicine or cytochalasin B.

RESULTS. Observations indicate that formation of the polar rings and meridional bands involve co-migration of mitochondria and of a network of F-actin towards the sites of egg deformation. The cytoskeleton of the polar rings seems to consist mostly of circumferentially-oriented actin filaments. The meridional bands, however, include an apparently more complex cytoskeleton of longitudinally-oriented microtubules and actin filaments. Hence, the polar rings and the meridional bands represent sites of condensation of the actin-lattice. As a result of the conjugated displacement of the polar rings and meridians towards their respective poles, mitochondria and a network of actin filaments agglomerate at the top of the animal hemisphere and at the bottom of the vegetal hemisphere. Meanwhile, the ectoplasmic framework of microtubules, that grew from the sperm centrosome, has been greatly dismantled. This process in clearly related to the disassembly of an interphase microtubular cytoskeleton and the assembly of the cleavage spindle. Deformation movements are disturbed or blocked in eggs treated with high concentrations of colchicine or cytochalasin B. Under these conditions the teloplasms present abnormal configurations or fail to be formed.

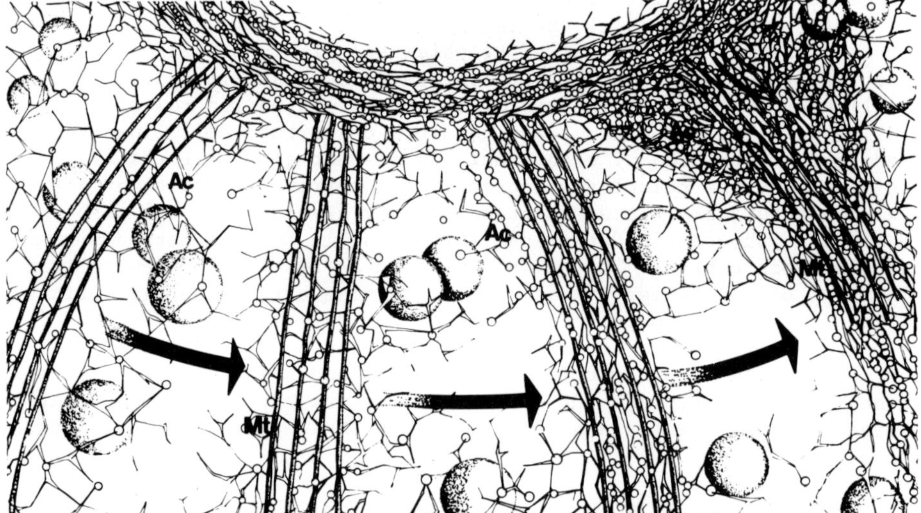

Fig. 1. Model of a *T. rude* egg that illustrates how the organization of the cytoskeleton appears to be modified during formation and displacement of the polar rings and meridional bands of contraction (direction of arrows). Notice that progressive condensation of the ectoplasmic actin lattice (Ac), together with mitochondria (small circles), is accompanied by depolymerization of microtubules (Mt). Large circles represent yolk platelets.

CONCLUSIONS. The cytoskeleton plays a very important role in the redistribution of cytoplasm destined to form the teloplasms. The ectoplasmic framework of microtubules, and its associated actin-lattice, appear to constitute a force-generating system capable of directing bipolar translocation of organelles. Changes in the state of aggregation of the actin filaments and in the dynamic equilibrium of the microtubules seem to be key features in the operation of the system. A model of the egg based on such a system is depicted in Fig. 1.

(Supported by Grant B-1987-8635 from the University of Chile and Grant 1218 from Fondo Nacional de Investigación Científica y Tecnológica).

REFERENCES

(1) Vacquier, V.D. (1981) Develop. Biol., 84: 1-26.
(2) Jeffery, W.R. and Meier, S. (1983) Develop. Biol., 96, 125-143.
(3) Dictus, W.J.A.G., van Zoelen, E.J.J., Tetteroo, P.A.T., Tertoolen, L.G.J., de Laat, S.W. and Bluemink, J.G. (1984) Develop. Biol., 101, 201-211.
(4) Speksnijder, J.E., Mulder, M.M., Dohmen, M.R., Hage, W.J. and Bluemink, J.G. (1985) Develop. Biol., 108, 38-48.
(5) Vincent, J.P., Oster, G.F. and Gerhart, J.C. (1986) Develop. Biol., 113, 484-500.
(6) Shimizu, T. (1982) in Developmental Biology of Freshwater Invertebrates (Harrison, F.W. and Cowden, R.R., eds.) pp. 283-316. Alan R. Liss Inc. New York.
(7) Fernández, J. (1980) Develop. Biol., 76, 245-262.
(8) Fernández, J., and Olea, N. (1982) in Developmental Biology of Freshwater Invertebrates (Harrison, F.W. and Cowden, R.R. eds) pp. 317-361. Alan R. Liss Inc. New York.
(9) Fernández, J. and Stent, G.S. (1980) Develop. Biol., 78, 407-434.
(10) Fernández, J., Olea, N. and Matte, C. (1987) Development. In press.

MACRONUCLEAR MICROTUBULES DURING EARLY EXCYSTMENT OF *ONYCHODROMUS ACUMINATUS* (CILIOPHORA,HYPOTRICHIA)

M.A. JAREÑO
Centro de Investigaciones Biológicas,C.S.I.C.
Velazquez 144,Madrid 28006, Spain.

In Protozoa has been confirmed that intranuclear micro tubules directly participate in macronuclear division.They are involved in the regular distribution of macronuclear material into the daughter macronuclei (1),but it has not been suggested a possible role of the microtubules in the nucleolar material transport, even thaught that microtubules coated by granules have been shown (2).

In most hypotrichs, among them *Onychodromus acuminatus* ciliary organelles as well as cytoplasmic microtubules disintegrate during excystment. So, during excystment an indifferentiate cell must be transformed into an organism with complex organelles. Positioning fibers (3) which arise from a small body, probably a microtubule-organizing center (4) have been exposed at light microscope level during the excystment of this species. They are responsible for the organization of the kinetosomes which appear following the orientation of the fibers during early excystment. The existence of these superficial structures brings forward many problems that remain to be resolved. Our published data also suggest that the macronucleus could be directly implicate in the initiation of the cortical morphogenesis in the differentiation process of excystment.

In this paper intramacronuclear microtubules are described and their possible role in facilitating the flow of nucleolar material is discussed.

MATERIALS AND METHODS

The strain of *O. acuminatus* and the techniques for culturing and induction of excystment have been already published (5). Processing of ciliates for electron microscope was carried out as previously described (6).

RESULTS AND DISCUSSION

During early excystment the macronucleus of this species form a sharpened point that reaches the cortex. Microtubules have been observed at an ultrastructural level inside the macronuclear point and in morphological relation to nucleoli. These support structural changes during the process (results in press) losing their granular component. Many of them become remnant nucleoli and finally disappear. Afterwards microtubules disassemble.

Fig. 1 shows a fragment of macronucleus(M) one hour after the start of the excystment. The compact nucleolus (N) is formed by granular and fibrillar components intermingled,and microtubules (arrows) are around it. Half an hour after, many nucleoli have a ring-like appearance and frequently they are also in close relation to microtubules (Fig.2,arrows). The ring-shape nucleoli are formed by a light core where dispersed granular material is observed, and an electron-dense cortex with their components organized as in compact nucleoli.

The morphological relation between microtubules and nucleoli during this differentiation process, suggest that microtubules are not only involved in the stretching of the macronuclear point but also they could facilitate the flow of ribosomal particles from the nucleoli into the cytoplasm. Some facts support this idea: 1) There is no macronuclear division in early excystment, so, they cannot be used to propulse nucleoli or chromatin into the daughter macronuclei. 2) Nucleoli lose their granular material. 3) Microtubules are in close relation to nucleoli during these phases. 4) Microtubules disassemble in the last phases of this nucleolar process. 5) At the same time as the nucleolar events, the macronuclear membrane increases the number of pores suggesting a high level of interchange between macronucleus and cytoplasm.

REFERENCES

(1) Raikov I.B.(1982).Cell Biol. Monographs 9
(2) Tucker J. et al.(1980) J.Cell Sci.44, 135-151
(3) Jareño M.A. & Tuffrau M. (1979). Protistologica XV , 597-605
(4) Jareño M.A. (1984). J. Protozool. 31 , 489-492
(5) " (1977). Protistologica XIII ,187-194
(6) " (1985). Protistologica XXI , 313-321

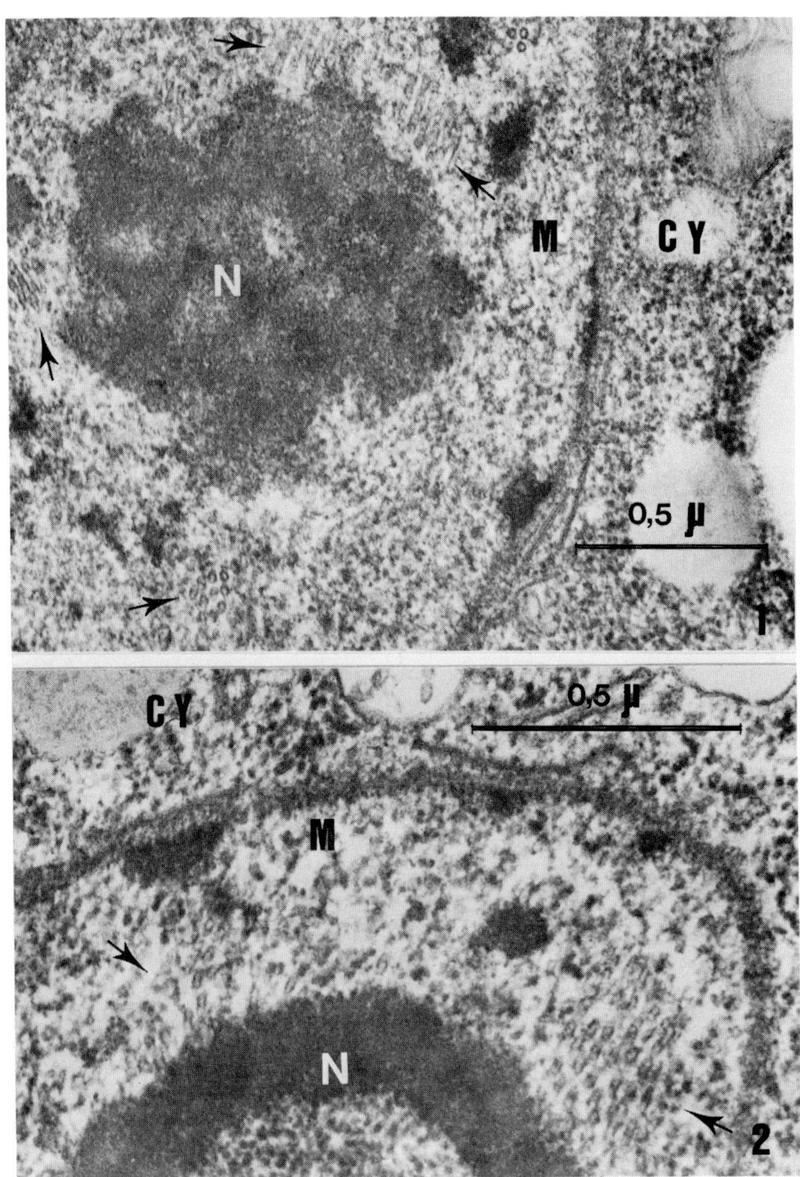
N
M
CY
0,5 µ
1
CY
0,5 µ
M
N
2

OBSERVATIONS ON THE CYTOSKELETAL ELEMENTS OF UROCENTRUM TURBO (PROTOZOA, CILIOPHORA)

R. Gil* and A. Guinea**

* Centro de Investigaciones Biológicas, C.S.I.C., 28006 Madrid, Spain

** Departamento de Microbiología, Facultad de Biología, Universidad Complutense, 28040 Madrid, Spain

INTRODUCTION. Several studies have been made on Urocentrum turbo, but only Didier (1) describes the ultrastructure, and Guinea & col. (3) determine the extent and position of the microfibrillar nets constituting the structure of its buccal apparatus. The aim of this work is to compare the results obtained using light and electron microscopies with those observed in permeabilized cells.

MATERIAL AND METHODS. With light microscopy, the staining method used was the pyridinated silver carbonate technique (2). Control cells were fixed for 20 min in a mixture of equal parts of glutaraldehyde (2%) and O_sO_4 (1%) in 0.05M phosphate buffer (pH 7.4). In order to obtain the permeabilization, the cells were treated with 5% Triton X-100 in PHEM, and then fixed with glutaraldehyde and O_sO_4. In both cases, the specimens were embedded in Epon, sections were cut with a Reichert-Jung ultramicrotome, stained with uranylacetat followed by lead citrate, and viewed with a Philips E.M. 300 (60-80 KV).

RESULTS. We have distinguished two parts in the cortex of Urocentrum turbo: somatic and buccal. After the treatment with Triton X-100, the main components of the somatic cytoskeleton remain. The somatic kinetosomes show kinetodesmal, transverse and postciliary fibers, while the cellular membrane (plasmalemma) disappears and the epiplasm is conserved. The microfibrillar nets that form the cytoskeleton of the buccal cavity also rest. The microfibrillar bundles surrounding the buccal cavity resist treatment of permeabilization, and they appear formed by microfibrils of 12 nm in diameter. We can also distinguish the microfibrillar net with condensation knots located beneath the ribbed wall, which is composed by microfibrils of 8 nm in diameter associated with knots of 24 nm in a triangular pattern

DISCUSSION. The fibers and microfibrillar elements observed in the somatic and buccal cortex of *Urocentrum turbo* after treatment with Triton X-100, coincide with those described by the other authors previously mentioned (1) and (3). The fibers observed in the somatic cortex are the three classic fibrillar systems associated to the somatic kinetosomes. In the buccal cavity, two types of microfibrils have been observed. The microfibrils of the bundles that surround the buccal cavity are similar in diameter to those described in vertebrate cells as "intermediate filaments" (4). On the other hand, the microfibrils of the net with condensation knots have a similar diameter to that described for "actin filaments" of vertebrate cells (4).

REFERENCES

(1) Didier, P. (1970) Ann. Stat. Biol. Besse-en-Chandesse 5, 1-274.
(2) Fernández-Galiano, D. (1976) Trans. Am. Microsc. Soc. 95, 557-560.
(3) Guinea, A., Gil, R. and Fernández-Galiano, D. (1987) Trans. Am. Microsc. Soc. 106, 53-62.
(4) Schliwa, M. and Blerkom, J. (1981) J. Cell. Biol. 90, 222-235.

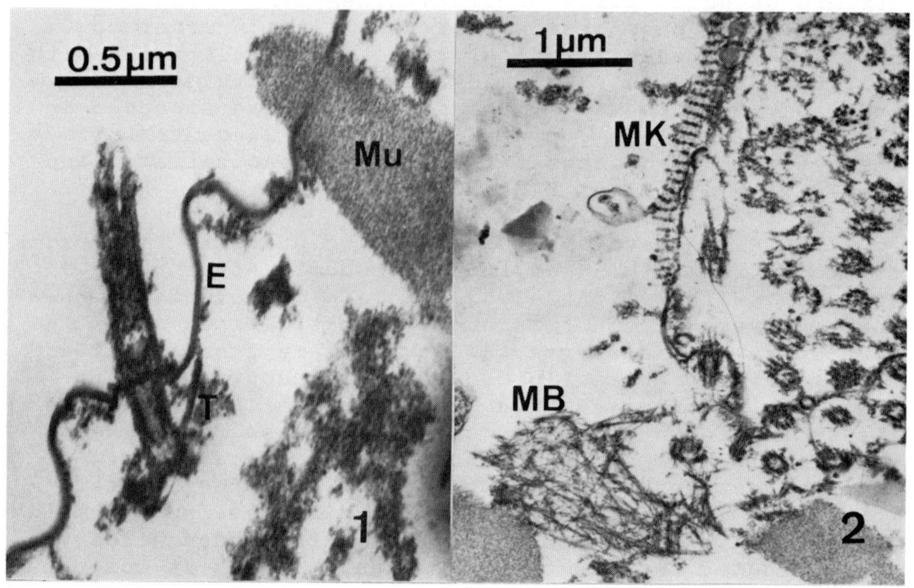

FIGS.1 and 2. Permeabilized cells of *Urocentrum turbo*. Fig.1. Somatic cortex. E, epiplasm; Mu, mucocyst; T, transverse fiber. Fig.2. Buccal cavity. MB, microfibrillar bundle; MK, microfibrillar net with condensation knots.

CLOSING REMARKS

Leslie Wilson, Department of Biological Sciences, University of California, Santa Barbara, California 93106, USA

It is a privilege and a pleasure to provide the closing remarks at this important conference; the First International Symposium on The Cytoskeleton in Cell Differentiation and Development. It was little more than ten years ago when we began to realize that the cytoplasm of eucaryotic cells is highly ordered, and that the order is conferred by an elaborate three-dimensional network of fibrous structures. It did not take long to learn that there exists along with the three principle filaments of the cytoskeleton, a large number of associated molecules that mediate the interactions of the fibrous elements with one another and with other cytoplasmic components. And it also did not take long to realize that the cytoskeletal network is highly dynamic and changes continually in response to the needs of the cell.

In thinking about the presentations of the past several days, I was especially impressed by the data emerging on the heterogeneity of the backbone proteins of the principle fibrous elements, and by the enormous diversity of fiber-associated proteins that have been identified. It is clear that our understanding of the cytoskeleton has come a long way in a short time, but the conclusion that we barely have scratched the surface is inescapable. We now realize that the cytoskeleton is remarkably complex, much more so than anyone could have imagined a few short years ago. The number of participating proteins is large, and the interactions among them will be challenging to unravel.

A strong impression I obtained while thinking about the many fine presentations of the last several days, is that we are about to enter a new era in cytoskeletal research. In retrospect, it seems clear that we can divide research on the cytoskeleton into two distinct phases, a "descriptive" phase, and a "mechanistic phase". Most of the research of the past decade has been descriptive, and we now appear on the verge of entering a mechanistic phase, in which the functions of the cytoskeletal elements at the molecular level will be elucidated. Our description of the cytoskeleton is far from complete. Many participating components remain to be discovered, and we still have much to learn about how cytoskeletal elements are organized and in the broad sense how they function in specific cells and tissues.

As we have heard during these proceedings, we are continuing to develop powerful tools to identify and characterize the organization of cytoskeletal components. The tools include sophisticated light and electron microscope techniques and molecular probes such as synthetic peptides and antibodies for analysis of cytoskeletal structure and organization in cells and for the molecular characterization of cytoskeletal proteins. We are also continuing at a rapid pace to identify and characterize new cytoskeletal components.

Without mentioning names or attempting to be comprehensive, I want to comment briefly on a few of the highlights of this symposium. Several talks and a number of posters were presented on assembly dynamics of cytoskeletal elements. The assembly dynamics of microtubules (MTs) continues to generate considerable interest as the complexity of MT assembly behavior becomes more apparent. With the aid of temperature-jump and other sophisticated approaches to the study of MT assembly dynamics in work described here and elsewhere we have learned that MTs can exhibit a diverse array of assembly behaviors that seem to be controlled by events occurring at both MT ends. We are also learning that MT-associated proteins (MAPs) rather than the tubulin backbone seem to be responsible for determining MT assembly behavior. Presumably the cell can utilize the various assembly behaviors to perform specific functions.

The technique of limited proteolysis and the use of synthetic peptides to create site-directed antibodies are permitting us to map functional domains of the alpha and beta chains of tubulin. Such studies surely will continue to increase our understanding of the relationship between the structure and function of the tubulins. We also are continuing to gain insight regarding post-translational modifications of tubulin such as the detyrosination and retyrosination at the C-terminus of alpha tubulin. Using specific antibodies, we are obtaining a sophisticated picture of the distribution of these tubulin forms in cells, but the functional significance of tyrosination, or of other post-translational modifications of tubulin or any other cytoskeletal element remains obscure.

There were significant new data presented on the tissue and cell specific expression of alpha and beta tubulins, and on the question of the function of tubulin diversity. For example, by transfecting cDNAs for specific tubulin isotypes into cells that do not normally have those isotypes expressed, it was found that the introduced isotypes were incorporated into the MTs without detectably affecting MT function. Thus, the functional significance of tubulin diversity remains unknown. These findings support the idea that MAPs rather than the tubulin backbone of MTs are responsible for the functional and organizational diversity of MTs.

A powerful advance has been the significant increase in production of specific antibodies directed toward cytoskeletal proteins. Many new antibodies have been described at this meeting. These antibodies are critical for identifying and localizing new cytoskeletal components and for determining their function. For example, with antibodies to specific intermediate filament (IF) proteins, we are developing a sophisticated description of the tissue specific organization of IFs during development and differentiation. From a practical point of view, these data will result in significant improvement in diagnosis of malignant disease. These studies have also revealed the remarkable chemical complexity of the IF network in cells. The question of the functional significance of the complexity, however, remains unanswered. As another example, the use of human autoantibodies that recognize cytoskeletal elements offers significant potential for identifying new cytoskeletal proteins. Linked to the acquisition of new antibodies and the identification of new cytoskeletal proteins, has been the establishment of computer-based protein databases that promise to be useful in the continued identification and cataloging of the properties and functions of the many cytoskeletal elements present in various cells and tissues.

Other significant advances described at this symposium include the development of new model systems for identifying and determining the functions of cytoskeletal proteins during development and differentiation. We learned about new model systems for probing cytoskeletal function during development in Drosophila and Xenopus embryos, and during differentiation in neuroblastoma, melanoma, and neural hybrid cell lines.

In the future one of our most significant challenges will be to determine the functions of individual cytoskeletal components on a molecular level. A most difficult area of research has concerned the function of IFs. Extensive studies including the use of microinjected antibodies to a variety of cytoskeletal elements to perturb IF function have failed to provide any mechanistic insight into IF function. Our only clue is that IF function may be linked to differentiation. Attempts to utilize the powerful tools of genetics in organisms such as Drosophila to examine IF function have proven frustratingly negative despite heroic efforts.

The directions for future research are clear. We must continue to identify and characterize new cytoskeletal elements as well as determine the function of the components at the molecular level. Despite the complexity of the cytoskeleton, we should continue to make progress at a rapid pace.

In closing, I know that all of the participants at this outstanding symposium join me in expressing our sincere gratitude to all of the organizers and supporters of the symposium, and especially, to Drs. Ricardo Maccioni and Juan Aréchaga, who worked extremely hard to organize the symposium and to bring us together in this exciting city. Until we meet again at the second International Symposium on the Cytoskeleton in Development and Differentiation, hasta la vista.

AUTHOR INDEX

SUBJECT INDEX